AN INTRODUCTION TO FLAPPING WING AERODYNAMICS

This is an ideal book for graduate students and researchers interested in the aerodynamics, structural dynamics, and flight dynamics of small birds, bats, and insects, as well as of micro air vehicles (MAVs), which present some of the richest problems intersecting science and engineering. The agility and spectacular flight performance of natural flyers – made possible by their flexible, deformable wing structures as well as outstanding wing, tail, and body coordination – are particularly significant. To design and build MAVs with performance comparable to natural flyers, it is essential to understand natural flyers' combined flexible structural dynamics and aerodynamics. The primary focus of this book is to address recent developments in flapping wing aerodynamics. This book extends the work presented in *Aerodynamics of Low Reynolds Number Flyers* (Shyy et al. 2008).

Dr. Wei Shyy is the Provost of the Hong Kong University of Science and Technology and former Clarence L. "Kelly" Johnson Collegiate Professor and Department Chair of Aerospace Engineering at the University of Michigan. Shyy is the author or co-author of four books and numerous journal and conference articles dealing with a broad range of topics related to aerial and space flight vehicles. He is Editor of the Cambridge Aerospace Series with Vigor Yang (Georgia Tech) and Co-Editor-in-Chief of the nine-volume *Encyclopedia of Aerospace Engineering* (2010). He received the 2003 AIAA Pendray Aerospace Literature Award and the ASME 2005 Heat Transfer Memorial Award. He has led multi-university centers under the sponsorship of NASA, the U.S. Air Force Research Laboratory, and industry. His professional views have been quoted in various news media, including the *New York Times* and *USA Today*.

Dr. Hikaru Aono is a Research Scientist at the Institute of Space and Astronautical Science, Japan Aerospace Exploration Agency. He has made contributions to biological aerodynamics and related fluid-structure interaction issues.

Dr. Chang-kwon Kang is a Postdoctoral Research Fellow at the University of Michigan. His expertise includes analytical and computational modeling of the performance of flapping wings for micro air vehicles, aeroelastic dynamics of flapping wings, and other complex systems.

Dr. Hao Liu is a Professor of Biomechanical Engineering at Chiba University in Japan. He is well known for his contributions to biological, flapping-flight research, including numerous publications on insect aerodynamics simulations and physical realization of MAVs.

CAMBRIDGE AEROSPACE SERIES

Editors
Wei Shyy and Vigor Yang

1. J. M. Rolfe and K. J. Staples (eds.): *Flight Simulation*
2. P. Berlin: *The Geostationary Applications Satellite*
3. M. J. T. Smith: *Aircraft Noise*
4. N. X. Vinh: *Flight Mechanics of High-Performance Aircraft*
5. W. A. Mair and D. L. Birdsall: *Aircraft Performance*
6. M. J. Abzug and E. E. Larrabee: *Airplane Stability and Control*
7. M. J. Sidi: *Spacecraft Dynamics and Control*
8. J. D. Anderson: *A History of Aerodynamics*
9. A. M. Cruise, J. A. Bowles, C. V. Goodall, and T. J. Patrick: *Principles of Space Instrument Design*
10. G. A. Khoury (ed.): *Airship Technology*, Second Edition
11. J. P. Fielding: *Introduction to Aircraft Design*
12. J. G. Leishman: *Principles of Helicopter Aerodynamics*, Second Edition
13. J. Katz and A. Plotkin: *Low-Speed Aerodynamics*, Second Edition
14. M. J. Abzug and E. E. Larrabee: *Airplane Stability and Control: A History of the Technologies that Made Aviation Possible*, Second Edition
15. D. H. Hodges and G. A. Pierce: *Introduction to Structural Dynamics and Aeroelasticity*, Second Edition
16. W. Fehse: *Automatic Rendezvous and Docking of Spacecraft*
17. R. D. Flack: *Fundamentals of Jet Propulsion with Applications*
18. E. A. Baskharone: *Principles of Turbomachinery in Air-Breathing Engines*
19. D. D. Knight: *Numerical Methods for High-Speed Flows*
20. C. A. Wagner, T. Hüttl, and P. Sagaut (eds.): *Large-Eddy Simulation for Acoustics*
21. D. D. Joseph, T. Funada, and J. Wang: *Potential Flows of Viscous and Viscoelastic Fluids*
22. W. Shyy, Y. Lian, H. Liu, J. Tang, and D. Viieru: *Aerodynamics of Low Reynolds Number Flyers*
23. J. H. Saleh: *Analyses for Durability and System Design Lifetime*
24. B. K. Donaldson: *Analysis of Aircraft Structures*, Second Edition
25. C. Segal: *The Scramjet Engine: Processes and Characteristics*
26. J. F. Doyle: *Guided Explorations of the Mechanics of Solids and Structures*
27. A. K. Kundu: *Aircraft Design*
28. M. I. Friswell, J. E. T. Penny, S. D. Garvey, and A. W. Lees: *Dynamics of Rotating Machines*
29. B. A. Conway (ed): *Spacecraft Trajectory Optimization*
30. R. J. Adrian and J. Westerweel: *Particle Image Velocimetry*
31. G. A. Flandro, H. M. McMahon, and R. L. Roach: *Basic Aerodynamics*
32. H. Babinsky and J. K. Harvey: *Shock Wave–Boundary-Layer Interactions*
33. C. K. W. Tam: *Computational Aeroacoustics: A Wave Number Approach*
34. A. Filippone: *Advanced Aircraft Flight Performance*
35. I. Chopra and J. Sirohi: *Smart Structures Theory*
36. W. Johnson: *Rotorcraft Aeromechanics*
37. W. Shyy, H. Aono, C. K. Kang, and H. Liu: *An Introduction to Flapping Wing Aerodynamics*
38. T. C. Lieuwen and V. Yang: *Gas Turbine Engines*

An Introduction to Flapping Wing Aerodynamics

Wei Shyy
Hong Kong University of Science and Technology

Hikaru Aono
Japan Aerospace Exploration Agency

Chang-kwon Kang
University of Michigan

Hao Liu
Chiba University

CAMBRIDGE
UNIVERSITY PRESS

CAMBRIDGE
UNIVERSITY PRESS

University Printing House, Cambridge CB2 8BS, United Kingdom

One Liberty Plaza, 20th Floor, New York, NY 10006, USA

477 Williamstown Road, Port Melbourne, VIC 3207, Australia

4843/24, 2nd Floor, Ansari Road, Daryaganj, Delhi - 110002, India

79 Anson Road, #06-04/06, Singapore 079906

Cambridge University Press is part of the University of Cambridge.

It furthers the University's mission by disseminating knowledge in the pursuit of education, learning and research at the highest international levels of excellence.

www.cambridge.org
Information on this title: www.cambridge.org/9781107037267

First published 2013

A catalogue record for this publication is available from the British Library

Library of Congress Cataloging in Publication data
Shyy, W. (Wei)
An introduction to flapping wing aerodynamics /Wei Shyy, Hikaru Aono, Chang-kwon Kang, Hao Liu.
pages cm. – (Cambridge aerospace series)
Includes bibliographical references and index.
ISBN 978-1-107-03726-7 (hardback) – ISBN 978-1-107-64035-1 (paperback)
1. Aerodynamics. 2. Airplanes –Wings. 3. Micro air vehicles.
4. Wings (Anatomy) 5. Animal flight. I. Aono, Hikaru, 1981–
II. Kang, Chang-kwon, 1978– III. Liu, Hao, Ph.D. IV. Title.
TL573.S46 2013
6229.132′38–dc23 2012047764

ISBN 978-1-107-03726-7 Hardback
ISBN 978-1-107-64035-1 Paperback

明 吕纪 残荷鹰鹭图 故宫博物院藏
Hawk preying on an egret, Chi Lu (1477–?), Ming Dynasty, Palace Museum, Beijing

Contents

Preface

This book is about flapping wing aerodynamics. It presents various aspects of the aerodynamics of natural flyers, such as birds, bats, and insects, and of human-engineered micro air vehicles (MAVs) for both rigid and flexible wing structures. This edition focuses on the many recent developments since the publication of our earlier book titled *Aerodynamics of Low Reynolds Number Flyers*. We have substantially expanded Chapter 1 to offer a general and comprehensive introduction to low Reynolds number flight vehicles for both biological flyers and human-made MAVs. In particular, we summarize the scaling laws to relate the aerodynamics and various flight characteristics to a flyer's size, weight, and speed on the basis of simple geometric and dynamics analyses. In Chapter 2, closely following the previous edition, we discuss the aerodynamics of fixed rigid wings. It considers both two- and three-dimensional airfoils with typically low aspect ratio wings. Both Chapters 3 and 4 have been significantly expanded and updated. Chapter 3 examines the interplay between flapping kinematics and key dimensionless parameters such as the Reynolds number, Strouhal number, and reduced frequency for rigid wings. The various unsteady lift enhancement mechanisms are addressed, including leading-edge vortex, rapid pitch-up and rotational circulation, wake capture, tip vortices, and clap-and-fling. It also discusses both detailed time-dependent and simplified quasi-steady analyses along with experimental observations. Efforts have been made to contrast fixed and flapping wing aerodynamics in the context of geometry and tip, as well as of stall margins. Chapter 3 presents individual and varied objectives in regard to maximizing lift, mitigating drag, and minimizing power associated with flapping wings.

Chapter 4 addresses the role of structural flexibility of low Reynolds number wing aerodynamics. Due to the interplay between structural and fluid dynamics, additional dimensionless parameters appear, resulting in multiple time and length scales. For fixed wings, structural flexibility can further enhance stall margin and flight stability; for flapping wings, passive control can complement and possibly replace active pitching to make the flight more robust and more power efficient. Chapter 4 also discusses the airfoil shape, the time-dependent fluid and structural dynamics, and the spanwise versus chordwise flexibility of a wing. The scaling laws linking lift and power with fluid and structural parameters are of fundamental interest and offer insight into low Reynolds number flight sciences while providing guidelines for vehicle development. Finally, recent advances and future perspectives are summarized and presented in Chapter 5.

As in the previous edition, we have benefited from collaborations and interactions with many colleagues. In addition to those colleagues named in the previous edition, we would like to acknowledge the generous intellectual and financial support provided by the U.S. Air Force Research Laboratory, in particular the Flight Vehicle Directorate (now Aerospace Systems Directorate) and the Office of Scientific Research.

We feel sure that significant advancements in both scientific and engineering endeavors of flapping wing aerodynamics will continue to be achieved, and we enthusiastically await these new breakthroughs and developments.

Wei Shyy, Hikaru Aono, Chang-kwon Kang, and Hao Liu

Preface of the First Edition (*Aerodynamics of Low Reynolds Number Flyers*)

Low Reynolds number aerodynamics is important for a number of natural and man-made flyers. Birds, bats, and insects have been of interest to biologists for years, and active study in the aerospace engineering community has been increasing rapidly. Part of the reason is the advent of micro air vehicles (MAVs). With a maximal dimension of 15 cm and nominal flight speeds around 10 m/s, MAVs are capable of performing missions such as environmental monitoring, surveillance, and assessment in hostile environments. In contrast to civilian transport and many military flight vehicles, these small flyers operate in the low Reynolds number regime of 10^5 or lower. It is well established that the aerodynamic characteristics, such as the lift-to-drag ratio of a flight vehicle, change considerably between the low and high Reynolds number regimes. In particular, flow separation and laminar-turbulent transition can result in substantial change in effective airfoil shape and reduce aerodynamic performance. Since these flyers are lightweight and operate at low speeds, they are sensitive to wind gusts. Furthermore, their wing structures are flexible and tend to deform during flight. Consequently, the aero/fluid and structural dynamics of these flyers are closely linked to each other, making the entire flight vehicle difficult to analyze.

The primary focus of this book is on the aerodynamics associated with fixed and flapping wings. Chapter 1 offers a general introduction to low Reynolds flight vehicles, including both biological flyers and MAVs, followed by a summary of the scaling laws that relate the aerodynamics and flight characteristics to a flyer's sizing on the basis of simple geometric and dynamics analyses. Chapter 2 examines the aerodynamics of fixed, rigid wings. Both two- and three-dimensional airfoils with typically low aspect ratio wings are considered. Chapter 3 examines structural flexibility within the context of fixed wing aerodynamics. The implications of laminar-turbulent transition, multiple time scales, airfoil shapes, angles-of-attack, stall margin, structural flexibility, and time-dependent fluid and structural dynamics are highlighted.

Unsteady flapping wing aerodynamics is presented in Chapter 4. In particular, the interplay between flapping kinematics and key dimensionless parameters such as the Reynolds number, Strouhal number, and reduced frequency is examined. The various unsteady lift enhancement mechanisms are also addressed, including leading-edge vortex, rapid pitch-up and rotational circulation, wake capture, and clap-and-fling.

The materials presented in this book are based on our own research, existing literature, and communications with colleagues. At different stages, we have benefited from collaborations and interactions with colleagues: Drs. Peter Ifju, David Jenkins, Rick Lind, Raphael Haftka, Roberto Albertani, and Bruce Carroll of the University of Florida; Drs. Luis Bernal, Carlos Cesnik, and Peretz Friedmann of the University of Michigan; Drs. Michael Ol, Miguel Visbal, and Gregg Abate, and Mr. Johnny Evers of the Air Force Research Laboratory; Dr. Ismet Gursul of the University of Bath; Dr. Charles Ellington of Cambridge University; Dr. Keiji Kawachi of the University of Tokyo; Mr. Hikaru Aono of Chiba University; Dr. Mao Sun of the Beijing University of Aeronautics and Astronautics. In particular, we have followed the flight vehicle development efforts of Dr. Peter Ifju and his group and enjoyed the synergy between us.

MAV and biological flight is now an active and well-integrated research area, attracting participation from a wide range of talents and specialties. The complementary perspectives of researchers with different training and backgrounds enable us to develop new biological insight, mathematical models, physical interpretation, experimental techniques, and design concepts.

Thinking back to the time we started our own endeavor a little more than ten years ago, substantial progress has taken place, and there is every expectation that significantly more will occur in the foreseeable future. We look forward to it!

Wei Shyy, Yongsheng Lian, and Jian Tang Dragos Viieru
Ann Arbor, Michigan, U.S.A.

Hao Liu
Chiba, Japan
December 31, 2006

List of Abbreviations

Abbreviation	*Definition*
2D	two-dimensional
3D	three-dimensional
AoA	angle of attack
DNS	direct numerical simulation
LES	large-eddy simulation
LEV	leading-edge vortex
LSB	laminar separation bubble
MAV	micro air vehicle
PIV	particle image velocimetry
RANS	Reynolds-averaged Navier-Stokes
TEV	trailing-edge vortex
TiV	tip vortex
UAV	unmanned air vehicle

Nomenclature

AR	aspect ratio	Eq. (1–7)
b	wingspan	Eq. (1–7)
c	chord length	Eq. (1–19)
$\mathbf{c}_3$	unit vector in the direction from the leading edge to the trailing edge	Eq. (4–28)
C_D	drag coefficient	Eq. (2–22)
$C_{D,F}$	drag coefficient due to skin friction	Eq. (2–22)
$C_{D,P}$	drag coefficient due to pressure	Eq. (2–22)
C_F	force coefficient	Eq. (3–35)
C_L	lift coefficient	Eq. (1–1)
C_T	tension coefficient, thrust coefficient	Eqs. (3–23) and (4–2)
D_{aero}	aerodynamic drag	Eq. (1–29)
D_{ind}	induced drag	Eq. (1–29)
D_{par}	parasite drag (drag on the body)	Eq. (1–29)
D_{pro}	profile drag	Eq. (1–29)
D_w	drag on a finite wing	Eq. (1–28)
e	span efficiency factor	Eq. (2–22)
E	elastic modulus	Eq. (4–1)
f	flapping (wing-beat) frequency	Eq. (1–12)
f_{ext}	distributed external force per unit	Eq. (4–1)
f_n	natural frequency	Eq. (1–21)
g	gravitational acceleration	Eq. (1–3)
h_a	flapping amplitude	Eq. (3–4)
h_s	thickness of wing, thickness of membrane	Eqs. (4–1) and (4–8)
$h(t)$	time-dependent flapping displacement	Eq. (3–4)
H	shape factor	Eq. (2–2)
I	moment of inertia	Eq. (1–10)
J	advance ratio	Eq. (3–14)
J_T	torque	Eq. (1–9)
k	reduced frequency, turbulent kinetic energy	Eqs. (1–19) and (2–6)
l	characteristic length	Eq. (1–4)
L	lift, length of membrane after deformation	Eqs. (1–1) and (4–10)

L_0	unstrained membrane length	Eq. (4–3)
L/D	lift-to-drag ratio or glide ratio	Eq. (2–20)
m	flyer's total mass	Eq. (1–3)
$\tilde{n}$	amplification factor	Eq. (2–12)
N	threshold value that triggers turbulent flow in e^N method	Eq. (2–17)
p	static pressure	Eq. (2–5)
P_{aero}	total aerodynamic power	Eq. (1–30)
P_{ind}	induced power	Eq. (1–32)
P_{iner}	inertial power	Eq. (1–33)
P_{par}	parasite power	Eq. (1–32)
P_{pro}	profile power	Eq. (1–32)
P_{tot}	total power required for flight	Eq. (1–33)
q_∞	far field dynamic pressure	Eq. (4–13)
R	wing length	Eq. (3–24)
Re	Reynolds number	
Re_{f2}	Reynolds number for 2D flapping motion	Eqs. (3–8a) and (3–8b)
Re_{f3}	Reynolds number for 3D flapping motion	Eq. (3–7)
Re_T	turbulent Reynolds number	Eq. (2–10)
Re_θ	momentum thickness Reynolds number	Eq. (2–12)
S	wing area	Eq. (1–1)
S^0	membrane prestress	Eq. (4–8)
St	Strouhal number	Eq. (3–9)
t	time	Eq. (2–5)
T	wing stroke time scale, thrust	Eqs. (1–12) and (1–31)
u_i	velocity vector in Cartesian coordinates	Eq. (2–4)
U	forward flight velocity (free-stream velocity)	Eq. (1–1)
U_f	flapping velocity	Eq. (1–20)
U_{mp}	velocity for minimum power (forward flight)	Eq. (1–35)
U_{Mr}	velocity for maximum range (forward flight)	Eq. (1–35)
U_r	relative flow velocity	Eq. (1–20)
U_{ref}	reference velocity	Eq. (1–19)
w	vertical velocity in the far wake, transverse deflection	Eqs. (3–26) and (4–1)
w_i	downwash (induced) velocity	Eq. (1–14)
W	weight	Eq. (1–1)
W/S	wing loading	Eq. (1–2)
x_i	spatial coordinates vector	Eq. (2–4)
α	angle of attack, feathering angle (pitch angle) of a flapping wing	Eqs. (3–3) and (3–5)
β	stroke plane angle	Eq. (3–25)
δ^*	boundary-layer displacement thickness	Eq. (2–3)
ϕ	positional angle of a flapping wing	Eq. (3–15)
Φ	stroke angular amplitude	Eq. (3–7)

μ	Coefficient of dynamic viscosity	Table 4.2
γ	membrane tension, non-dimensional tip, deformation parameter	Eqs. (4–4) and (4–34)
Γ	circulation	Eq. (2–23)
ϕ	phase difference between plunging and pitching motion	Eq. (3–5)
ν	kinematic viscosity, Poisson's ratio	Eqs. (2–5) and (4–23)
ν_{Te}	effective eddy viscosity	Eq. (2–18)
ν_T	turbulent eddy viscosity	Eq. (2–6)
Π_0	effective inertia	Eq. (4–32)
Π_1	effective stiffness	Eq.(4–15)
$\Pi_{1,\text{pret}}$	effective pretension	Eq. (4–17)
Π_2	effective rotational inertia	Eq. (4–25)
θ	gliding angle, boundary-layer momentum thickness, elevation angle of a flapping wing	Eqs. (1–17), (2–5), and Eq. (3–2)
ρ	fluid density	Eq. (1–1)
ρ_s	structural density	Eq. (4–1)
τ_{ij}	Reynolds-stress tensor	Eq. (2–6)
ω	dissipation rate for k-ω turbulence model	Eq. (2–7)
ω_n	natural angular frequency of the beam model	Eq. (4–33)
$\dot{\omega}$	angular acceleration	Eq. (1–11)
η	propulsive efficiency for forward flight	Eq. (4–41)
Ψ	the bending angle	Eq. (4–29)
$()^*$	non-dimensional quantity	
$\langle\rangle$	time-averaged quantity	

1 Introduction

Bird, bat, and insect flight has fascinated humans for many centuries. As enthusiastically observed by Dial [1], most species of animals fly. Based on his acute observation of how birds fly, Leonardo da Vinci conceptualized flying machines, which can be seen in documents such as the *Codex on the Flight of Birds*, published circa 1505 [2]; some illustrations of his work are shown in Figure 1.1. Otto Lilienthal was among the most dedicated and successful creators of flying machines at the dawn of human flight. He designed and demonstrated many hang gliders (see Fig. 1.2). Unfortunately, Lilienthal lacked sufficient knowledge of the science of flight and was killed in a fatal fall. For those who wish to explore in greater detail the history and the technology of early flight, John Anderson's *Inventing Flight* [3] offers interesting and well-documented information. Of course, there are ample records of humankind's interest in natural flyers from the artistic angle. Figure 1.3 shows four examples: (Figure 1.3a) decorative art done about 2,500 years ago, in China's Warring Period; (Figure 1.3b) a bronze crane model uncovered from the First Emperor's grave, who died in 210 BC; a pair of bas-reliefs (Figure 1.3c,d) uncovered from the Assyrian palace in today's Iraq, dated back to the 8th century BC; (Figure 1.3e) a stone sculpture of a standing owl from the Shang Dynasty, China, created in the 12th century BC or earlier!

There are nearly a million species of flying insects, and of the non-insects, another 13,000 warm-blooded vertebrate species (including mammals, about 9,000 species of birds, and 1,000 species of bats) take to the skies. In their ability to maneuver a body efficiently through space, birds, bats, and insects represent one of nature's finest locomotion experiments. Although aeronautical technology has advanced rapidly over the past 100 years, nature's flying machines, which have evolved over 150 million years, are still impressive. Considering that humans move at top speeds of 3–4 body lengths per second, a race horse runs approximately 7 body lengths per second, a cheetah accomplishes 18 body lengths per second [4], and a supersonic aircraft such as the SR-71 "Blackbird" traveling near Mach 3 (~900 m/s) covers about 32 body lengths per second, it is remarkable that a Common Pigeon (*Columba livia*) frequently attains speeds of 22.4 m/s, which converts to 75 body lengths per second. A European Starling (*Sturnus vulgaris*) is capable of flying at 120 body lengths per second, and various species of swifts are even faster, flying more than 140 body lengths per second. Whereas the roll rate of highly aerobatic aircraft (e.g., A-4 Skyhawk) is approximately 720°/s, a Barn Swallow (*Hirundo rustics*) has a roll

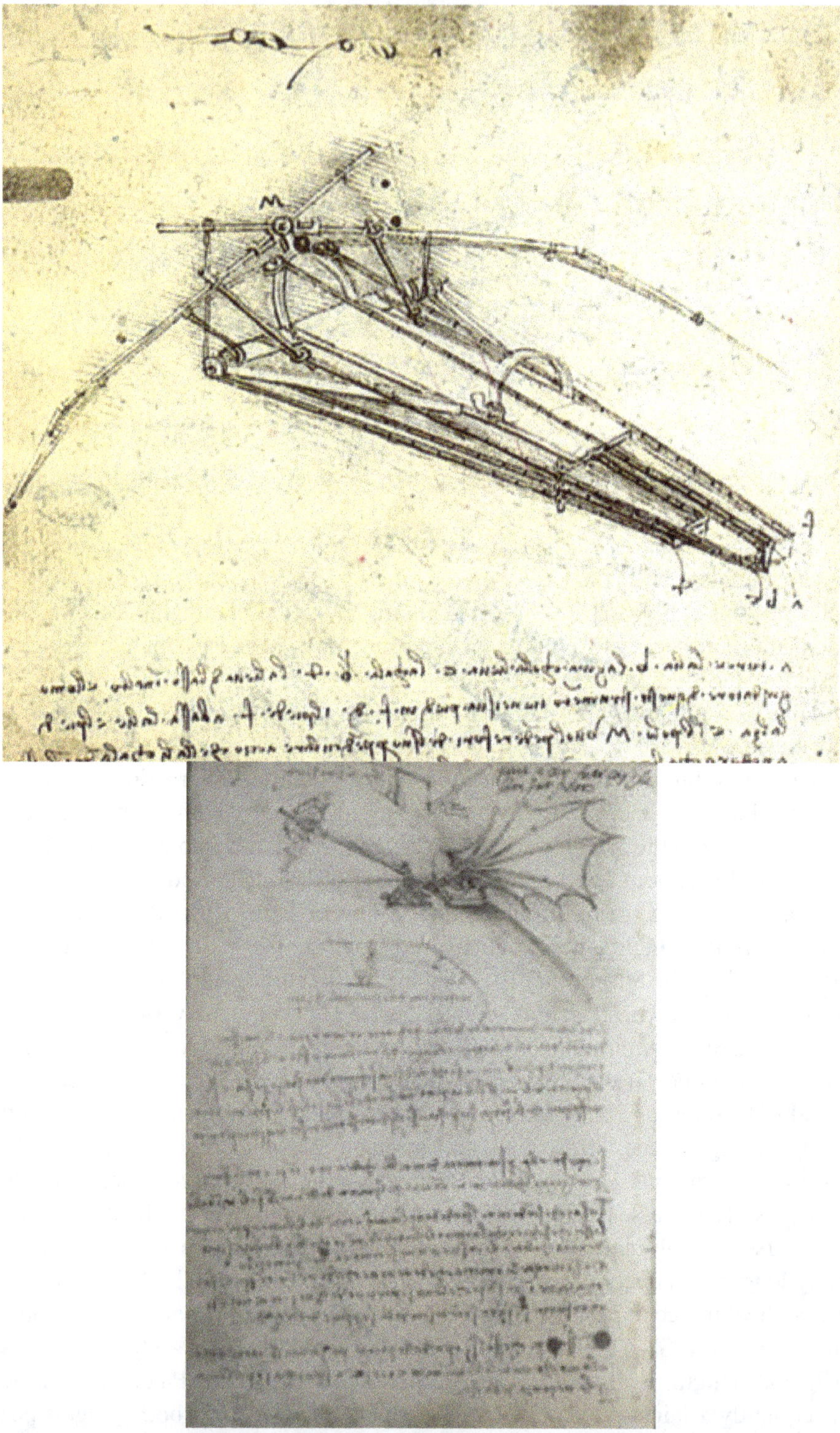

Figure 1.1. A drawing of a design for a flying machine by Leonardo da Vinci (c. 1488). This machine was an ornithopter, with flapping wings similar to a bird, first presented in his *Codex on the Flight of Birds* circa 1505 [2].

Figure 1.2. German engineer Otto Lilienthal flies his hang glider some 2,000 times during 1891–1896 before a fatal fall [3].

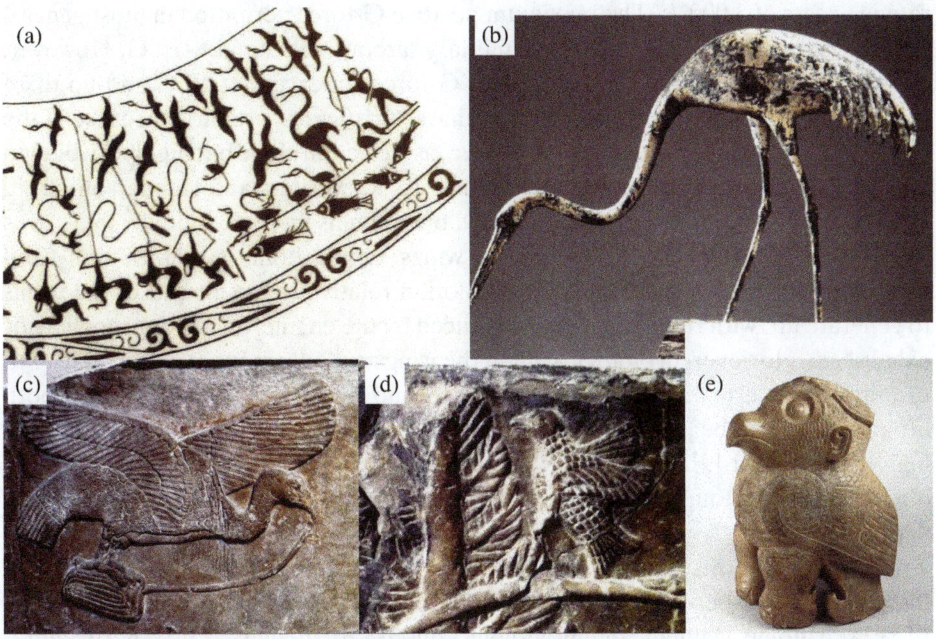

Figure 1.3. Birds recorded in early human history: (a) design of a wine vessel, early Warring period (475-early fourth century BC), China (Shanghai Museum, Shanghai); (b) a bronze crane-eating fish, uncovered inside the First Emperor's grave site, Xian, China (Museum of Emperor QinShihuang, Xian); (c,d) Assyrian bas-reliefs, circa eighth century BC (British Museum, London); (e) Stone sculpture of a standing owl, Shang Dynasty, China, around the 12th century BC or earlier (Academia Sinica, Taipei).

Figure 1.4. According to the simplified static aerodynamics, bumblebees were proclaimed to be unfit to fly.

rate in excess of 5,000°/s. The maximum positive G-force permitted in most general aviation aircraft is 4–5 G, and select military aircraft withstand 8–10 G. However, many birds routinely experience positive G-forces in excess of 10 G and up to 14 G. Such superior maneuvering and flight characteristics are primarily because of the "scaling laws" with respect to a vehicle's size, as well as intuitive but highly developed sensing, navigation, and control capabilities. As McMasters and Henderson put it, humans fly commercially or recreationally, but animals fly professionally [5].

Compared to vehicles with flapping wings, conventional airplanes with fixed wings are relatively simple; the forward motion relative to the air causes the wings to generate lift, with the thrust being produced by the engine via either propellers or exhaust gas. However, in biological flight the wings not only move forward relative to the air but they also flap up and down, plunge, and sweep [1] [4] [6]–[8], so that both lift and thrust can be generated and balanced in accordance with the instantaneous flight task. Aymar [9] and Storer [10] provide early photographs and some general observations. Although in the early days of flight studies, much of the analysis of flapping wing aerodynamics was based on an analogy to the fixed-wing counterpart, it was known that this approach encountered qualitative difficulties, especially when the size of a flyer became smaller, as in small birds, small bats, and insect regimes. In 1934, Antoine Magnan, an entomologist, discussed an analysis by Andre Sainte-Lague, an engineer, that whereas the lift generated by wings can adequately support an aircraft to stay aloft, the same is not true at equivalent speeds of a bee [11]. In other words, an airplane the size of a bee, moving as slowly as a bee, should not be able to fly [12]. Yet, of course, bumblebees, shown in Figure 1.4, can fly.

This example illustrates in simple fashion the implied conclusion – that the theory of fixed-wing aerodynamics cannot explain certain critical aspects of

Figure 1.5. The instantaneous flapping wing patterns can sometimes look reasonable to be analogous to stationary wings under a series of quasi-statically defined conditions.

flapping wing aerodynamics. The aforementioned framework of fixed wings essentially considers flapping wing dynamics as a series of "snapshots" (as illustrated by Fig. 1.5), neglecting the influence of the aerodynamics and wing motion at an earlier moment on the aerodynamics at a later time, based on the so-called quasi-steady approach. In reality, a small flyer can often benefit from manipulating unsteady fluid flows using flapping wing aerodynamics. A rich variety of natural flyers, as depicted in Figure 1.6, can be observed to characterize the instantaneous flapping wing motions. Depending on the real-time flight requirements, these complex motions and wing shapes generate the desirable lift and thrust in different flight environments.

As another example, Franco et al. [13] investigated fluid-structure interactions (FSIs) seen in jellyfish. They reported aperiodic flow despite the relative simplicity of a jellyfish's body shape and motion, as shown in Figure 1.7. Muscle contraction reduces the volume of the subumbrellar cavity (i.e., the region underneath its umbrella-shaped body), resulting in a net downward flux of fluid. The motion of the lower margin of the bell generates vortex rings of opposite rotational sense during the contraction and relaxation phases of the swimming cycle. Franco et al. [13] observed that these vortices act to entrain fluid from above the animal into the subumbrellar cavity, where the feeding and sensory apparatuses of the animal are located. Furthermore, despite the approximate periodicity of the swimming motion, inspection of the flow created by the animal indicates that it is indeed aperiodic in time. Because the animal does not swim at a constant velocity, a periodic flow cannot be constructed by any Galilean transformation of frame. Instantaneous streamlines of the flow field, measured by Franco et al. [13] using digital particle image velocimetry,

Figure 1.6. In reality, the instantaneous flapping wing patterns are very complicated. Depending on the real-time flight requirements, necessary lift and thrust are generated by dynamic mechanisms resulting from unsteady wing movement and shape changes.

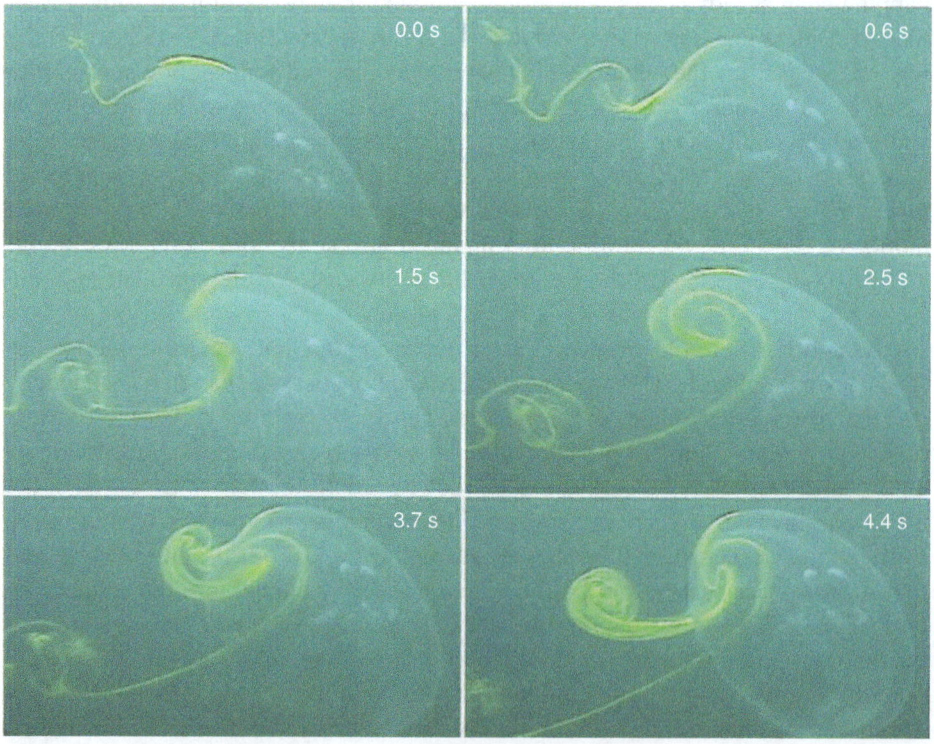

Figure 1.7. Dye visualization of jellyfish vortex wake. Time series shows vortices of clockwise and counterclockwise rotation sense generated during the contraction and relaxation phases of the swimming cycle, respectively. Bell diameter is 10 cm. Images from Franco et al. [13].

Figure 1.8. Instantaneous streamlines of flow around a jellyfish as it swims vertically. Left: End of relaxation phase of swimming cycle. Right: End of contraction phase of swimming cycle. Bell diameter is 10 cm. Images from Franco et al. [13].

indicate local entrainment of fluid from above the animal into the subumbrellar cavity during the entire swimming cycle. Simultaneously, as shown in Figure 1.8, a net downward momentum flux propels the animal forward. This study illustrates that (i) the flow field and force generation processes are highly time dependent and cannot be described based on a simple quasi-steady framework without substantial corrections, and that (ii) the movement and shape deformation of the body influence animal locomotion in major ways.

In addition to generating the aerodynamic forces, flapping wings can also significantly enhance the maneuverability of a flyer. Figure 1.9 illustrates several maneuvering characteristics of biological flyers; these capabilities are difficult to mimic by human-made machines. By combining flapping motion, wing deformation, body contour, and tail adjustment, natural flyers can track targets precisely at amazing speeds. Another issue of interest is the weight of the wings relative to the total animal weight. As summarized in Table 1.1, bat wings, depending on the species, are close to 20 percent of the wing-to-total bat weight. The wings of other natural flyers, as shown in Table 1.2, including large birds such as osprey and vultures, account for 20 percent or more of their body weight. Many butterflies (such as Scarce Swallowtail, Large White) also have relatively heavy wings, around 15 to 25 percent of body weight. The wings of a Small Heath, a small, yellow-orange butterfly that flies close to the ground, are about 5 percent of its body weight. Bees, wasps, flies, and the like all have very light wings, typically less than 1 percent of their body weight.

Heavier wings require more energy to flap; nevertheless their larger inertia enables the flyers to make turns within one or two flapping periods. Those with a higher wing-to-body mass ratio and moment of inertia, such as bats and butterflies, are more maneuverable, capable of making abrupt changes of trajectories within a time comparable to that of a flapping cycle. However, they pay a penalty for this ability because a heavier wing consumes more energy while flapping. Many small

Figure 1.9. Maneuvering capabilities of natural flyers: (a) Canadian Geese's response to wind gust; (b) speed control and target tracking of a seagull; (c) precision touchdown of a finch; (d) a hummingbird defending itself against a bee; (e) the asymmetric movement of wings and tail of a Black Kite while hunting.

flyers such as hummingbirds and insects (with many butterflies as a noticeable exception) tend to have much faster flapping time scales than their bodies' response time scale. For the higher wing-to-body mass ratio group, with the flapping and body response time scales being comparable, the flyer's flight dynamics and control need to be closely linked to the instantaneous aerodynamics, because the time history of the flapping wing aerodynamics directly affects a flyer's performance characteristics. For the lower wing-to-body mass ratio group, whose flapping time scales are significantly shorter than their bodies' response time scale, the lift, drag, and thrust variations during the flapping cycle tend to be smoothed out over the entire flight flapping cycles.

Table 1.1. *Mass of wing and body, and wing dimensions for eight species of bats. With relatively heavier wings, bats can maneuver and make a turn within a stroke or two. On the other hand, they have to work harder to flap*

Species	m_b (kg)	m_w (kg)	m_w/m_b (%)	b (m)	S (m²)	AR (–)	W/S (N/m²)
Egyptian Fruit Bat (*Rousettus aegyptiacus*)	8.34×10^{-2}	2.06×10^{-2}	24.70	5.30×10^{-1}	4.65×10^{-2}	6.4	22
Minor Epauletted Fruit Bat (*Epomophorus anurus*)	4.16×10^{-2}	8.76×10^{-3}	21.03	4.00×10^{-1}	2.90×10^{-2}	5.8	18
Common Pipistrelle (*Pipistrellus pipistrellus*)	4.57×10^{-3}	7.34×10^{-4}	16.08	2.09×10^{-1}	6.50×10^{-3}	6.7	8.0
Common Noctule (*Nyctalus noctula)*	2.35×10^{-2}	3.00×10^{-3}	12.76	3.44×10^{-1}	1.61×10^{-2}	7.4	16
Northem Bat (*Eptesicus nilssonii*)	8.20×10^{-3}	1.70×10^{-3}	20.73	2.77×10^{-1}	1.15×10^{-2}	6.7	8.4
Particoloured Bat (*Vespertilio murinus*)	1.24×10^{-2}	1.72×10^{-3}	13.87	2.98×10^{-1}	1.22×10^{-2}	7.3	11
Brown Long-Eared Bat (*Plecotus auritus*)	7.83×10^{-3}	1.17×10^{-3}	14.94	2.70×10^{-1}	1.23×10^{-2}	5.9	7.2
Large-Eared Free-Tailed Bat (*Otomops martiensseni*)	3.01×10^{-2}	5.48×10^{-3}	18.19	4.49×10^{-1}	2.17×10^{-2}	9.3	16

Notes: m_b is body mass, m_w is wing mass (total), b is wingspan, S is wing area (total), and AR is aspect ratio. Note the body mass is computed by the subtraction of the total wing mass from the total mass.
Source: [47].

However, this feature does not mean that the flapping wing aerodynamics of small flyers can simply be considered as quasi-steady. As is discussed in detail in Chapter 3, the time history of the wing motion is often important to flapping wing aerodynamics. Figure 1.10 shows hummingbirds conducting highly difficult and precise flight control. As illustrated in Figure 1.11, in which several photos highlight the flapping pattern along with a flow field illustration from Warrick et al. [14], hummingbird wing motion exhibits a figure-eight pattern and is highly adaptive to accommodate the challenges posed by wind gust, target tracking, and mitigation of potential interference and threat.

Natural flyers synchronize their wings, body, legs, and tail to perform many tasks. As shown in Figure 1.12, they can take off on water, from land, and off a tree, exhibiting varied and sophisticated patterns. While gliding, as shown in Figure 1.13, they flex their wings to control their speed and direction. On landing, as depicted in Figure 1.14, birds use wing-tail combinations to correct flight trajectory and to adjust for the location of the available landing area. If they need to slow down and adjust the detailed flight trajectory, they fully expand their wings to increase drag and reduce speed; otherwise, they simply fold their wings to reduce lift without slowing down.

Table 1.2. *Mass of wing and body, and wing dimensions for some insects [48] and birds [49]. With lighter wings, which results in time scale differences between the wing movement and the body movement, flyers such as bumblebees need to flap a number of times before being able to make turns*

Flyers/Species	m_b (kg)	m_w (kg)	m_w/m_b (%)	R (m)	S(m^2)	AR	W/S (N/m^2)
Birds							
Griffon Vulture (*Gyps fulvus*)	7.27×10^{0}	1.60×10^{-3}	22.00	6.98×10^{-1}	1.05×10^{0}	1.85	67.59
Osprey (*Pandion haliaetus*)	1.11×10^{0}	3.10×10^{-1}	28.05	4.96×10^{-1}	2.92×10^{-1}	3.37	37.07
Pallid Harrier (*Circus macrourus*)	3.86×10^{-1}	7.51×10^{-2}	19.46	3.57×10^{-1}	1.41×10^{-1}	3.61	26.77
Hummingbrid (*Lampornis clemenciae*)	8.4×10^{-3}	6.00×10^{-4}	7.14	8.5×10^{-2}	3.50×10^{-3}	8.26	25.2
Insects							
Giant Peacock Moth (*Saurnia pyri*)	1.89×10^{-3}	3.00×10^{-4}	15.87	7.00×10^{-2}	1.20×10^{-2}	1.85	1.54
Butterflies							
Scarce Swallowtail (*Papilio podalirius*)	3.00×10^{-4}	8.00×10^{-5}	26.67	3.70×10^{-2}	3.60×10^{-3}	1.52	0.82
Large White (*Pieris brassicae*)	1.27×10^{-4}	2.10×10^{-5}	16.54	3.10×10^{-2}	1.84×10^{-3}	2.09	0.68
Small Heath (*Coenonympha pamphilus*)	4.60×10^{-5}	3.50×10^{-6}	7.61	1.61×10^{-2}	4.80×10^{-4}	2.13	0.94
Moths, Bees, and Other Insects							
Death's-Head Hawkmoth (*Acheronia atropos*)	1.60×10^{-3}	6.70×10^{-5}	4.19	5.10×10^{-2}	2.05×10^{-3}	5.08	7.65
Hummingbird Hawkmoth (*Macroglossum stellatarum L.*)	2.82×10^{-4}	9.18×10^{-6}	3.26	2.13×10^{-2}	3.79×10^{-4}	4.79	7.29
Hawkmoth (*Manduca sexta*)	1.60×10^{-3}	9.00×10^{-5}	5.63	4.85×10^{-2}	1.80×10^{-3}	5.23	9.20
German Wasp (*Vespula germanica F.*)	2.40×10^{-4}	1.39×10^{-6}	0.58	1.62×10^{-2}	1.33×10^{-4}	7.89	17.68
European Hornet (*Vespa crabro L.*)	5.97×10^{-4}	5.68×10^{-6}	0.95	2.43×10^{-2}	3.04×10^{-4}	6.08	19.25
European Hoverfly (*Eristalis tenax*)	1.29×10^{-4}	1.13×10^{-6}	0.88	1.27×10^{-2}	8.26×10^{-5}	7.78	15.31
Honeybee (*Apis mellifica L.*)	9.75×10^{-5}	4.25×10^{-7}	0.44	9.95×10^{-3}	5.98×10^{-5}	6.62	15.98
Red-Tailed Bumblebee (*Bombus lapidaries*)	4.95×10^{-4}	3.10×10^{-6}	0.63	1.65×10^{-2}	1.65×10^{-4}	6.60	29.40
Buff-Tailed Bumblebee (*Bombus terrestris*)	3.88×10^{-4}	2.50×10^{-6}	0.64	1.60×10^{-2}	1.42×10^{-4}	7.21	26.78
Blue Fly (*Caliphora erythorocephala*)	6.10×10^{-5}	5.27×10^{-7}	0.86	1.04×10^{-2}	6.33×10^{-5}	6.83	9.44

Source: Insects [48] and birds [49].

Figure 1.10. Natural flyers can track a target precisely and instantaneously. Shown here are hummingbirds using flapping wings, contoured body, and tail adjustment to conduct flight control.

Since the late 1990s, micro air vehicles (MAVs) have attracted substantial and growing interest in the engineering and science community. An MAV is defined as a vehicle with a maximal dimension of 15 cm or less, which is comparable to the size of small birds or bats, and with a flight speed of 10–20 m/s [15]. Equipped with a video camera or a sensor, MAVs can perform surveillance and reconnaissance, targeting, and biochemical sensing at remote or otherwise hazardous locations. With the rapid progress made in structural and material technologies, miniaturization of power plants, communication, visualization, and control devices, numerous groups have developed successful MAVs. In the process, many MAV concepts, based on the fixed wing, rotary wing, and flapping wing, have been investigated [16]. Figure 1.15a shows an inventive 15-cm MAV design by Ifju et al. [17] that uses a fixed, flexible wing concept. Figure 1.15b shows a rotary wing MAV with 8.5-cm rotary diameter designed by Muren (http://www.proxflyer.com). Figure 1.15c shows a biplane MAV designed by Jones and Platzer [18], which uses a hybrid flapping-fixed-wing design, with the flapping wing generating thrust and the fixed wing producing necessary lift. Figure 1.15d shows an MAV by Kawamura et al. [19] that relies on a flapping wing to generate both lift and thrust and possesses some flight control capabilities. Figure 1.15e shows a 16.5-cm MAV recently developed by AeroVironment (http://www.avinc.com/), which uses a hummingbird-like flapping wing design. It can hover for eight minutes without external power, fly forward with a speed of 4.9 m/s, return to hover, and then withstand a 2 m/s wind gust from the side without drifting more than 1 meter. Figure 1.15f shows a recently developed 10-cm and 3-gram flapping wing MAV (http://www.delfly.nl/?site=diii&menu=home&lang=en). It can carry a small camera and 1-gram battery and fly up to 5 m/s for 3 minutes. As

Figure 1.11. Flapping pattern and flow field of a hummingbird. Agile maneuvering in varied and uncertain environments is critical. The bottom image is from Warrick et al. [14].

Figure 1.12. During takeoff, natural flyers display larger flapping amplitudes, wider and larger wing and tail surfaces, and springing. The level of coordination between wing, tail, and body is remarkable.

Figure 1.13. Birds such as seagulls glide while flexing their wings to adjust their speed as well as to control their direction.

Figure 1.14. While landing, birds use various options. They may fold their wings to reduce the lift if the flight speed is appropriate and there are no environmental disturbances such as wind gust. They also often deliberately stall their fully stretched wings on final approach, maximizing drag to obtain quick deceleration and not being concerned about lift during the final stage of flight.

demonstrated by the U.S. Defense Advanced Research Projects Agency (DARPA), there is substantial interest in developing and using MAVs with robust "perch and stare" capabilities that can take off without external assistance. There are a number of possible candidates, including the aforementioned flapping-wing-based vehicles and rotorcraft such as quad-rotors [20].

Figure 1.16 highlights detailed vehicle characteristics of flexible wing MAVs designed by Ifju and co-workers. The various international MAV competition venues, initiated by the University of Florida and now widespread globally, have substantially encouraged the development of MAVs. The MAVs operate in the low

Figure 1.15. Representative MAVs: (a) flexible fixed wing [17]; (b) small-sized rotary wing (http://www.proxflyer.com); (c) hybrid flapping-fixed wing, with the fixed wing used for lift and the flapping wing for thrust [18]; (d) flapping wing for both lift and thrust [19]; (e) hummingbird-inspired robotic Nano Air Vehicle (http://www.avinc.com/); and (f) a 5-cm wingspan flapping wing MAV (http://www.delfly.nl/?site=diii&menu=home&lang=en).

Reynolds number (characterizing the relative importance between inertia and viscous effects of the fluid) and, compared to large manned flight vehicles, experience unfavorable aerodynamic characteristics, such as low lift-to-drag ratio [21]. However, the MAVs' small geometric dimensions result in favorable scaling characteristics, such as reduced stall speed and better structural survivability.

General references offering broad accounts of biological flight, including geometric scaling laws, power, and morphology, as well as simplified modeling, can be found in Alexander [22] [23], Azuma [24], Biewener [25], Brodsky [26], Dudley [27], Grodnitsky [28], Norberg [4], Tennekes [29], Videler et al. [30], Vogel [31],

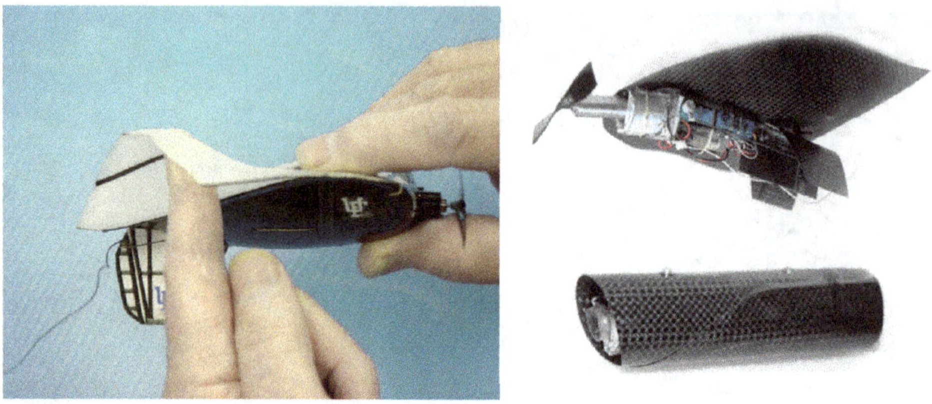

(a) (b)

Figure 1.16. The flexible-wing MAVs (a) can benefit from passive shape adaptation in accordance with instantaneous aerodynamic loading and (b) can be packed very easily based on need. (Courtesy Peter Ifju).

Ward-Smith [32], Dalton [33], Pringle [34], Henderson [35], and Floreano et al. [36]. The symposia volumes edited by Wu et al. [37], Pedley [38], and Maddock et al. [39] offer multiple aspects of biological flight as well as swimming. Lighthill [40], Wu [41], Childress [42], and Maxworthy [43] discuss swimming and flying primarily from analytical viewpoints with simplified problem definitions. Finally, the standard texts by Anderson [44], Katz and Plotkin [45], and Shevell [46] present basic knowledge of human-made aerodynamics relevant to both biological and human-made MAVs.

In the following chapters, we offer an introductory and simplified, concept-based framework on the lift- and thrust-generation mechanisms of flapping wing vehicles. Specifically, we present the scaling laws related to the mechanics and energetics of avian flight. Then, we use simple mechanics to describe gliding, forward, and hovering flight, as well as drag and power related to avian flight. The different power components are first presented separately and then summed together, giving the total power required for hovering and forward flight. The results of these different power calculations can be summarized in the form of power curves. The analyses are based on the quasi-steady approach, neglecting the time history of the wing motion. This approach, although incomplete in certain important aspects (such as coupled and multi-scale mechanics between wing and fluid and between wing and body and tail, shape deformation, and the time history of force generation), can offer useful insights into some of the important and basic issues related to flapping wing flight. More detailed and critical discussions accounting for unsteady fluid physics and fluid-structure interactions are presented in later chapters.

1.1 Flapping Flight in Nature

While flapping, birds systematically twist their wings to produce aerodynamic effects in ways similar to that of the ailerons on the wings of conventional airplanes. Specifically, one wing is twisted downward (pronated) to reduce the angle of attack (AoA)

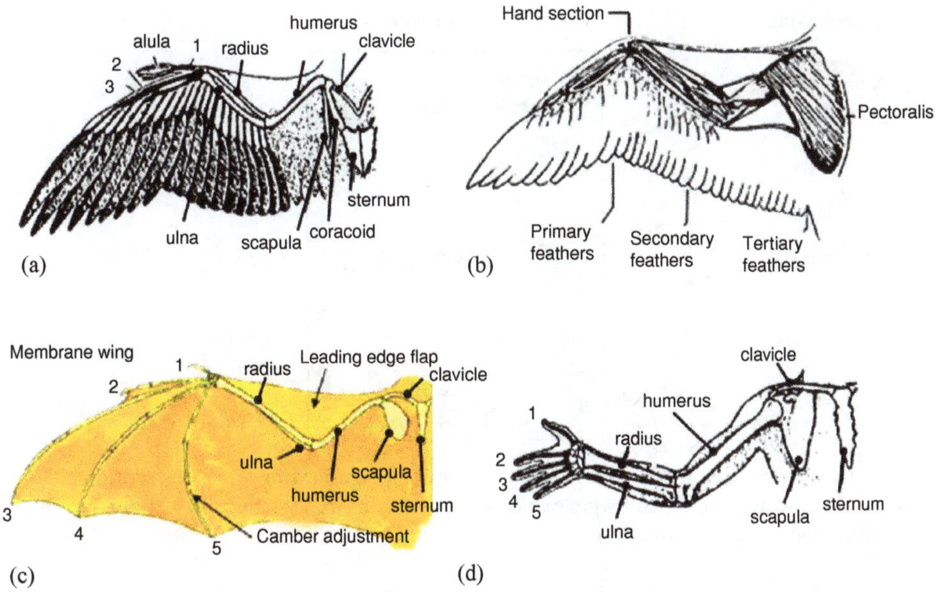

Figure 1.17. Schematics of (a,b) a bird wing, (c) a bat wing, and (d) a human arm. For birds, the upper arm, the "humerus," is proportionately shorter, and the "wrist" and "palm" bones are fused together for greater strength in supporting the primary flight feathers. For bats, the bone-membrane combination creates a leading-edge flap and allows passive camber adaptation in the membrane area. (a), (b), and (d) are modified from Dhawan [50]; (c) is from Anders [51].

and corresponding lift, while the other wing is twisted upward (supinated) to increase lift. With different degrees of twisting between wings, a bird is able to roll [1]. For a bird to be able to deform and twist its wings, an adaptation in the skeletal and muscular systems is required. The key features that seem desirable are modification of camber and flexing of the wing planform between upstroke and downstroke, twisting, area expansion and contraction, and transverse bending. To perform these functions, birds have a bone structure in their wings similar to the one in a human arm or a bat wing, as shown in Figure 1.17. Despite these similarities birds have more demanding muscle and bone movement in their wings during flight than a human arm is capable of making. Figure 1.18 compares the cross-sectional shapes of a pigeon wing and a conventional transport airplane wing. To a noticeable extent, the pigeon wing exhibits a greater number of variations in camber and thickness along the span.

1.2 Scaling

When studying natural flyers, it is insightful to assess the effects of different parameters, such as wing area and wing span, on the flight characteristics based on the dimensional analysis [4] [29] [52]–[54]; this analysis presents very interesting correlations that can be used to summarize the various scaling laws ranging from birds and insects to aircraft. Tennekes [29] considered the relations between cruising speed, weight, and wing loading and established *The Great Flight Diagram*. The diagram is shown in Figure 1.19.

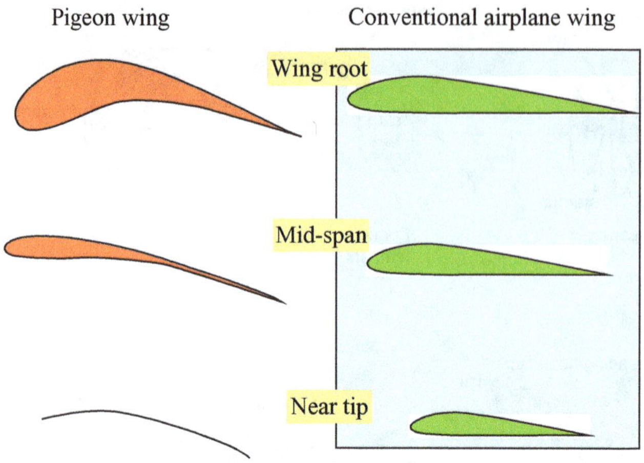

Figure 1.18. Comparison of cross-sectional shapes of a pigeon wing and a conventional transport airplane wing. The pigeon wing exhibits noticeably more variations in camber and thickness along the spanwise direction.

With technical advancement, the MAV dimensions, wing loading, and speed will continue to decrease, moving toward the lower left corner in Figure 1.19. From Figure 1.19 one can compare and correlate relations between species with a pronounced difference in size. For example, the small fruit fly (*Drosophila melanogaster*) can be compared to the Boeing 747, which weighs about 500 billion times more. By using scaling analysis, one can predict how a parameter, such as wingspan, varies with another parameter, such as the body mass, for natural flyers in general or specific animal groups.

As an illustration, consider the balance between lift (L) and weight (W) during steady-state flight,

$$L = W = \frac{1}{2}\rho U^2 S C_L. \tag{1–1}$$

From Eq. (1–1) it is possible to get an understanding of how wing area (S), air speed (U), density (ρ), and wing loading (W/S) are connected. *Wing area* (S): The wing area for a flight vehicle is often defined as the area projected when the wing is seen from above, and usually it includes the contribution from the "wing area" inside the fuselage. *Air speed* (U): The air speed is defined as the forward velocity for the flight vehicle. Given a particular AoA, a twofold increase in speed will result in a fourfold increase in lift. *Density* (ρ): For cases of interest to bird and insect flight, the density of the air is basically unchanged, because birds fly within a narrow altitude near sea level. In general, a decrease of density due to an increase in altitude will decrease the lift.

From Eq. (1–1), it is clear that the cruising speed depends on the wing loading (W/S):

$$\frac{W}{S} = \frac{1}{2}\rho U^2 C_L. \tag{1–2}$$

Equation (1–2) shows that the greater a flyer's wing loading, the faster it has to fly. Some of the relations between body mass and parameters connected to birds

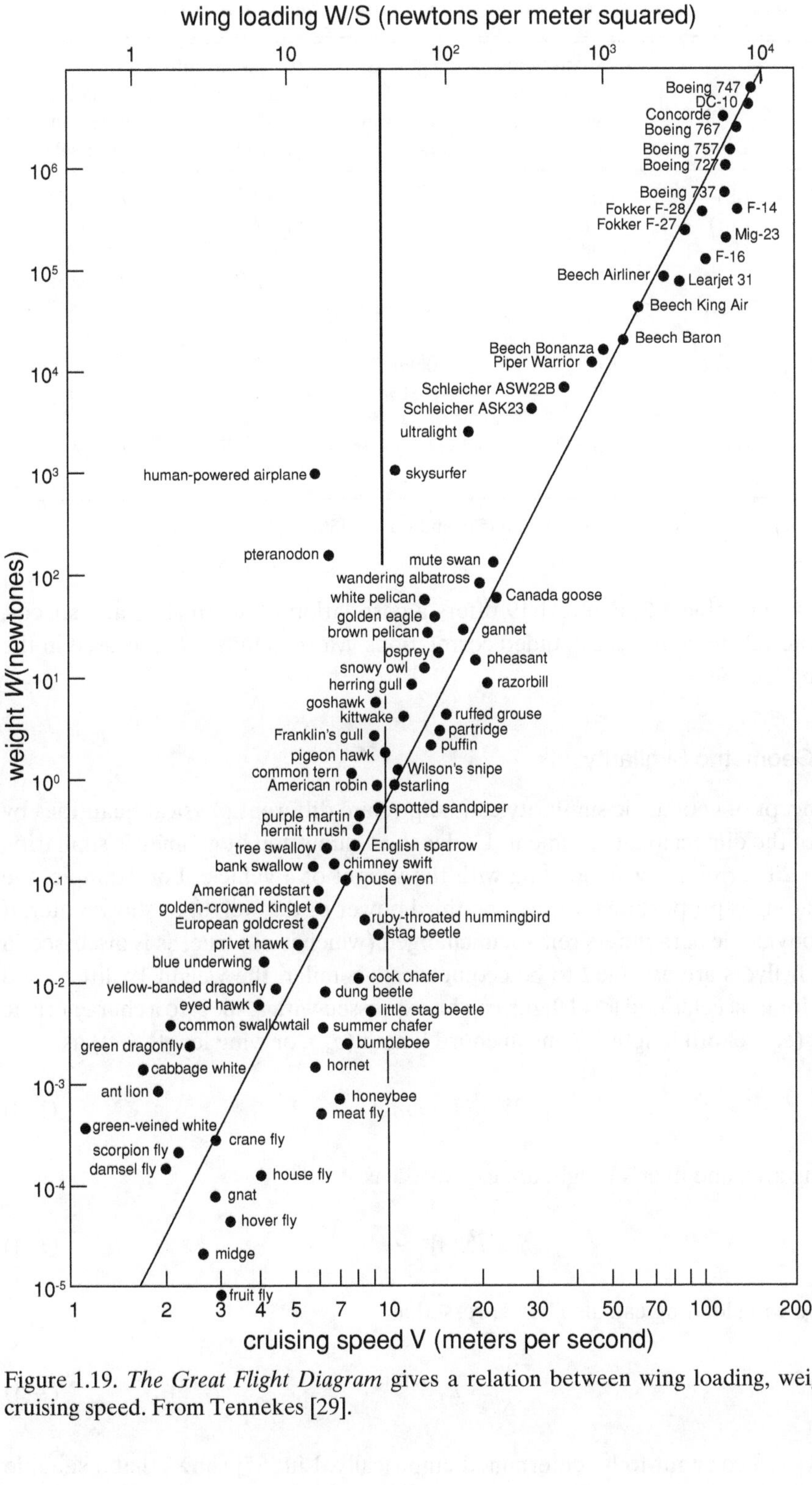

Figure 1.19. *The Great Flight Diagram* gives a relation between wing loading, weight, and cruising speed. From Tennekes [29].

Table 1.3. *Power functions of wing dimensions and flight parameters against body mass m*

Animal groups	Dimensions	Correlation (based on geometrical similarities)	All birds (based on empirical data)	All birds except Hummingbirds (based on empirical data)	Hummingbirds (based on empirical data)
Wingspan	(m)	$m^{0.33}$	–	$1.17m^{0.39}$	$2.24m^{0.53}$
Wing area	(m^2)	$m^{0.67}$	–	$0.16m^{0.72}$	$0.69m^{1.04}$
Wing loading	(N/m^2)	$m^{0.33}$	–	$62.2m^{0.28}$	$17.3m^{-0.04}$
Aspect ratio	–	0	–	$8.56m^{0.06}$	$7.28m^{0.02}$
Minimum power speed U_{mp}	(m/s)	$m^{0.17}$	$5.70m^{0.16}$	–	–
Minimum power P_{mp}	(W)	$m^{0.17}$	$10.9m^{0.19}$	–	–
Minimum cost of transport C_{min}	–	0	$0.21m^{-0.07}$	–	–
Wing-beat frequency f_w	(Hz)	$m^{-0.33}$	$3.87m^{-0.33}$	$3.98m^{-0.27}$	$1.32m^{0.60}$

Source: Originally compiled by Norberg[4], Greenewalt [55], and Rayner [56].

are shown in Table 1.3. Figure 1.19 offers a correlation between sizes and speeds, and Table 1.3 summarizes expanded correlations. More details are discussed in the following sections.

1.2.1 Geometric Similarity

The concept of geometric similarity can help relate different physical quantities by means of the dimensional argument. Under the assumption of geometric similarity, Figure 1.20 correlates wing loading with the weight of a vehicle. For example, the wing loading is proportional to the one-third power of the weight as shown later, if the aerodynamic parameters remain unchanged (which is not true, as is discussed in detail). If flyers are assumed to be geometrically similar, the weight W, lift L, and mass m for unaccelerated level flight, can be expressed with respect to a characteristic length l (e.g., chord length (c), mean chord length (c_{m}), or wing length (R)) as

$$W = L = mg. \tag{1–3}$$

The wing area and flyer's weight are expressed as

$$S \sim l^2, \ W \sim l^3. \tag{1–4}$$

Then the wing loading can also be expressed as

$$\frac{W}{S} = k_1 W^{1/3}, \tag{1–5}$$

where k_1 is a constant to be determined empirically. Liu [57] shows that a suitable value of k_1 is 53 and 30.6 for aircraft and birds, respectively. The correlation is shown in Figure 1.20.

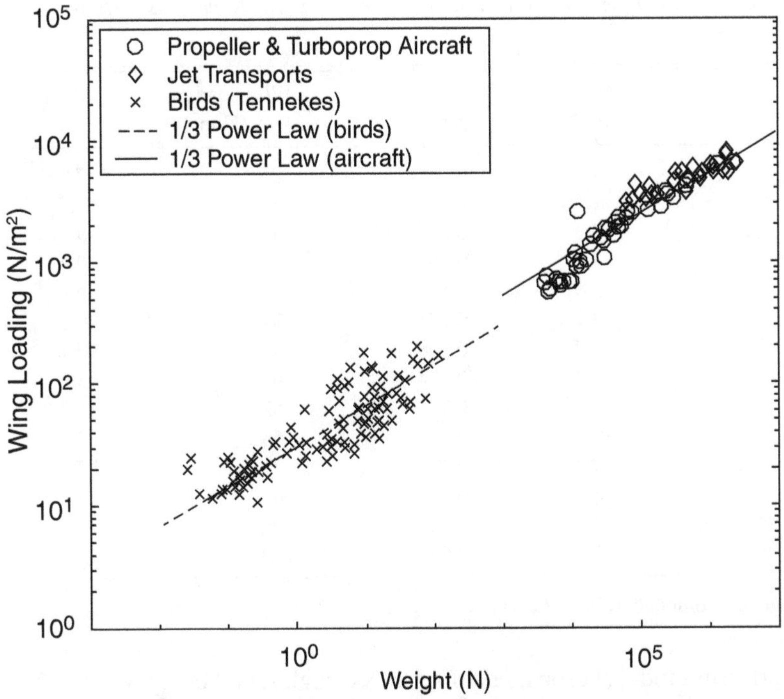

Figure 1.20. The relationship between weight and wing loading. From Liu [57].

1.2.2 Wingspan

When studying flapping animals, parameters of interest are often related to the body mass m of the animal. Using the dimensional argument method and assuming geometric similarity for the animals considered, one can determine a relation between the wingspan and mass. Collecting data from Tennekes [29] for birds ranging from a 0.026-N black-chinned hummingbird to a 116-N mute swan, and data for propeller/turboprop aircraft and jet transports published by Jackson [58] covering a broad spectrum of aircraft from a 1500-N ultralight to a 1800-kN Boeing 747–400, Liu [57] suggests that, over a large range of the weight, birds and aircraft basically follow the power law:

$$l = 1.654m^{1/3} \text{ (aircraft)};\ l = 1.704m^{1/3} \text{ (birds)}. \tag{1–6}$$

1.2.3 Wing Area

Norberg [4] reports that there are greater variations between groups of animals in wing area than in wingspan. The departure from the geometrical relation is obvious, which is shown in Table 1.3. As for wingspan, hummingbirds have the largest deviation from the geometrical relation. They seem to have a larger wing area for a given body mass than do birds in general. Based on the variation of the wing area for different groups of birds, Greenewalt [55] subdivides birds into different classes or "models." His model offers the following correlations:

Table 1.4. *Weight, wing area, wing loading, and airspeeds for various seabirds assumed to be geometrically similar*

Seabird	Weight W (N)	Wing area S (m^2)	Wing loading W/S (N/m^2)	Air speed (m/s)
Common Tern	1.15	0.05	23	7.8
Dove Prion	1.7	0.046	37	9.9
Black-Headed Gull	2.3	0.075	31	9
Black Skimmer	3	0.088	34	9.4
Common Gull	3.67	0.115	32	9.2
Kittiwake	3.9	0.101	39	10.1
Royal Tern	4.7	0.108	44	10.7
Fulmar	8.2	0.124	66	13.2
Herring Gull	9.4	0.181	52	11.7
Great Skua	13.5	0.214	63	12.9
Great Black-Backed Gull	19.2	0.272	71	13.6
Sooty Albatross	28	0.34	82	14.7
Black-Browed Albatross	38	0.36	106	16.7
Wandering Albatross	87	0.62	140	19.2

Source: Data originally compiled by Tennekes [29].

1. The Passeriform model (herons, falcons, hawks, eagles, and owls): $s \sim m^{0.78}$
2. The Shorebird model (doves, parrots, geese, swans, and albatross): $s \sim m^{0.71}$.
3. The Duck model (grebes, loons, and coots):$s \sim m^{0.78}$.

These relations are consistent with those presented in Table 1.3 for all birds other than hummingbirds.

1.2.4 Wing Loading

Regarding wing loading, although the overall correlation shown in Eq. (1–5) seems reasonable, Greenewalt [55] found that, in many cases, the relation between wing loading and mass increases more slowly than that indicated in Eq. (1–5). For example, the three families of birds (i.e., the Passeriforms, the Shorebirds, and the Ducks) do not follow the one-third law. As indicated in Table 1.3, for hummingbirds, wing loading is almost independent of body mass; hence different species can have the same wing loading. Tennekes [29] used the data collected by Greenewalt [55] and summarized the various scaling relations for seabirds, shown in Table 1.4. All gulls and their relatives have long, slender wings and streamlined bodies, so it was reasonable to assume geometric similarity. From Table 1.4 it is obvious that the wing loading and cruising speed generally increase with weight.

1.2.5 Aspect Ratio

As for aircraft, the aspect ratio (AR) can give an indication of the flight characteristics for flapping animals. The AR is a relation between the wingspan b and the wing area S:

$$AR = \frac{b^2}{S}. \tag{1–7}$$

In general, for both human-made and natural flyers, the agility and maneuverability improve with a smaller AR. High AR wing flyers, such as an albatross and the U-2, are excellent flyers for steady-state forward flight, but are not suitable for fast course changes.

Another consideration is that the induced drag, which is caused by the lift, tends to decrease with higher AR. Obviously, the minimum induced drag is obtained with an infinitely long wing. Similarly, for steady forward flight, with a large AR, the lift-to-drag ratio (L/D) or the so-called glide ratio increases with an increasing AR. The largest AR for birds is found among species that typically spend a substantial portion of their time in soaring instead of flapping flight. A typical example is the wandering albatross (*Diomeda exulans*), which has an aspect ratio of about 15. As noted by Tennekes [29], the glide ratio for the wandering albatross is around 20. Human engineering can now offer a sailplane with a glide ratio of around 60.

1.2.6 Wing-Beat Frequency

The main function of wing bones is to transmit force to the external environment during flight. This force cannot, however, be too high because of the risk for bone or muscle failure. These limitations, along with the amount of power available from flight muscles, settle the upper and lower limits of wing-beat frequency for flapping animals [59]–[61]. Based on the insight into the wing-beat or flapping frequency, it is possible to estimate the power output from a bird's flight muscles and achieve an estimation of the power required for flying. According to Pennycuick [62], this means we can estimate the maximum wing-beat frequency $f_{\mathrm{w,max}}$ for geometrically similar animals, as shown in the following discussion. Because the force F_{m} exerted by a muscle is assumed to be proportional to the cross-sectional area of its attachment, we get

$$F_{\mathrm{m}} \,\alpha\, S \sim l^2. \tag{1–8}$$

Pennycuick [62] assumes that the stresses in muscles and bones are constant and that the torque acting about the center of rotation of the proximal end of the limb can be expressed as

$$J_{\mathrm{T}} = F_{\mathrm{m}} l. \tag{1–9}$$

The moment of inertia of the limb can now be determined as follows:

$$I = m_{\mathrm{limb}} \left(\frac{l}{2}\right)^2 \sim l^5. \tag{1–10}$$

The mass of the limb is denoted by m_{limb} and it is assumed that the limb has a uniform density. The muscle in action has an angular acceleration, which can be determined as

$$\dot{\omega} = \frac{J_{\mathrm{T}}}{I} \sim \frac{l^3}{l^5} \sim l^{-2}. \tag{1–11}$$

From Eq. (1–11) it is easy to determine the stroke time scale T, and with the frequency $f \sim T^{-1}$, we get

$$T = \dot{\omega}^{-1/2} \Rightarrow f \propto \dot{\omega}^{1/2}. \tag{1–12}$$

A relation between the body mass m and the maximum wing-beat frequency $f_{w,\max}$ can also be derived:

$$f_{\mathrm{w,max}} \sim T^{-1} \sim l^{-1} \sim m^{-1/3}. \tag{1–13}$$

With the assumption of geometric similarity, this is the upper limit of the flapping frequency. For the lower flapping-frequency limit, which is the case for most birds in slow forward flight or hovering, the induced velocity w_i, – the airflow speed in the wake right beneath the animal – dominates. Still, the weight W of the flyer must be balanced by the lift L so that, referring to Eq. (1–1), we obtain the following relation for the induced velocity w_i:

$$W = L = \frac{1}{2}\rho w_i^2 S C_L \Rightarrow w_i = \sqrt{\frac{2mg}{\rho S C_L}}. \tag{1–14}$$

The angular velocity of the wings can be dimensionally expressed as

$$\omega \sim w_i/l. \tag{1–15}$$

With Eqs. (1–3), (1–4), and (1–15), we obtain the final expression for the lower flapping limit as

$$\begin{aligned} f_{\mathrm{w,min}} &= \frac{\omega_{\min}}{2\pi} \sim \frac{w_i}{l} \\ &= \frac{1}{l}\sqrt{\frac{2mg}{\rho S C_L}} \sim \sqrt{\frac{2mg}{\rho C_L l^4}} \sim \left(\frac{l^3}{l^4}\right)^{1/2} \sim l^{-1/2} \sim m^{-1/6}. \end{aligned} \tag{1–16}$$

Because of these two physical limits, animal flight has an upper and lower bound for the flapping frequency.

1.3 Simple Mechanics of Gliding, Forward, and Hovering Flight

1.3.1 Gliding and Soaring

Flying animals usually flap their wings to generate both lift and thrust. But if they stop flapping and keep their wings stretched out, their wings will actively produce lift, not thrust. When this happens, we call them gliders. Thrust can then be produced by gravitational force while an animal descends. In addition to bats and larger birds, gliders can also be found among fish, amphibians, reptiles, and mammals.

To maintain level flight, a flying animal must produce both lift and thrust to balance the gravity force, in the vertical direction, and the drag, in the horizontal direction, respectively. Because gliding occurs with no active thrust production, an animal always resorts to the gravity force to overcome the drag. In gliding, the animal tilts its direction of motion slightly downward relative to the air that it moves through. When the animal tilts downward, the resulting angle between the motion direction and the air becomes the gliding angle. The gliding angle directly controls the lift-to-drag ratio. The higher this ratio, the shallower the glide becomes. Recall from a basic fluid dynamics class that the lift-to-drag ratio increases with the Reynolds number, a parameter proportional to animal size and flight speed. Large flying animals fly at high Reynolds numbers and have a large lift-to-drag ratio. For example, a Wandering

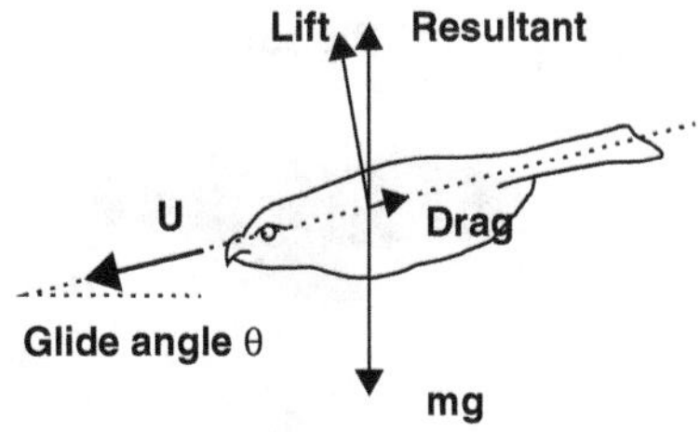

Figure 1.21. Force balance of gliding [63].

Albatross, with a wing span more than 3 m, has a reported lift-to-drag ratio of 19, whereas the fruit fly, which has a span of 6 mm, has a ratio of 1.8 [22]. If the animal has a low lift-to-drag ratio, it must glide (if it can) with a considerable increase in glide angle. For example, a lizard (the genus *Draco*) from Southeast Asia has a lift-to-drag ratio of 1.7 and glides at an angle of 30°, whereas a North American flying squirrel glides at the angle of about 18° to 26° with a lift-to-drag ratio of 2 or 3 [22].

Whereas gliding animals take a downward tilt to acquire gravity-powered flight, many birds can ascend without flapping their wings. This is called soaring. Instead of using gravity, soaring uses energy in the atmosphere, such as rising air currents [22].

Consider a gliding flyer at an angle θ relative to the free-stream (see Fig. 1.21). The lift and drag are perpendicular and parallel to the flyer's inclination axis, while the gravitational force is invariant for our purpose. It is known that good bird flyers often attain a maximum glide ratio of 10 to 15 (θ about 5°), whereas Wandering Albatrosses record 23 ($\theta = 2.5°$). Using the planform area S and the lift coefficient C_L,

$$L = \frac{1}{2}\rho S U^2 C_L = mg\cos\theta \approx mg\,(\text{as}\,\theta\,\text{is small}) \tag{1–17}$$

By measuring m, S, and U, the lift coefficient is found to be typically < 1. Maximum C_L (about 1.5) occurs at minimum glide speed, when the bird starts to flap its wings just before they stall. One can also estimate the glide speeds as a function of flyer sizes:

$$U \approx \sqrt{\frac{2W}{\rho S C_L}}. \tag{1–18}$$

Table 1.5 summarizes gliding characteristics of selected natural flyers, showing that the glide speed is indeed well correlated with the square root of the wing loading,

Table 1.5. *Glide characteristics of selected flyers, showing that the glide speed is generally proportional to (wing loading)*$^{1/2}$

	Wing loading (N/m^2)	Glide speed (m/s)
Mute Swan	170	21
Wandering Albatross	140	19
Ultralight	133	18
Graylag Goose	115	17
Common Gull	32	9

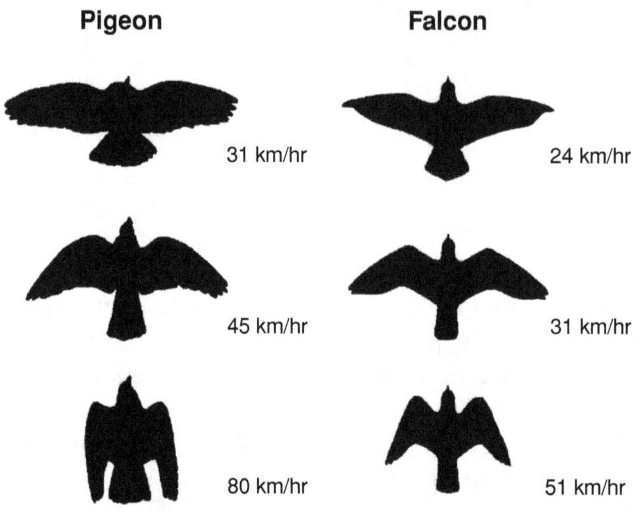

Figure 1.22. Variable span gliders: wing morphing to accommodate gliding speed [29].

W/S. Furthermore, statistically, as shown in Table 1.5, for most birds, except hummingbirds, the planform area is $S \propto m^{0.72}$ and the wing loading is $W/S \propto m^{0.28}$, which result in the glide speed: $U \propto m^{0.14}$.

Because lift (L) is proportional to flight speed squared, birds control their speed by varying S and hence the wing loading (W/S). This enables the bird to adjust the speed range so that the wings operate at an optimal AoA at all speeds. To complement the trends highlighted in Figure 1.13, Figure 1.22 illustrates such trends for a falcon and a pigeon under varied flight conditions.

1.3.2 Powered Flight: Flapping

An alternative method to gliding and soaring that is used by many biological flyers to produce lift and thrust is flapping wing flight. The similarities between the aerodynamics of a flapping wing and that of a rotary wing, although limited, illustrate a few key ideas. For example, the rotors of a helicopter rotate about the central shaft continuously, while the relative flow around the rotors produces lift. Likewise, a flapping wing rotates, swings in an arc around its shoulder joint, and reverses direction every half-stroke. Helicopters and biological flyers also use similar techniques to accelerate from hovering to forward flight. Helicopters tilt the rotational plane of rotors from horizontal to forward. The steeper the tilt of the rotor, the faster the helicopter accelerates. Biological flyers also tilt their flapping stroke plane: down and forward on the downstroke, and up and backward on the upstroke. To fly faster, biological flyers make the stroke more vertical by increasing the up-and-down amplitude of the movements. When biological flyers decrease their speed, they tend to flap their wings more horizontally, similar to the way helicopters change the angle of the rotors.

Birds, bats, and insects apply a variety of different flapping patterns in hovering and in forward flight to generate lift and thrust. Larger birds have relatively simple wingtip paths. For example, an oval tip path is often associated with albatrosses (see Fig. 1.23). Smaller flyers exhibit more complicated flapping patterns. Figure 1.23 illustrates the highly curved tip paths of a locust and a fruit fly, the figure-eight

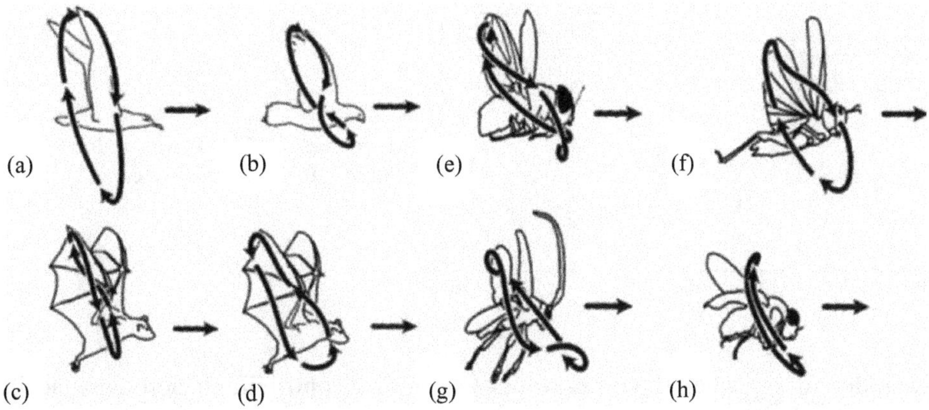

Figure 1.23. Wingtip paths relative to the body for a variety of flyers, as indicated by arrows. (a) albatross, fast gait; (b) pigeon, slow gait; (c) horseshoe bat, fast flight; (d) horseshoe bat, slow gait; (e) blow fly; (f) locust; (g) June Beetle; (h) fruit fly. From Alexander [22].

pattern of a pigeon (see Fig. 1.23b), and the more complicated paths of June Beetles and blowflies.

1.3.2.1 Forward Flight

When analyzing a natural flyer's aerodynamic performance, an important parameter is the ratio between the forward velocity and the flapping velocity, which is expressed in terms of the reduced frequency k,

$$k = \frac{2\pi fc}{2U_{\mathrm{ref}}}, \tag{1–19}$$

where f, c, and U_{ref} are, respectively, the flapping frequency, the mean chord length, and the reference velocity – in this case the flyer's forward flight velocity. Unsteady effects increase as flapping frequency decreases, and therefore, depending on the forward velocity, different techniques have been devised to calculate the forces acting on a specific species.

In slow forward flight, both reduced frequency and wing-beat amplitude tend to be high, resulting in highly unsteady flow structures. In accordance with the Lifting Line Theory [64], the lift on a wing is related to the strength of the bound vortex. The trailing vortices (the tip vortices) are of the same circulation magnitude as the bound vortex. At the beginning or end of the downstroke, when the flapping velocity changes direction, a transverse vortex (starting or stopping vortex) is produced at the trailing edge, and according to Kelvin's circulation theorem, these two transverse vortices connect the two tip vortices and result in the shedding of a vortex ring. Some flyers (for example, doves) make use of the clap-and-fling mechanism to generate the starting vortex and by this means reduce the delay in building up maximum lift during the first part of the downstroke. This mechanism is highlighted in Chapter 3.

In fast forward flight, the reduced frequency and the wing-beat amplitude tend to be low, and the wake consists of a pair of continuous undulating vortex tubes – or line vortices – approximately behind the wingtips. In such cases, it is not unusual for the outer part of the wing to be folded so that it aligns with the free-stream direction

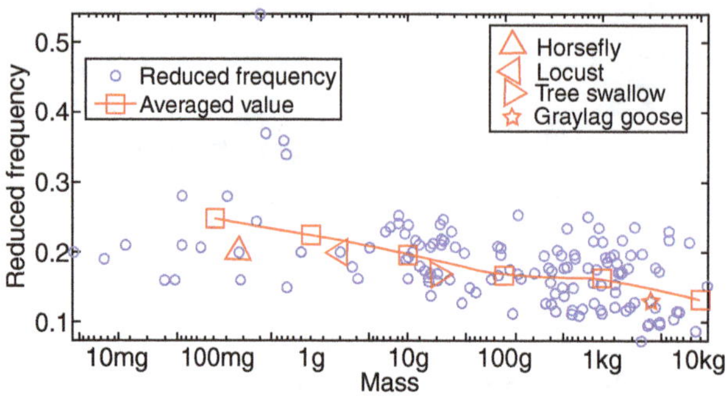

Figure 1.24. Mass versus reduced frequency for birds and insects.

(to reduce drag); when this occurs only the arm wing contributes to aerodynamic lift during the upstroke.

To evaluate flying animals' lift and thrust, either unsteady or quasi-steady methods can be used depending on the magnitude of the reduced frequency. Early work by Ellington [65] showed that quasi-steady analysis substantially under-predicts the aerodynamic force needed to sustain the insect weight. As discussed in Chapters 3 and 4, much recent flapping wing research has focused on the understanding of unsteady aerodynamic mechanisms resulting from the wing movement. Figure 1.24 shows the correlation between a flyer's mass and the reduced frequency. The data are based on those reported by Azuma [24] and Pennycuick [59], aided by the cruising-velocity estimate documented by Tennekes [29]. Overall, the reduced frequency decreases as the size and mass grow, indicating that small flyers are more unsteady in their flight than large flyers. Although this figure does not explain how they use this unsteadiness, it does indicate that unsteadiness plays a critical role in small flyers' movement.

To quantify the lift and thrust generated by the flapping motion requires more sophisticated tools. However, we can understand the role of the unsteady effects by examining the relationship between the forward velocity and the flapping velocity in terms of the reduced frequency. It is also noted that different sections of the wing function differently in force generation. This concept can be better understood by introducing the relative flow velocity (U_r), defined as

$$U_r = U + U_f + w_i, \tag{1–20}$$

where U is the forward velocity of the bird, U_f the flapping velocity, and w_i the downwash (induced) velocity. The relative velocity determines the aerodynamic forces on the wing. For fast forward flight, the downwash velocity is small and can be largely neglected. As the wingspan increases, U_f increases and changes direction; consequently, the magnitude and direction on U_r change as well. Since U_r determines the resulting aerodynamic force F acting on each wing section along the span, F also changes in magnitude and direction. The changes can be seen in Figure 1.25.

It is commonly held that during the downstroke the inner part of the wing produces lift and drag, while the outer part produces lift and thrust. The net aerodynamic force produced by the wings during the downstroke is directed upward and forward, providing both lift and thrust. To obtain this favorable force distribution, the wings have to be twisted. By twisting the wings an optimal relative velocity can be

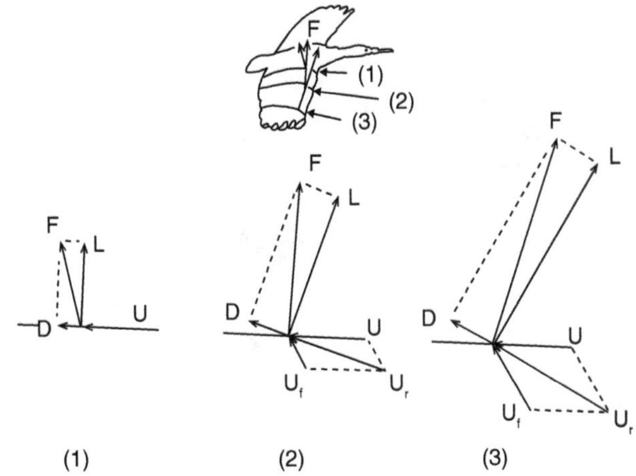

Figure 1.25. Velocity-vector diagram at different wingspan locations for fast forward flight. Here, the lift and drag are defined based on the effective velocity combining forward and local flapping velocities. For the entire vehicle, the lift is defined to be normal to the forward velocity (U) (i.e., in the vertical direction), and drag or thrust in the horizontal direction. According to the resulting force vector F illustrated here, drag of the vehicle is generated by the inner wing, and thrust of the vehicle is generated by the outer wing.

obtained at each wing section throughout the wing stroke. Since the relative velocity determines the direction of the resultant aerodynamic force, this force is directed backward at the wing root and gradually turned forward when moving along the wingspan. At the wingtip region, the resultant aerodynamic force points toward the forward direction, giving both lift and thrust.

Flying animals employ different mechanisms for various missions such as takeoff, landing, or gliding. Even during forward flight, they change their wing and body movements while flying through a range of speeds. Tobalske and Dial [8] analyzed high-speed (60 Hz) videotapes of black-billed magpies (*Pica pica*) flying at speeds of 4–14 m/s and of pigeons (*Columbia livia*) flying at 6–20 m/s in a wind tunnel. Pigeons have higher wing loading and higher aspect ratio wings compared with magpies. Both species alternate phases of steady-speed flight with phases of acceleration and deceleration, particularly at intermediate flight speeds. The birds modulate their wing-beat kinematics among these phases and frequently exhibit non-flapping phases while decelerating. During steady-speed flight, wing-beat frequency does not change appreciably with horizontal flight speed. Instead, with increasing flight speed the body angle relative to the horizontal decreases, thereby illustrating that the dominant function of wing flapping changes from weight support at slow speeds to positive thrust at fast speeds. Pigeons progressively flex their wings during glides as flight speed increases but never perform bounding. For magpies, the wingspan during glides does not vary with flight speed, but the percentage of bounding among non-flapping intervals increases with speed from 10–14 m/s. The use of non-flapping wing postures seems to be related to the gaits used during flapping and to the aspect ratio of the wings.

In general, the outer wing is mainly responsible for generating weight support and thrust, largely on the downstroke (see Fig. 1.26). For most birds, force generation is usually minimized on the upstroke to prevent excessive negative thrust. Furthermore, the shallow undulation of the inner wing areas cannot significantly contribute to thrust production (see Fig. 1.27a). However, the forces exerted on the inner wings can be used for weight support throughout the cycle without much penalty (see Fig. 1.27).

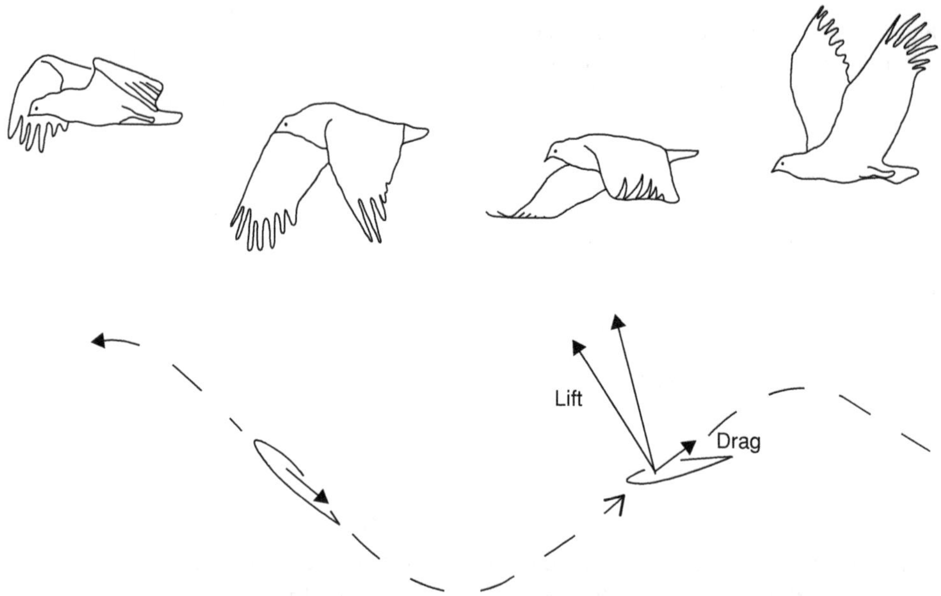

Figure 1.26. Condor in fast flapping flight [66]. (a) Schematic of resultant forces acting on an outer or an inner wing during up- and downstrokes. (b) Asymmetric strokes of a pigeon in slow flight [67].

1.3.2.2 Hovering Flight

Whether a flying animal can hover or not depends on its size, the moment of inertia of its wings, the degree of freedom in the movement of the wings, and the wing shape. As a result of these limitations, hovering is mainly performed by smaller birds and insects. Larger birds can hover only briefly. Although some larger birds, such as kestrels, seem to hover more regularly, in fact they use the incoming wind to generate enough lift. The two kinds of hovering, symmetric and asymmetric, are described by Norberg [4] and Weis-Fogh [68].

Because larger birds cannot rotate their wings between forward and backward strokes, they extend their wings to provide more lift during downstroke, whereas during the upstroke their wings are flexed backward to reduce drag. In general the flex is more pronounced in slow forward flight than in fast forward flight. This type of asymmetric hovering, usually called "avian stroke" [24], is illustrated in Figure 1.28. As shown, to avoid large drag forces and negative lift forces, these birds flex their wings during the upstroke by rotating the primaries (tip feathers) to let air through.

Symmetrical hovering, also called normal or true hovering, or "insect stroke," is performed by hummingbirds or insects that hover with fully extended wings during the entire wing-beat cycle. Lift is produced during the entire wing stroke, except at the reversal points. The wings are rotated and twisted during the backstroke so that the leading edge of the wing remains the same throughout the cycle, but the upper surface of the wing during the forward stroke becomes the lower surface during the backward stroke. The wing movements during downstroke and upstroke can be seen in Figure 1.29. As shown, hummingbirds can rotate their shoulder joint enough to flip the wing over on the upstroke. Like most insects, they hover symmetrically with

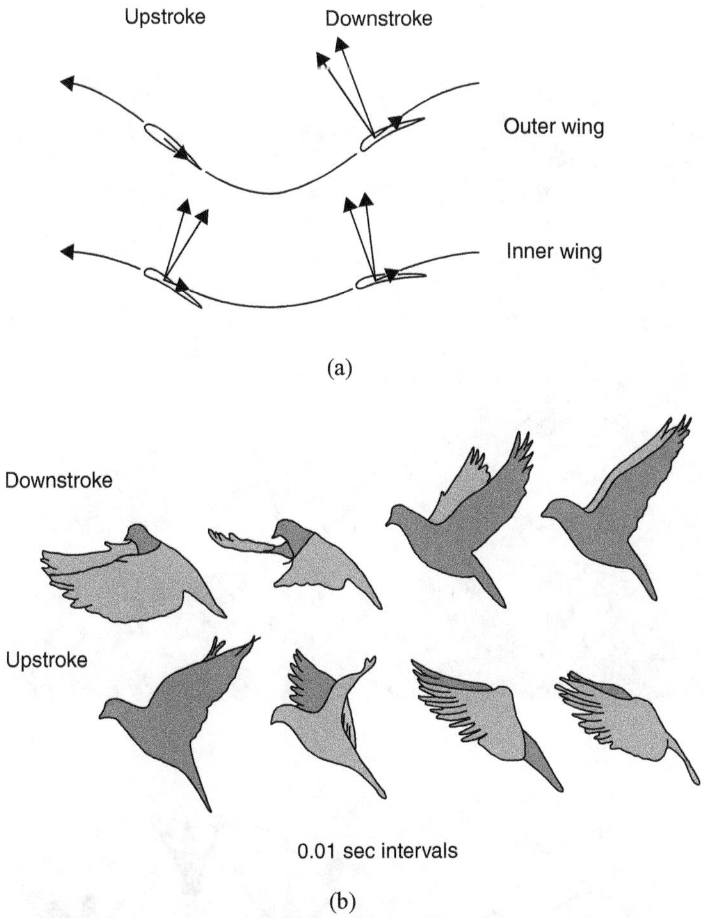

Figure 1.27. Schematic sketches of the roles of the outer and inner wing (a) and the up- and downstroke (b) for bird flight. The outer wing area is the main source of weight support and thrust generation during the downstroke. During the upstroke the forces are minimized. Upstroke is suitable for a small amount of thrust in slow flight, but not much weight support.

Figure 1.28. Selected seagull wing configurations during flapping, which show various stages of stroke. Note the wings are often flexed with their primaries rotated.

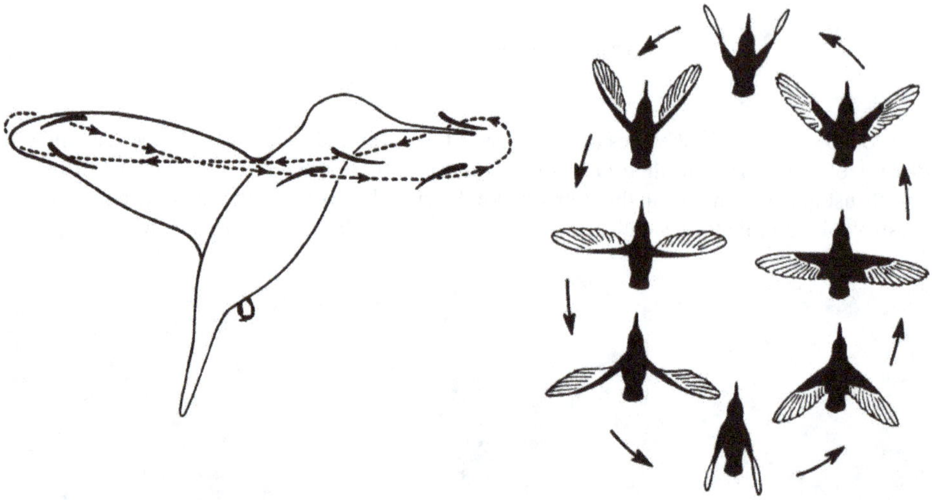

Figure 1.29. Illustration of the flapping wing patterns of a hummingbird: forward and back strokes, flexible and asymmetric wing motions, and a figure-eight pattern. (Bottom images are from Stolpe and Zimmer [69].) (a) Inclined hovering mode of a bat. The flapping motion is asymmetric, such that most of the required force is generated during the downstroke. (b) Resultant force in the downstroke in inclined hover is vertical [70].

a horizontal stroke plane. Their *supracoracoideus* (pull the wing down) is half the size of the *pectoralis* (pull the wing up). Note that, during hovering, the body axis is inclined at a desirable angle and the wing movements describe a figure-eight pattern lying in the vertical plane.

Most birds and bats hover with the stroke plane inclined about 30 degrees to the horizontal. As illustrated in Figure 1.30, the resultant downstroke force with a

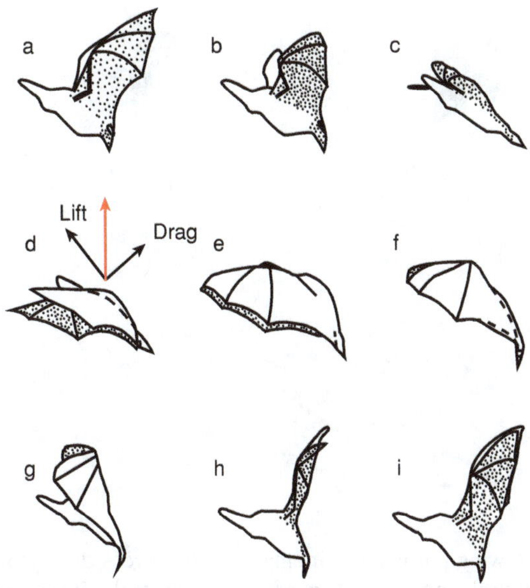

(a) Inclined hovering mode of a bat. The flapping motion is asymmetric, such that most of the required force is generated during the downstroke.

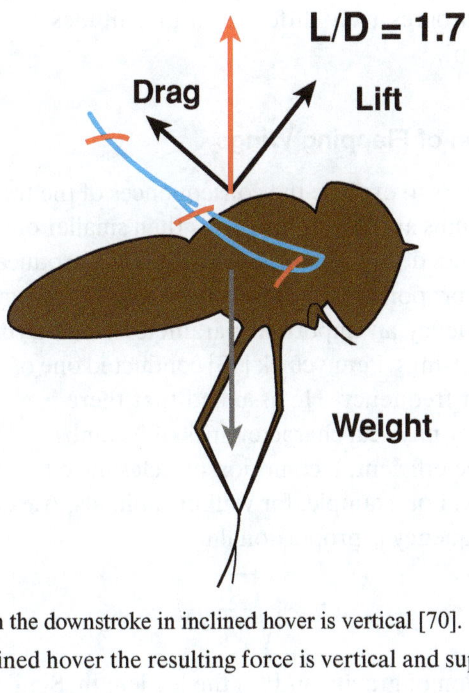

(b) Resultant force in the downstroke in inclined hover is vertical [70].

Figure 1.30. In inclined hover the resulting force is vertical and supports the weight.

lift-to-drag ratio of 1.7 is vertical. Except for small species, noticeably hummingbirds, the wing is typically flexed on the upstroke, minimizing forces on the wing. In inclined hovering flight, lift and drag on the downstroke support the weight.

Figure 1.31 illustrates the wing paths and associated kinematics at different speeds of a bumblebee flying at varying flight speed. The wings flap in a stroke plane

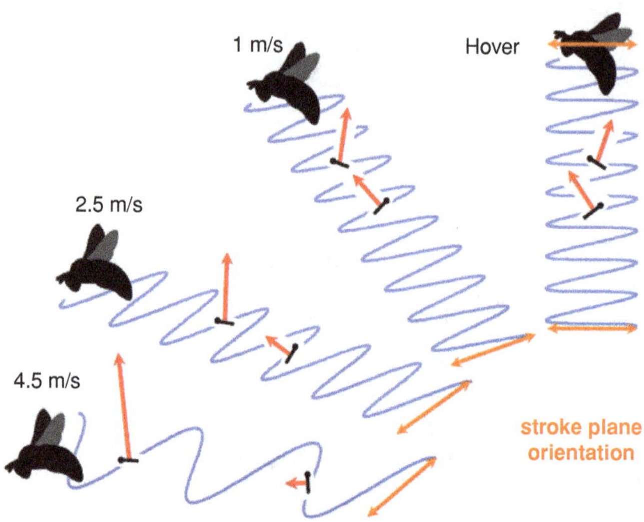

Figure 1.31. Bumblebee wing paths and kinematics. The forces during hovering are almost symmetric, but as the bumblebee moves forward the downstroke force dominates [71].

and flip over between half-strokes, resulting in forces that are nearly symmetrical in hovering. As the forward speed increases, the force generation between the up- and downstrokes becomes progressively different in magnitudes, with the downstroke becoming more dominant.

1.4 Power Implication of Flapping Wings

One of the first researchers to explore the consequences of the trend in which larger animals oscillate their limbs at lower frequencies than smaller ones of a similar type was Hill [72]. He concluded that the mechanical power produced by a particular flight muscle is directly proportional to the contraction frequency. This conclusion has made flapping frequency an important parameter when trying to describe the theories behind flapping wings. Pennycuick [73] conducted one of the most thorough studies of the wing-beat frequency. He assumed that there is a natural frequency, imposed on the animal by physical characteristics of its limbs and by the forces they have to overcome. To be efficient, locomotion muscles have to be adapted to work at a particular frequency. For example, for walking animals, Alexander [74] showed that the first natural frequency is proportional to

$$f_n \sim \sqrt{\frac{g}{l_{leg}}}, \tag{1–21}$$

where g is the acceleration of gravity and l_{leg} the leg length. Similar approaches can be used for flight.

Pennycuick [73] identified several physical variables that affect the wing-beat frequency:

b: Wing span (m)
S: Wing area (m^2)
I: Wing moment of inertia ($kg{\cdot}m^2$) $\sim m(b/2)^2$
ρ: Air density (kg/m^3)

Allowing these variables to vary independently and assuming that the wing moment of inertia is proportional to $m(b/2)^2$, Pennycuick [73] deduced the following correlation for the wing-beat frequency by incorporating data from 32 different species:

$$f = 1.08(m^{1/3}g^{1/2}b^{-1}S^{-1/4}\rho^{1/3}). \tag{1–22}$$

In an updated study Pennycuick [59] added 15 species and made a more detailed analysis. This leads to the following expression:

$$f = m^{3/8}g^{1/2}b^{-23/24}S^{-1/3}\rho^{3/8}. \tag{1–23}$$

Equation (1–21) can be used to predict the wing-beat frequency of species whose mass, wing span, and wing area are known. As mentioned earlier, the moment of inertia I for the wing is dependent on both the span b and the body mass m; hence, a change in any of these variables will result in a change in the moment of inertia. This effect may not be appropriate, given that a change in, for instance, body mass does not necessarily affect the number of the moment of inertia I. So, if we intend to predict these effects on the wing-beat frequency, it is more suitable to include I as an independent parameter [59]:

$$f = (mg)^{1/2}b^{-17/24}S^{-1/3}I^{-1/8}\rho^{3/8}. \tag{1–24}$$

Another relation observed by Pennycuick et al. [75] is the effect of air speed on wing-beat frequency when body mass changes. The data for the wing-beat frequency and the air speed, U, are fitted with a least squares method:

$$f = k_2 + k_3/U + k_4U^3, \tag{1–25}$$

where k_2, k_3, and k_4 are proportional constants.

1.4.1 Upper and Lower Limits

Can scaling arguments provide any information about limits on the size of flapping flyers capable of sustained flight? As mentioned earlier, large *pterosaurs* once flew long ago; some of these species were much larger than the birds of today. There are discussions about whether they were able to flap or only to soar [4]. There are many parameters to consider when flapping flight is studied, but limitations to this kind of flight mainly depend on the power available and structural limits.

These limitations are intimately connected, because flapping frequency affects both power and structural limits. To generate the power required for flight, most birds and other flapping animals have well-developed flight muscles. For birds these muscles are the pectoral muscles, which power the downstroke of the wings, and the *supracoracoideus* muscles powering the upstroke. Much effort has been made to determine power output and frequency levels and to compare the masses of these muscles with the mass of the whole specimen. According to Rayner [56], relations between body mass m and the mass of the pectoral and supracoracoideus muscles, m_p and m_s, respectively, can be expressed as

$$m_p = 0.15\,m^{0.99}, \tag{1–26}$$

$$m_s = 0.016\,m^{1.01}. \tag{1–27}$$

This means that the flight muscles constitute approximately 17 percent of the total weight. In comparison, the muscle of human arms accounts for about 5 percent of total body weight, according to Collins and Graham [76]. The power output from bird muscles and "fast" human muscles is about the same, 150 W/kg. Because the wings are often flexed during the upstroke and therefore not exposed to the same aerodynamic force or moment of inertia as during the downstroke, the weight of the supracoracoideus muscle is generally low compared with the weight of the pectoral muscle. Hummingbirds are different, having an aerodynamically active upstroke (producing lift). In their case, the weight of the supracoracoideus is higher; according to Norberg [4], this muscle group can constitute up to 12 percent of the body weight. The smallest supracoracoideus muscles are found among species with large wingspan, where the muscle mass is about 6 percent of the total mass. This value is comparable to that of the human body, and hence, these species have difficulties taking off without a headwind, running start, or a slope-start from a height. However, species with a long aspect ratio are usually able to soar, so the duration of the flapping flight mode can therefore be decreased.

Pennycuick [62] [77] [78] defined the power margin as the ratio of the power available from the flight muscles to that required for horizontal flight at the *minimum power speed*. As already mentioned, the power available depends on the flapping frequency, which determines the upper and lower limits of the size of flying animals. Pennycuick [78] concluded that the upper limit for flapping flight, based on actual sizes of the largest birds with powered flight, is a body mass of about 12 to 15 kg. Larger birds do not have the possibility of beating their wings fast enough to generate lift to sustain horizontal flight. Smaller birds have the advantage of being able to use different flapping frequencies, but for animals with a weight of about 1 gram, there is another upper limitation. Their muscles need time to reset the contractile mechanism after each contraction [4]. For insects with wing-beat frequencies up to 400 Hz, this problem is solved with special fibrillar muscles capable of contracting and resetting at very high frequencies. This limitation results in a minimum mass for birds of 1.5 grams and, for bats, 1.9 grams.

The upper and lower wing-beat frequencies are also restricted because of the structural limits. Bones, tendons, and muscles are not capable of performing wing motions above a certain wing-beat frequency. Wing bones that have to transmit forces to the external environment during flight must be strong enough to not fail under the imposed loads. This means that the bones have to be stiff and strong and at the same time not too heavy. Kirkpatrick [61] investigated the scaling relationships between body size and several morphological variables of bird and bat wings in order to estimate the stress levels in their wings. He also estimated the bending, shearing, and breaking stresses in the wing bones during flight. He suggested that the breaking stress for a bat humerus bone is around 75 MPa and for birds 125 MPa. This structural limit helps explain why no bat weighs more than 1.5 kg. Kirkpatrick [61] found that no relationship exists between either bending or shearing stresses and wingspan during gliding flight and during the downstroke in hovering flight. In general, the safety factors are greater for birds than for bats. Hence birds are more capable of withstanding higher wing loading. A final conclusion by Kirkpatrick [61] is that the stresses examined are scale independent.

1.4.2 Drag and Power

Like an aircraft, a natural flyer has to generate power to produce lift and to overcome drag during the flight. When soaring or gliding without flapping, the flyer produces much of the power required by converting potential energy to kinetic energy, and vice versa. When the flyer flaps, the power is the rate at which work is produced by the flight muscles. For basic aerodynamic concepts discussed in this section, please refer to the standard textbooks such as Anderson [44] and Shevell [46].

The total aerodynamic drag (D_{aero}), acting on a flight vehicle, is a result of the resistance to the motion through the air. It can be divided into two components. The two drag components acting on a wing in steady flight are the induced drag (D_{ind}), which is the drag that is due to lift, and the profile drag (D_{pro}), which is associated with form and friction drag on the wing. The drag on a finite wing (D_{w}) is the sum of these two components:

$$D_{\text{w}} = D_{\text{ind}} + D_{\text{pro}}. \tag{1–28}$$

The parasite drag (D_{par}), which is defined as the drag on the body and *only* on the body, also contributes to the total drag on the bird. This drag component is caused by the form and friction drag of the "non-lifting" body (it is true that, if the body is tilted at an angle to the free-stream, it will contribute to lift, but this contribution is very small and is neglected). If the drags of the wing and body are summed, the total aerodynamic drag (D_{aero}) of the bird can be expressed as

$$D_{\text{aero}} = D_{\text{ind}} + D_{\text{pro}} + D_{\text{par}}. \tag{1–29}$$

The different powers presented in this section are defined as the powers needed to overcome specific drags at a certain velocity. The total aerodynamic power required for steady forward flight is obtained by multiplying the drag force by the forward velocity (U_{ref}):

$$P_{\text{aero}} = D_{\text{aero}} U_{\text{ref}}. \tag{1–30}$$

The primary goal of this section is to describe the different methods used to determine the total power required (P_{tot}) for flight. The power components are calculated in different ways depending on the forward velocity. Because there is a clear difference between flight at zero velocity (hovering flight) and forward flight, these two cases are dealt with separately when calculating the power components.

For hovering flight, as discussed in Section 1.2.6, the resulting velocity is essentially the same as the induced velocity (w_{i}), which is the air speed in the wake right beneath the flyer, because of negligible forward velocity. In this case, the lift is equal to the thrust T, namely, the weight, and the total aerodynamic power required for hovering flight is

$$P_{\text{aero}} = T w_{\text{i}}. \tag{1–31}$$

For forward flight, there exist three different power components corresponding to the three drag components in Eq. (1–27). The three components are the induced power (P_{ind}), which is the rate of work required to generate a vortex wake whose reaction generates lift and thrust; the profile power (P_{pro}), which is the rate of work needed to overcome form and friction drag of the wings; and the parasite power

(P_{par}), which is the rate of work needed to overcome form and friction drag of the body. As is the case for the drag components, the power components are added together to produce the total aerodynamic power required for horizontal flight:

$$P_{aero} = P_{ind} + P_{pro} + P_{par}, \tag{1–32}$$

where P_{ind} is the power needed for lift production during flight and decreases with increasing flight velocity. In the theory developed by Rayner [79], the upstroke is assumed not to contribute to any useful aerodynamic forces and is therefore not included. The wings are considered only to move in the stroke plane (i.e., no forward or backward movement). The induced power is calculated from the kinetic energy increment in the wake from a single stroke. The shed vortex rings are elliptical and inclined at an angle to the horizontal. The kinetic energy has two components, the self-energy of the newly generated ring and the mutual energy of the new ring, with each of the existing rings in the wake. The mutual energy contribution decreases with higher forward velocities and can be neglected for velocities above the minimum power-required velocity. With this method, the induced power can be calculated as a function of the forward velocity, from the total energy increment divided by the stroke period.

Depending on the forward velocity, different methods are used to estimate the power components in Eq. (1–32). If the forward velocity is high, the unsteady effects are small and quasi-steady assumptions can give good approximations. For slow forward velocities the vortex theory is more accurate, especially when one is estimating the induced power.

In addition to the components previously introduced, the inertial power (P_{iner}) refers to the power needed to move the wings and only the wings. The most important parameter when calculating this power is the wing's moment of inertia. Two main ways exist to obtain a low moment of inertia: to keep the mass of the wing as low as possible and to concentrate the mass as much as possible near the axis of rotation. The inertial power is typically small under medium to fast forward flight conditions and can be neglected [80]. However, for slow or hovering flight, this power must be accounted for.

The total power required (P_{tot}) for flight is the sum of the total aerodynamic power and the inertial power:

$$P_{tot} = P_{aero} + P_{iner} = P_{ind} + P_{pro} + P_{par} + P_{iner}. \tag{1–33}$$

Note that this is only the power required for flight and is not the same as the power input [81]. Since the flight muscles are limited by their own mechanical efficiency, and all living animals are regulated by their own metabolism, the power input needed is higher than the total power required in Eq. (1–33).

The power required (P) is strongly connected to the forward flight speed. A common way of describing this relationship is by means of a power curve. For a fixed-wing air vehicle, the induced power is proportional to U^{-1}, the profile and parasite powers to U^3, and the power required is given by

$$P \sim k_5 U^{-1} + k_6 U^3, \tag{1–34}$$

where k_5 and k_6 are constants.

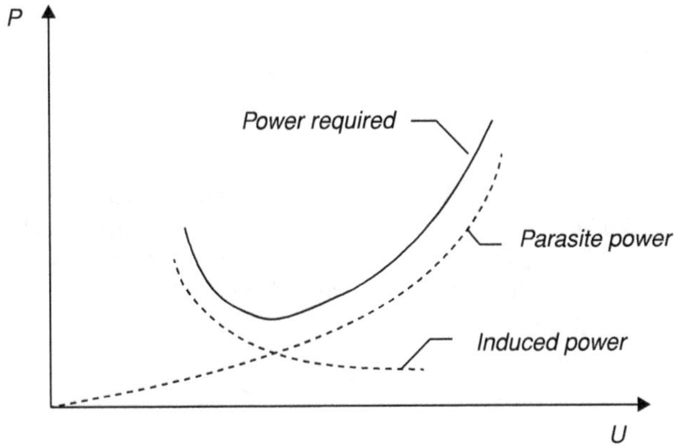

Figure 1.32. The two power components for a fixed-wing air vehicle and the power required, as a result of adding these two components together. The parasite power curve represents the function $P = f(U^3)$ and the induced power curve $P = f(U^{-1})$.

By expressing each power component as a function of velocity, $P = f(U^3)$ and $P = f(U^{-1})$, respectively, two curves can be plotted (see Fig. 1.32). The solid curve in Figure 1.32 represents the power required for steady flight of a fixed-wing aircraft.

As suggested in Figure 1.32, the most common powered flight speed curve is the U-shaped curve (further illustrated in Fig. 1.33), in which there exists a particular speed U_{mp}, where the required power becomes a minimum value. The straight dashed line in Figure 1.33 starts at the origin and intersects the U-curve at a certain point and has the same slope. The velocity at this point is the velocity for maximum range, U_{Mr}. When flyers migrate, they need to cover a long distance for a given amount of energy and therefore tend to fly at this velocity.

For a fixed-wing flyer, as shown by Lighthill [52], U_{mp} and U_{Mr}, are related as follows:

$$U_{\mathrm{Mr}} = 1.32\, U_{\mathrm{mp}}. \tag{1–35}$$

For birds, the power curve is not necessarily U-shaped. Different researchers in the area of avian flight have come up with different shapes of the power curve [82], depending on the power components and muscle efficiency measures they studied.

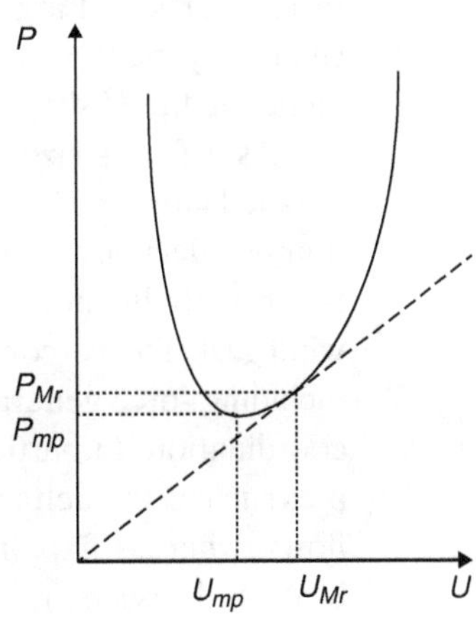

Figure 1.33. The U-shaped power curve for a fixed-wing aircraft. U_{mp} is the velocity for minimum power (P_{mp}) and U_{Mr} is the velocity for maximum range.

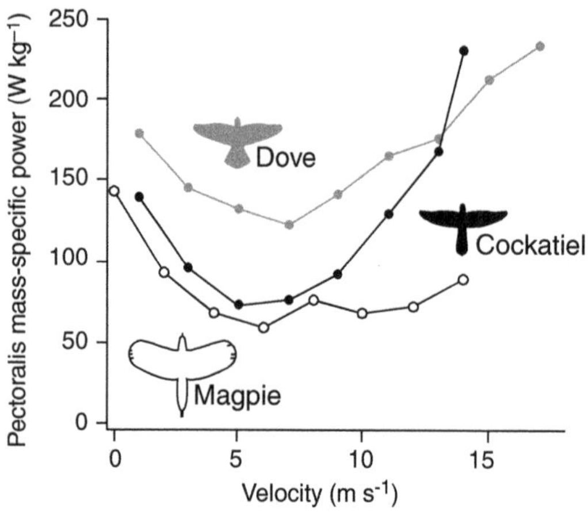

Figure 1.34. Comparative mass-specific pectoralis power as a function of flight velocity in cockatiels, doves, and magpies. Bird silhouettes are shown to scale digitized from video [83].

Nevertheless, as shown in Figure 1.34, the U-shaped power-flight speed curve is indeed observed in natural flyers.

1.5 Concluding Remarks

In this chapter, we have offered an overview of the various low Reynolds number flyers, discussing simple mechanics for several flight modes and highlighting flight characteristics and scaling laws related to wingspan, wing area, wing loading, wing-beat frequency, and vehicle size and weight.

The scaling laws indicate that, as a flyer's size reduces, it has to flap faster to stay in the air, experiences lower wing loading, is capable of cruising slower, has a lower stall speed, and, consequently, can survive much better in a crash landing. In addition, as a flyer becomes smaller, its weight shrinks at a much faster rate, meaning that it can carry very little "fuel" and has to resupply frequently. Birds, bats, and insects apply different flapping patterns in hovering and forward flight to generate lift and thrust. Typically, in slow forward flight the reduced frequency and wing-beat amplitude tend to be high, resulting in highly unsteady flow structures. In fast forward flight the reduced frequency and the wing-beat amplitude tend to be low, and the wake often consists of a pair of continuous undulating vortex tubes or line vortices. Larger birds have relatively simple wingtip paths in comparison to smaller flyers. We have also discussed the power requirement associated with flight, including the U-shaped curve between specific power and flight speed.

As a flyer's size reduces, its operating Reynolds number becomes lower. As detailed later, the lift-to-drag ratio of a stationary wing deteriorates as the Reynolds number decreases. Due to the low weight and slower flight speed, a small flyer is substantially more influenced than large flyers by the flight environment such as wind gust. To overcome these challenges, natural flyers improve flight performance, including force generation and maneuverability, using flapping wings and wing-tail coordination. In particular, the fixed-wing design (other than the noticeable examples such as the delta wing) tends to favor flow patterns with no large scale vortical flows, whereas flapping wing flyers, especially smaller species such as hummingbirds and insects, generate coherent, large-scale vortical flows with low-pressure regions. As discussed in detail in Chapter 3, several fluid flow mechanisms, including a leading-edge vortex caused by varying the AoA under dynamic pitch (induced either actively, by arm and muscle, or passively, by structural shape deformation) and spanwise flows associated with modest aspect ratios, among others, contribute to generating necessary lift and thrust. Of course, the movement of the wing, the substantial impact of wind gust and environmental uncertainty, and the associated structural deformation of the wing during flapping make flapping wing aerodynamics more complex and challenging to analyze.

In the next chapter, we summarize the rigid fixed-wing aerodynamics that shares fundamental aspects of low Reynolds aerodynamics such as flow separation, laminar-to-turbulent transition, and so on. In Chapter 3, we discuss rigid flapping wing aerodynamics and focus on the wing kinematics and the resulting *unsteady* flow features involving flapping wings. In Chapter 4, we discuss aeroelasticity of flexible flapping wings including both active and passive mechanisms and implications.

2 Rigid Fixed-Wing Aerodynamics

As already mentioned, there are several prominent features of MAV flight: (i) low Reynolds numbers (10^4 to 10^5), resulting in degraded aerodynamic performance; (ii) small physical dimensions, resulting in much reduced payload capabilities and some favorable scaling characteristics including structural strength, reduced stall speed, and impact tolerance; and (iii) low flight speed, resulting in an order one effect of the flight environment such as wind gust, as well as intrinsically unsteady flight characteristics. The preferred low Reynolds number airfoil shapes are different in thickness, camber, and aspect ratio from those typically employed for manned aircraft. In this chapter, we discuss the low Reynolds number aerodynamics associated with rigid fixed wings, including the implications of airfoil shapes, laminar-turbulent transition, and unsteady free-stream on the performance outcome.

Schmitz [84] was among the first to investigate the aerodynamics for model airplanes in Germany, and he published his research in 1942. His work is often considered to be the first reported low-speed wind-tunnel research. However, experimental investigations of low Reynolds number aerodynamics were conducted earlier by Brown [85] and by Weiss [86], in the two (and only) issues of *The Journal of International Aeromodeling*. Brown's experiments focused on curved-plate airfoils, made by using two circular arcs meeting at a maximum camber point of 8 percent, but at varying locations. The wings' test sections were all 12.7 cm × 76.2 cm, giving an AR of 6. In all cases, the tests were conducted at a free-stream velocity of 94 cm/s. The Reynolds number, although not mentioned in Brown's study [85], is estimated to be about 8×10^3.

It is hard to judge the quality of the measurements reported in these early works (see Fig. 2.1). Nevertheless, these publications have clearly demonstrated that work with model airplanes offers much to scientific investigation and continues to generate enthusiastic inquiries about many aspects related to low Reynolds number aerodynamics. Representative figures from Brown's experiments [85] are included here for us to gain a historical perspective.

Many published papers have improved our understanding and experimental database and have provided airfoil design guidance in the lower Reynolds number regime. For example, valuable insight has been offered by Liebeck [87], Selig et al. [88]–[90], and Hsiao et al. [91]. Liebeck [87] has addressed the laminar separation and airfoil design issues for the Reynolds number between 2×10^5 and 2×10^6, and

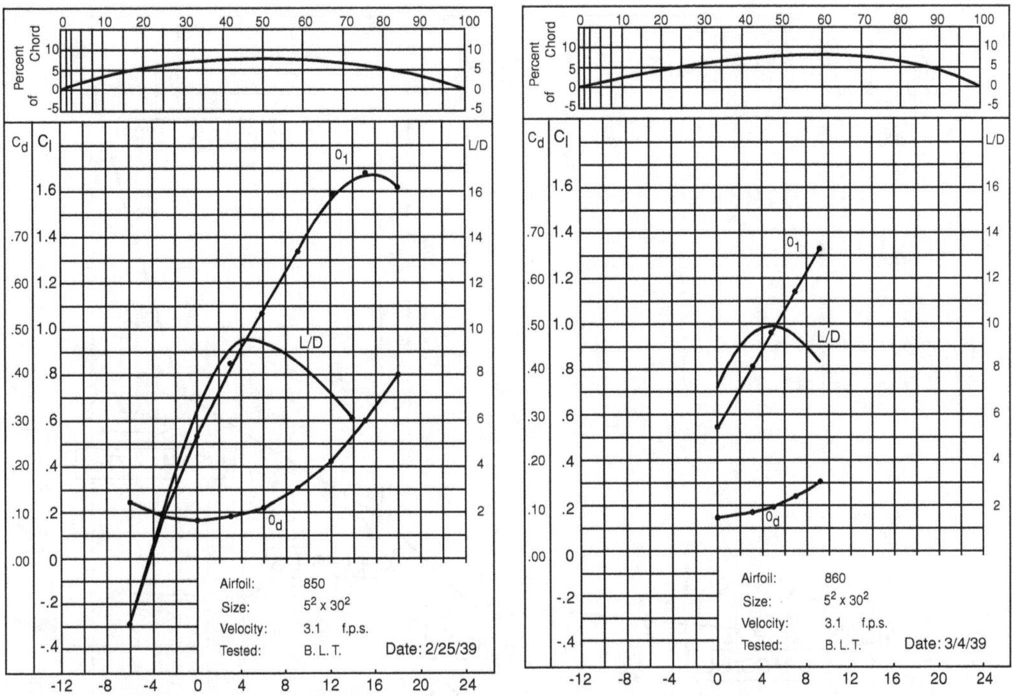

Figure 2.1. Low-speed aerodynamic tests reported by Brown for two airfoils [85]. The chord was 12.7 cm and the free-stream velocity was 94 cm/s.

Hsiao et al. [91] have investigated the aerodynamic and flow structure of an airfoil, NACA 63_3–018, for the Reynolds number between 3×10^5 and 7.74×10^5. Selig et al. have reported on a wide variety of airfoils with basic aerodynamic data for the Reynolds number between 6×10^4 and 3×10^5 [88] [90] and for the Reynolds number between 4×10^4 and 3×10^5 [89]. In the following sections, we discuss the various aerodynamics characteristics and fluid physics for the Reynolds number between 10^2 and 10^6, with a focus on issues related to the Reynolds number of 10^5 or lower.

2.1 Laminar Separation and Transition to Turbulence

Figure 2.2 illustrates the aerodynamic performance and shapes of several representative airfoils under steady-state free-stream. A substantial reduction in lift-to-drag ratio is observed as the Reynolds number becomes lower. The observed aerodynamic characteristics are associated with the laminar-turbulent transition process. With conventional manned aircraft wings whose Reynolds numbers exceed 10^6, the flows surrounding them are typically turbulent, with the near-wall fluid capable of strengthening its momentum via energetic "mixing" with the free-stream. Consequently, flow separation is not encountered until the AoA becomes high. With low Reynolds number aerodynamics, the flow is initially laminar and is prone to separate even under a mild adverse pressure gradient. Under certain circumstances, as discussed next, the separated flow reattaches and forms a laminar separation bubble (LSB) while transitioning from a laminar to a turbulent state. Laminar separation can modify the effective shape of an airfoil, and so it consequently influences the aerodynamic performance.

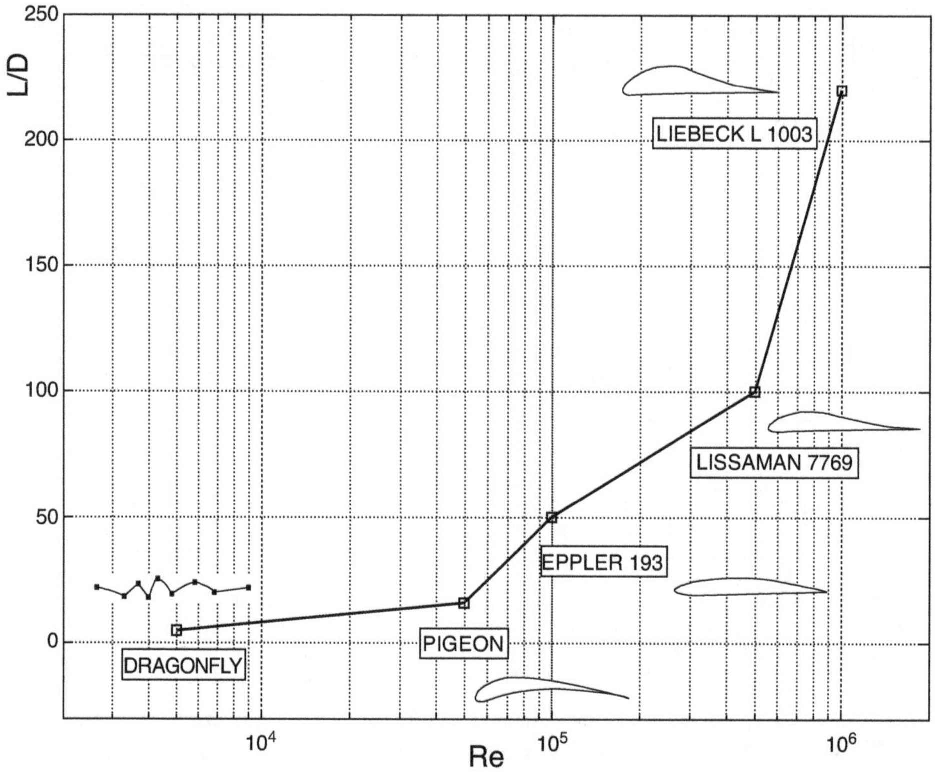

Figure 2.2. Aerodynamic characteristics of representative airfoils. Figure plotted based on the data from Lissaman [21]. Note *Re* indicates the Reynolds number and L/D indicates the lift-to-drag ratio.

The first documented experimental observation of an LSB was reported by Jones [92]. In general, under an adverse pressure gradient of sufficient magnitude, the laminar fluid flow tends to separate before becoming turbulent. After separation, the flow structure becomes increasingly irregular, and beyond a certain threshold, it undergoes a transition from laminar to turbulent. The turbulent mixing process brings high-momentum fluid from the free-stream to the near-wall region, which can overcome the adverse pressure gradient, causing the flow to reattach.

The main features of an LSB are illustrated in Figure 2.3a. After separation, the laminar flow forms a free-shear layer, which is contained between outer edge S"T" of the viscous region and the mean dividing streamline ST'. Downstream of the transition point T, turbulence can entrain a significant amount of high-momentum fluid through diffusion [93], which enables the separated flow to reattach to the wall and form a turbulent free-shear layer. The turbulent free-shear layer is contained between lines T"R" and T'R. The recirculation zone is bounded by the ST'R and STR.

Just downstream of the separation point, there is a "dead-fluid" region, where the recirculation velocity is significantly lower than the free-stream velocity and the flow can be considered almost stationary. Because the free-shear layer is laminar and is less effective in mixing than in turbulent flow, the flow velocity between the separation and transition is virtually constant [93]. That the velocity is almost

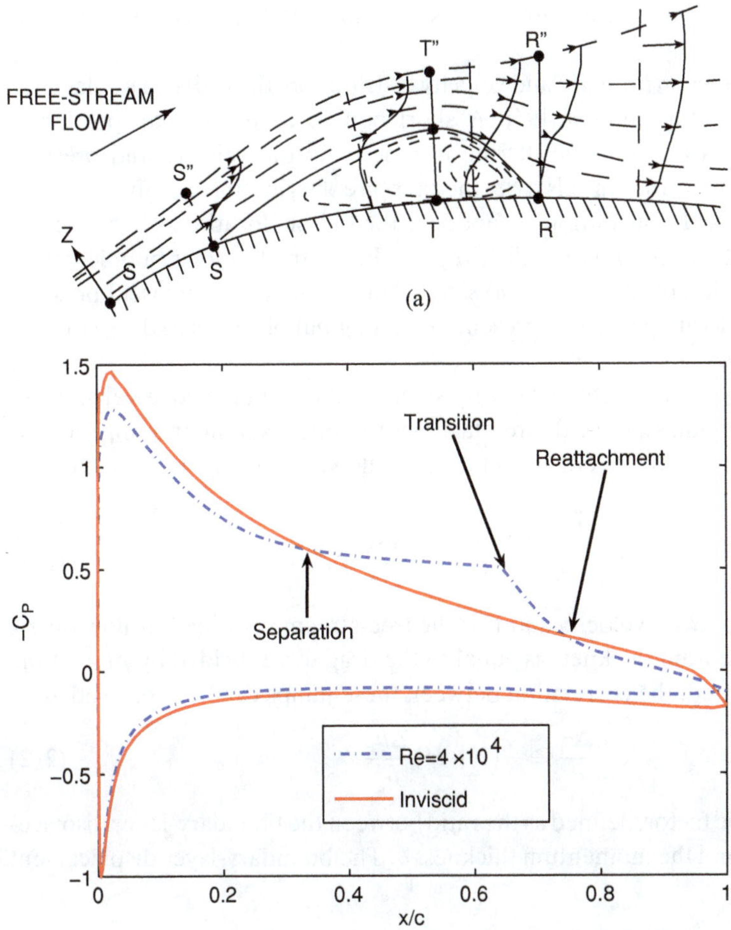

Figure 2.3. (a) Schematic flow structures illustrating the laminar-turbulent transition [93] (copyright by AIAA). (b) Pressure distribution over an SD7003 airfoil, as predicted by XFOIL [96].

constant is also reflected in the pressure distribution in Figure 2.3b. The pressure "plateau" is a typical feature of the laminar part of the separated flow.

The dynamics of an LSB depends on the Reynolds number, pressure distribution, geometry, surface roughness, and free-stream turbulence. An empirical rule given by Carmichael [94] says that the Reynolds number, based on the free-stream velocity and the distance from the separation point to the reattachment point, is approximately 5×10^4. It suggests that, if the Reynolds number is less than 5×10^4, an airfoil will experience separation without reattachment; in contrast, if the Reynolds number is slightly higher than 5×10^4, a long separation bubble will occur. This rule provides general guidance to predict the reattachment, but should be used with caution. As we discuss later, the transition and the reattachment process are too complicated to be described by the Reynolds number alone.

As the Reynolds number decreases, the viscous damping effect increases, and it tends to suppress the transition process or to delay reattachment. The flow will not reattach if the Reynolds number is sufficiently low to enable the flow to completely remain laminar or the pressure gradient is too strong for the flow to reattach.

Thus, without reattachment, a bubble does not form and the flow is then fully separated.

Based on its effect on pressure and velocity distribution, the LSB can be classified as either a short or long bubble [95]. A short bubble covers a small portion of the airfoil and plays an insignificant rule in modifying the velocity and pressure distributions over an airfoil. In this case, the pressure distribution closely follows its corresponding inviscid distribution except near the bubble location, where there is a slight deviation from the inviscid distribution. In contrast, a long bubble covers a considerable portion of the airfoil and significantly modifies the inviscid pressure distribution and velocity peak. The presence of a long bubble leads to decreased lift and increased drag.

Typically, a separation bubble has very steep gradients in the edge velocity, u_e, and momentum thickness, θ, at the reattachment point, resulting in jumps in Δu_e and $\Delta\theta$ over a short distance. For incompressible flow, the momentum thickness is defined as

$$\theta = \int_0^{\xi\to\infty} \frac{u}{U}\left(1-\frac{u}{U}\right)d\xi, \tag{2–1}$$

where u is the streamwise velocity and U is the free-stream velocity. For flow over a flat plate the momentum thickness is equal to the drag force divided by ρU^2. If the skin friction is omitted, the correlation between these jumps can be expressed as

$$\frac{\Delta\theta}{\theta} \cong -(2+H)\frac{\Delta u_e}{u_e}, \tag{2–2}$$

where H is the shape factor, defined as the ratio between the boundary-layer displacement thickness δ^* and the momentum thickness θ. The boundary-layer displacement is defined as

$$\delta^* = \int_0^{\xi\to\infty} \left(1-\frac{u}{U}\right)d\xi. \tag{2–3}$$

Due to the change in flow structures, the shape factor H increases rapidly downstream of the separation point. Hence, according to Eq. (2–2), the momentum thickness jump is sensitive to the location of the transition point in the separation bubble. Furthermore, because airfoil drag is directly affected by a momentum thickness jump, an accurate laminar-turbulent transition model is important for drag prediction.

Figure 2.4 illustrates the behavior of an LSB in response to variations in the Reynolds number. The analyses are based on the XFOIL code [96], which uses the thin-layer fluid flow model, assuming that the transverse length scale is much smaller than the streamwise length scale. At a fixed AoA, four flow regimes can be identified as the Reynolds number varies. As indicated in Figure 2.4, at the Reynolds number $Re = 10^6$, on the upper surface there exists a short LSB, which affects the velocity distribution locally. At an intermediate Reynolds number (e.g., $Re = 4\times10^4$), the short bubble bursts to form a long bubble. The peak velocity is substantially lower than that of the inviscid flow. As the Reynolds number decreases to, for example, $Re = 2\times10^4$, the velocity peak and circulation decrease further, reducing the pressure gradient after the suction peak. A weaker pressure gradient attenuates the amplification of disturbance in the laminar boundary layer, which

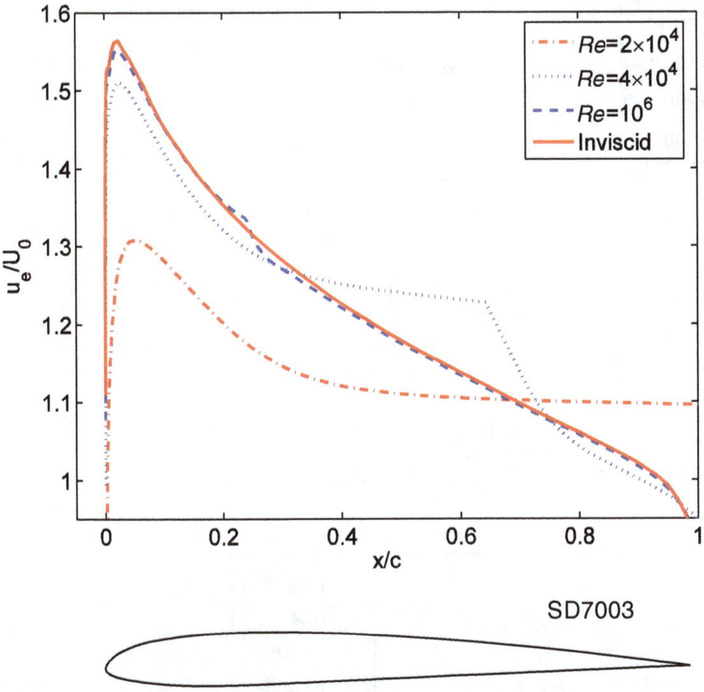

Figure 2.4. Streamwise velocity profiles over the upper surface of an SD7003 airfoil with varying Reynolds numbers, at the AoA of 4°. Inviscid as well as viscous flow solutions are shown. At the Reynolds number, $Re = 10^6$, a short bubble is observed; otherwise, the velocity distribution largely matches that of the inviscid model. At a lower Reynolds number, $Re = 4 \times 10^4$, a long bubble after bursting is observed, causing significant impact on the velocity distribution. Finally, at the Reynolds number, $Re = 2 \times 10^4$, a complete separation with no reattachment is noticed. u_e is the velocity at the boundary-layer edge parallel to the airfoil surface, and U_0 is the free-stream velocity. The results are based on computations made with XFOIL [96].

delays the transition and elongates the free-shear layer. At this Reynolds number, the separated flow no longer reattaches to the airfoil surface, and the main structures are no longer sensitive to the exact value of the Reynolds number.

For a fixed Reynolds number, varying the AoA changes the pressure gradient aft of the suction peak and therefore changes the LSB. In this aspect, varying the AoA has the same effect on the LSB as changing the Reynolds number. Figure 2.5 illustrates that, at a fixed Reynolds number of $6.01 \times 10^4$for the Eppler E374, a zigzag pattern appears in the lift-drag polar:

1. At a lower AoA, for example, 2.75°, there is a long bubble on the airfoil surface, which leads to a large drag.
2. When the AoA is increased (from 4.03° to 7.82°), the adverse pressure gradient on the upper surface grows, which intensifies the Tollmien-Schlichting (TS) wave, resulting in an expedited laminar-turbulent transition process. A shorter LSB leads to more airfoil surface being covered by the attached turbulent boundary-layer flow, resulting in a lower drag. This corresponds to the lift-drag polar's left turn.

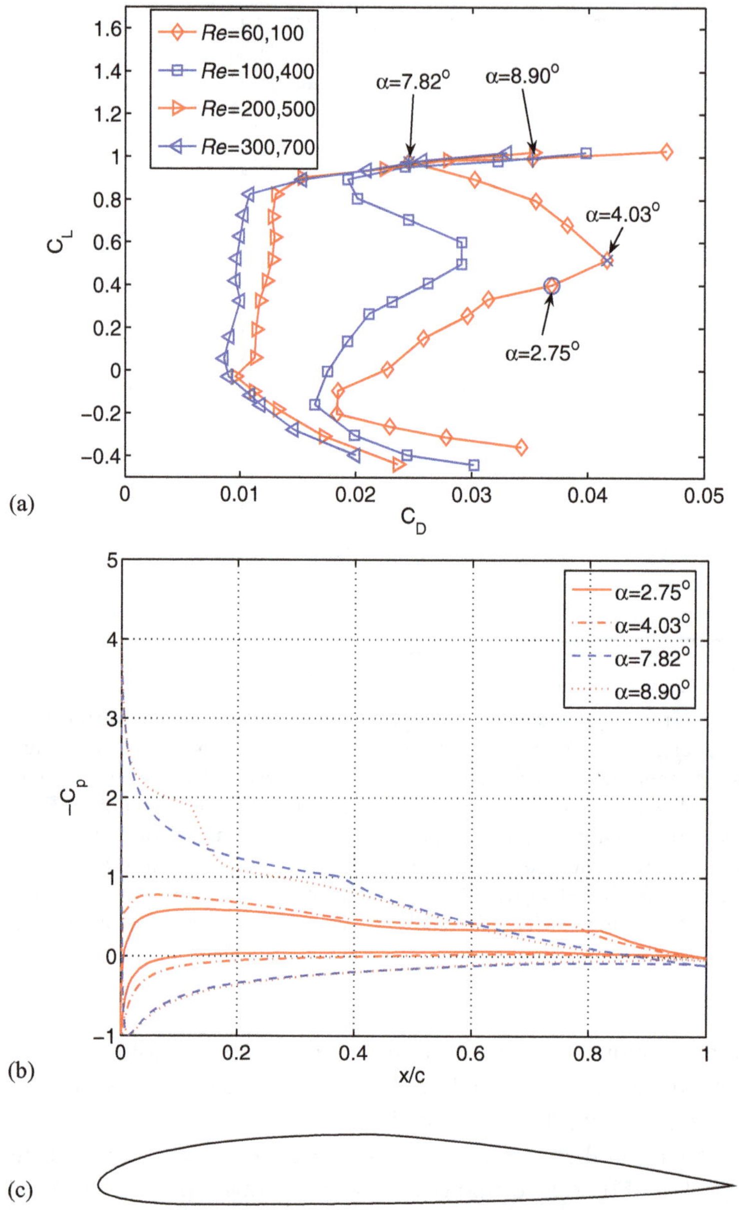

Figure 2.5. Lift-to-drag polars of the Eppler E374 airfoil at different chord Reynolds numbers. (a) Lift-to-drag polars at different Reynolds numbers; (b) pressure coefficient distribution at different AoAs for the Reynolds number, $Re = 6.01 \times 10^4$; (c) shape of the Eppler E374 airfoil. The results are computed with XFOIL [96].

3. When the AoA is further increased (beyond 7.82°), the separated flow quickly experiences transition; however, with a massive separation, the turbulent diffusion can no longer make the flow reattach, and the drag increases substantially with little changes in lift.

The previously described zigzag pattern of the lift-drag polar is a noticeable feature of low Reynolds number aerodynamics. As illustrated in Figure 2.5, at a sufficiently high Reynolds number, the polar exhibits the familiar C-shape.

Earlier experimental investigations on low Reynolds number aerodynamics were reviewed by Young and Horton [97]. Carmichael [94] further reviewed theoretical and experimental results of various airfoils with Reynolds numbers spanning from 10^2 to 10^9. In particular, many investigators studied the near-surface flow and aerodynamic loads of a wing at Reynolds numbers in the range of 10^4 to 10^6. Crabtree [98] studied the formation of short and long LSBs on thin airfoils. Consistent with the preceding discussion of the two types of separation bubbles, he suggested that the long bubble directly influences aerodynamic characteristics, whereas the short one serves as an agent for initiating a turbulent boundary layer.

In the last two decades, numerous investigations have been reported on the interplay between the near-wall flow structures and aerodynamic performance. For example, Huang et al. [99] studied the aerodynamic performance versus the surface-flow mode at different Reynolds numbers. Hillier and Cherry [100] and Kiya and Sasaki [101] studied the influence of the free-stream turbulence on the separation bubble along the side of a blunt plate with right-angled corners, finding that the bubble length, sizes of vortices in the separating region, and level of the suction peak pressure can all be well correlated with the turbulence outside the shear layer and near the separation point.

2.1.1 Navier-Stokes Equation and the Transition Model

The constant property Navier-Stokes equations adequately model the fluid physics needed to perform practical laminar- and turbulent-flow computations in the Reynolds number range typically used by the low Reynolds number flyers:

$$\frac{\partial u_i}{\partial x_i} = 0, \tag{2–4}$$

$$\frac{\partial u_i}{\partial t} + \frac{\partial}{\partial x_j}(u_i u_j) = -\frac{1}{\rho}\frac{\partial p}{\partial x_i} + \nu\frac{\partial^2}{\partial x_j^2}(u_i), \tag{2–5}$$

where u_i are the mean flow velocities and ν is the kinematic viscosity. For turbulent flows, turbulent closures are needed if one is solving the ensemble-averaged form of the Navier-Stokes equations. Numerous closure models have been proposed in the literature [102]. Here we present the two-equation $k-\omega$ turbulence model [102] as an example. For clarity, the turbulence model is written in Cartesian coordinates as follows:

$$\frac{\partial k}{\partial t} + \frac{\partial (u_j k)}{\partial x_j} = \tau_{ij}\frac{\partial u_i}{\partial x_j} - \beta^* \omega k + \frac{\partial}{\partial x_j}\left[(\nu + \sigma^* \nu_{\mathrm{T}})\frac{\partial k}{\partial x_j}\right], \tag{2–6}$$

$$\frac{\partial \omega}{\partial t} + \frac{\partial (u_j \omega)}{\partial x_j} = \frac{\alpha\omega}{k}\tau_{ij}\frac{\partial u_i}{\partial x_j} - \beta\omega^2 + \frac{\partial}{\partial x_j}\left[(\nu + \sigma \nu_{\mathrm{T}})\frac{\partial \omega}{\partial x_j}\right], \tag{2–7}$$

where

$$\nu_{\mathrm{T}} = \frac{\alpha^*\omega}{k}, \quad \tau_{ij} = 2\nu_{\mathrm{T}} S_{ij} - \frac{2}{3}k\delta_{ij}, \quad S_{ij} = \frac{1}{2}\left(\frac{\partial u_i}{\partial x_j} + \frac{\partial u_j}{\partial x_i}\right), \tag{2–8}$$

$$\alpha^* = \frac{\alpha_0^* + Re_T/R_k}{1 + Re_T/R_k}, \quad \alpha = \frac{13}{25}\frac{\alpha_0 + Re_T/R_k}{1 + Re_T/R_k}\frac{1}{\alpha^*}, \tag{2–9}$$

$$\beta^* = \frac{9}{100}\frac{4/15 + (Re_T/R_\beta)^4}{1 + (Re_T/R_\beta)^4}, \quad Re_T = \frac{k}{\omega\nu},$$

$$\beta = \frac{9}{125}, \quad \sigma^* = \sigma = \frac{1}{2}, \quad \alpha_0^* = \frac{1}{3}\beta, \quad \alpha_0 = \frac{1}{9}, \tag{2–10}$$

$$R_\beta = 8, \quad R_k = 6, \quad R_\omega = 2.95. \tag{2–11}$$

For the preceding equations, k is the turbulent kinetic energy, ω is the dissipation rate, ν_T is the turbulent kinematic eddy viscosity, Re_T is the turbulent Reynolds number, and α_0, β, R_β, R_k, and R_ω are model constants. To solve for the transition from laminar to turbulent flow, the incompressible Navier-Stokes equations are coupled with a transition model.

The onset of laminar-turbulent transition is sensitive to a wide variety of disturbances, as produced by the pressure gradient, wall roughness, free-stream turbulence, acoustic noise, and thermal environment. A comprehensive transition model considering all these factors currently does not exist. Even if we limit our focus to free-stream turbulence, it is still a challenge to provide an accurate mathematical description. Overall, approaches to transition prediction can be categorized as (i) empirical methods and those based on linear stability analysis, such as the e^N method [96]; (ii) linear or non-linear parabolized stability equations [103]; and (iii) large-eddy simulation (LES) [104] or direct numerical simulation (DNS) methods [105].

Empirical methods have also been proposed to predict transition in a separation bubble. For example, Roberts [93] and Volino and Bohl [106] developed models based on local turbulence levels; Mayle [107], Praisner and Clark [108], and Roberts and Yaras [109] tested concepts by using the local Reynolds number based on the momentum thickness. These models use only one or two local parameters to predict the transition points and hence often oversimplify the downstream factors such as the pressure gradient, surface geometry, and surface roughness. For attached flow Wazzan et al. [110] proposed a model based on the shape factor H. Their model gives a unified correlation between the transition point and Reynolds number for a variety of problems. For separated flow, however, no similar models exist, in part because of the difficulty in estimating the shape factor.

Among the approaches employing linear stability analysis, the e^N method has been widely adopted [111] [112]. It solves the Orr-Sommerfeld equation to evaluate the local growth rate of unstable waves based on velocity and temperature profiles over a solid surface. Its successful application is exemplified in the popularity of airfoil analysis software such as XFOIL [96]. XFOIL uses the steady Euler equations to represent the inviscid flows, a two-equation integral formulation based on dissipation closure to represent boundary layers and wakes, and the e^N method to tackle transition. Coupling the Reynolds-averaged Navier-Stokes (RANS) solver with the e^N method to predict transition has been practiced by Radespiel et al. [113], Stock and Haase [114], and He et al. [115]. An application of this approach for low Reynolds number applications can be found in the work of Yuan et al. [116] and Lian and Shyy [117].

The e^N method is based on the following assumptions: (i) the velocity and temperature profiles are essentially 2D and steady, (ii) the initial disturbance is infinitesimal, and (iii) the boundary layer is thin. Even though in practice the e^N method has been extended to study 3D flow, strictly speaking, such flows do not meet the preceding conditions. Furthermore, even in 2D flow, not all these assumptions can be satisfied [118]. Nevertheless, the e^N method remains a practical and useful approach for engineering applications.

Advancements in turbulence modeling have made possible alternative approaches for transition prediction. For example, Wilcox devised a low Reynolds number $k - \omega$ turbulence model to predict transition [119]. One of his objectives is to match the minimum critical Reynolds number beyond which the TS wave begins forming in the Blasius boundary-layer context. However, this model fails if the separation-induced transition occurs before the minimum Reynolds number is reached, as frequently occurs in the separation-induced transition. Holloway et al. [120] used unsteady RANS equations to study the flow separation over a blunt body for the Reynolds number range of 10^4 to 10^7. It has been observed that the predicted transition point can be too early even for a flat-plate flow case, as illustrated by Dick and Steelant [121]. In addition, Dick and Steelant [122] and Suzen and Huang [123] incorporated the concept of an intermittency factor to model transitional flows. One can model this either by using conditional-averaged Navier-Stokes equations or by multiplying the eddy viscosity by the intermittency factor. In either approach, the intermittency factor is solved based on a transport equation, aided by empirical correlations. Mary and Sagaut [124] studied the near-stall phenomena around an airfoil using LES, and Yuan et al. [116] studied transition over a low Reynolds number airfoil using LES.

2.1.2 The e^N Method

In this section we offer a more detailed presentation of the e^N method, because it forms the basis for low Reynolds number aerodynamics predictions and has proven to be useful for engineering applications. As already mentioned, the e^N method is based on linear stability analysis, which states that transition occurs when the most unstable TS wave in the boundary layer has been amplified by a certain factor. Given a velocity profile, one can determine the local disturbance growth rate by solving the Orr-Sommerfeld eigenvalue equations. Then, the amplification factor is calculated by integrating the growth rate, usually the spatial growth rate, starting from the point of neutral stability. The *Transition Analysis Program System* (TAPS) by Wazzan and co-workers [125] and the *COSAL* program by Malik [126] can be used to compute the growth rate for a given velocity profile. Schrauf also developed a program called *Coast3* [127]. However, it is very time consuming to solve the eigenvalue equations. An alternative approach was proposed by Gleyzes et al. [128], who found that the integrated amplification factor $\tilde{n}$ can be approximated by an empirical formula as follows:

$$\tilde{n} = \frac{d\tilde{n}}{dRe_\theta}(H)[Re_\theta - Re_{\theta_o}(H)], \tag{2–12}$$

where Re_θ is the momentum thickness Reynolds number, Re_{θ_o} is the critical Reynolds number that we define later, and H is the shape factor previously discussed. With this approach, one can approximate the amplification factor with a reasonably good accuracy without solving the eigenvalue equations. For similar flows, the amplification factor $\tilde{n}$ is determined by the following empirical formula:

$$\frac{d\tilde{n}}{dRe_\theta} = 0.01\{[2.4H - 3.7 + 2.5\tanh(1.5H - 4.65)]^2 + 0.25\}^{1/2}. \tag{2–13}$$

For non-similar flows (i.e., those that cannot be treated by similarity variables using the Falkner-Skan profile family [129]) the amplification factor with respect to the spatial coordinate ξ is expressed as

$$\frac{d\tilde{n}}{d\xi} = \frac{d\tilde{n}}{dRe_\theta}\frac{1}{2}\left(\frac{\xi}{u_e}\frac{du_e}{d\xi} + 1\right)\frac{\rho u_e \theta^2}{u_e \xi}\frac{1}{\theta}. \tag{2–14}$$

An explicit expression for the integrated amplification factor then becomes

$$\tilde{n}(\xi) = \int_{\xi_o}^{\xi} \frac{d\tilde{n}}{d\xi} d\xi, \tag{2–15}$$

where ξ_0 is the point where $Re_\theta = Re_{\theta_o}$, and the critical Reynolds number is expressed by the following empirical formulas:

$$\log_{10} Re_{\theta_o} = \left(\frac{1.415}{H-1} - 0.489\right)\tanh\left(\frac{20}{H-1} - 12.9\right) + \frac{3.295}{H-1} + 0.44. \tag{2–16}$$

Once the integrated growth rate reaches the threshold N, flow becomes turbulent. To incorporate the free-stream turbulence level effect, Mack [130] proposed the following correlation between the free-stream intensity T_i and the threshold N:

$$N = -8.43 - 2.4\ln(T_i), \quad 0.0007 \leq T_i \leq 0.0298. \tag{2–17}$$

However, care should be taken in using such a correlation. The free-stream turbulence level itself is not sufficient to describe the disturbance, and other information, such as the distribution across the frequency spectrum, should also be considered. The so-called receptivity – how the initial disturbances within the boundary layer are related to the outside disturbances – is a critically important issue. Actually, we can only determine the N factor if we know the "effective T_i," which can be defined only through a comparison of the measured transition position with calculated amplification ratios [131].

A typical procedure to predict the transition point using coupled RANS equations and the e^N method is as follows. First, solve the Navier-Stokes equations together with a turbulence model without invoking the turbulent production terms, for which the flow is essentially laminar; then integrate the amplification factor $\tilde{n}$ based on Eq. (2–12) along the streamwise direction; once the value reaches the threshold N, the production terms are activated for the post-transition computations. After the transition point, flow does not immediately become fully turbulent;

instead, the movement toward full turbulence is a gradual process. This process can be described with an intermittency function, allowing the flow to be represented by a combination of laminar and turbulent structures. With the intermittency function, an *effective eddy viscosity* is used in the turbulence model and can be expressed as follows:

$$\nu_{\mathrm{Te}} = \gamma \nu_{\mathrm{T}}, \tag{2–18}$$

where γ is the intermittency function and ν_{Te} is the effective eddy viscosity.

In the literature a variety of intermittency distribution functions have been proposed. For example, Cebeci [132] presented such a function by improving a model previously proposed by Chen and Thyson [133] for the Reynolds number range of 2.4×10^5 to 2×10^6 with an LSB. However, no model is available when the Reynolds number is lower than 10^5. Lian and Shyy [117] suggested that, for separation-induced transition at such a low Reynolds number regime, the intermittency distribution is largely determined by the distance from the separation point to the transition point: the shorter the distance, the quicker the flow becomes turbulent. In addition, previous work suggested that the flow property at the transition point is also important. From the available experimental data and our simulations, Lian and Shyy [117] proposed the following model:

$$\gamma = \begin{cases} 1 - \exp\left(-\left[\exp\left(\frac{\max(H_T - 2.21, 0)}{20}\right)^2 - 1\right]\left(\frac{x - x_T}{x_T - x_S}\right) Re_{\theta T}\right) & (x \geq x_T) \\ 0 & (x < x_T) \end{cases}, \tag{2–19}$$

where x_T is the transition onset position, x_S is the separation position, H_T is the shape factor at the transition point, and $Re_{\theta T}$ is the Reynolds number based on the momentum thickness at the transition point.

2.1.3 Case Study: SD7003

Lian and Shyy [117] studied the Reynolds number effect with a Navier-Stokes equation solver augmented with the e^N method. The lift and drag coefficients of the SD7003 airfoil against AoA are plotted in Figure 2.6. This figure vividly shows the good agreement between the numerical results [117] and the experimental measurements by Ol et al. [134] and Selig et al. [88]. Both the simulation and the measurement by Ol et al. [134] predict that the maximum lift coefficient happens at the AoA = 11°. Close to the stall AoA, the simulations over-predict the lift coefficients.

As the AoA increases, as illustrated in Figure 2.7, the adverse pressure gradient downstream of the point of suction peak becomes stronger and the separation point moves toward the leading edge. The strong pressure gradient amplifies the disturbance in the separation zone and prompts transition. As the turbulence develops, the increased entrainment causes reattachment. At an AoA of 2°, the separation position is at around 37 p of the chord length, and transition occurs at 75 percent of the chord length. A long LSB forms. The plateau of the pressure distribution shown in Figure 2.7a is characteristic of an LSB. As seen in Figure 2.7b the bubble length decreases with an increase in the AoA.

Lian and Shyy [117] compared computed shear stress with experimental measurements by Radespiel et al. [135], using the low-turbulence wind tunnel and the

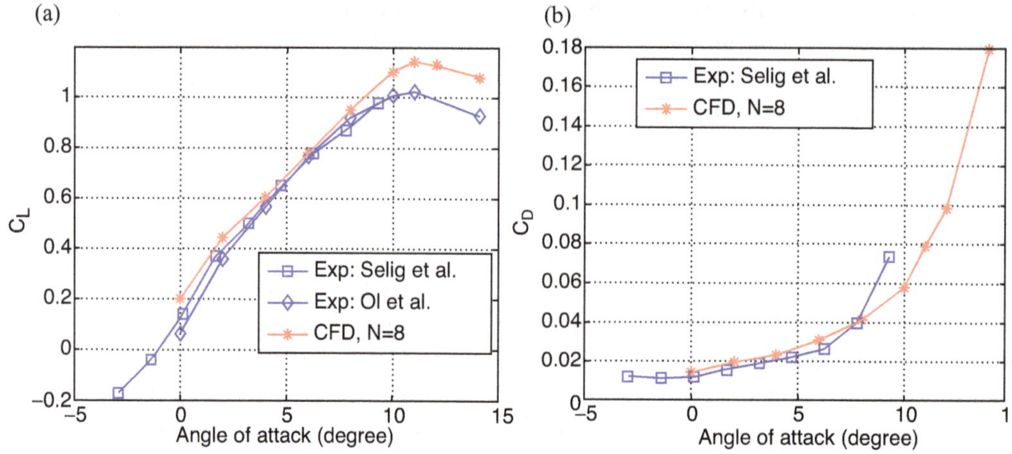

Figure 2.6. (a) Lift and (b) drag coefficients against the AoA for an SD7003 airfoil at the Reynolds number, $Re = 6 \times 10^4$ [117]. CFD refers to computational fluid dynamics.

water tunnel because of its low-turbulent nature. Radespiel et al. [135] suggested that large values of critical N factor should be appropriate. As shown in Figure 2.8, the simulation by Lian and Shyy [117] with $N = 8$ shows good agreement with measurements in terms of the transition position, reattachment position, and vortex core position. It should be noted that, in the experiment, the transition location is defined as the point where the normalized Reynolds shear stress reaches 0.1 percent and demonstrates a clearly visible rise. However, the transition point in the simulation is defined as the point where the most unstable TS wave has amplified over a factor of e^N. This definitional discrepancy may cause some problems when we compare the transition position. In any event, simulations typically predict noticeably lower shear-stress magnitude than the experimental measurement. Recently, LES simulations of an SD7003 rectangular wing at the Reynolds number of 6×10^4 were performed by Galbraith and Visbal [136] and Uranga et al. [137]; they discussed transition mechanisms based on instantaneous and phase-averaged flow features and

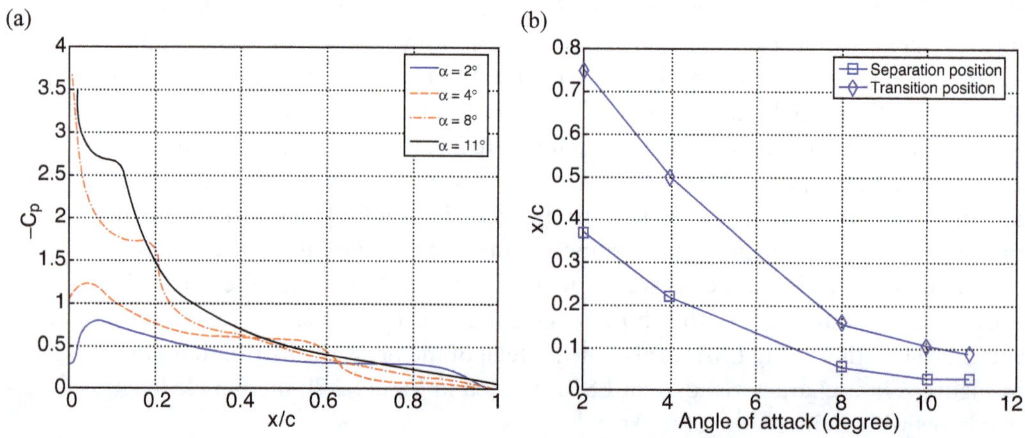

Figure 2.7. (a) Pressure coefficients and (b) separation and transition position against the AoA for an SD7003 airfoil at Reynolds number, $Re = 6 \times 10^4$ [117].

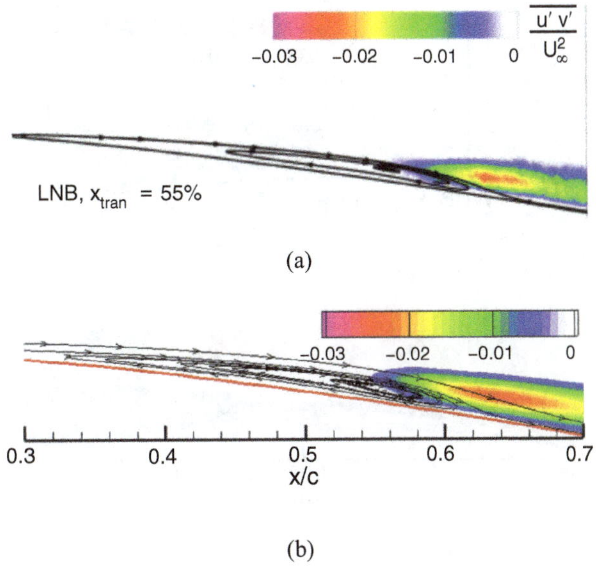

Figure 2.8. Streamlines and turbulent shear stress for the AoA = 4°. (a) Experimental measurement by Radespiel et al. [135] and (b) numerical simulation with $N = 8$ by Lian and Shyy [117].

turbulence statics. Moreover, using time-resolved particle image velocimetry (PIV), Hain et al. [138] studied the transition mechanism in the flow around an SD7003 wing at Reynolds number of 6.6×10^4; they claimed that the TS waves trigger the amplification of the KH waves, which explains why the size of the separation bubble is affected by the magnitude of the TS waves at separation.

As the AoA increases, both the separation and the transition positions move upstream, and the bubble shrinks. The measurements at the AoAs of 8° and 11° are performed in the water tunnel with a measured free-stream turbulence intensity of 0.8 percent. At the AoA = 8° the simulation by Lian and Shyy [117] predicts that the flow goes though transition at 15 percent of the chord, which is close to the experimental measurement of 14 percent. The bubble covers approximately 8 percent of the airfoil upper surface. The computational and experimental results for the AoA of 8° are shown in Figure 2.9. With the AoA of 11°, the airfoil is close to stall. The separated flow requires a greater pressure recovery in the laminar bubble for reattachment. Lian and Shyy [117] predicted that flow separates at 5 percent of the chord and that the separated flow quickly reattaches after it experiences transition at the 7.5 percent chord position, whereas in the experiment transition actually occurred at 8.3 percent. This quick reattachment generally represents the transition-forcing mechanism. Comparison shows that the computed Reynolds shear stress matches the experimental measurement well (see Fig. 2.10).

For low Reynolds number airfoils, the chord Reynolds number is a key parameter used to characterize the overall aerodynamics. Between the separation position and the transition position, as shown in Figure 2.11a, the shape factor H and the momentum-thickness-based Reynolds number increase with the chord Reynolds number. As shown in Figure 2.11b the effective airfoil shape, which is the superimposition of the airfoil and the boundary-layer displacement thickness, has the largest camber at the Reynolds number $Re = 4 \times 10^4$. This helps explain why the largest lift coefficient is obtained at that Reynolds number (see Fig. 2.11c). The camber decreases significantly when the Reynolds number increases from 4×10^4 to 6×10^4,

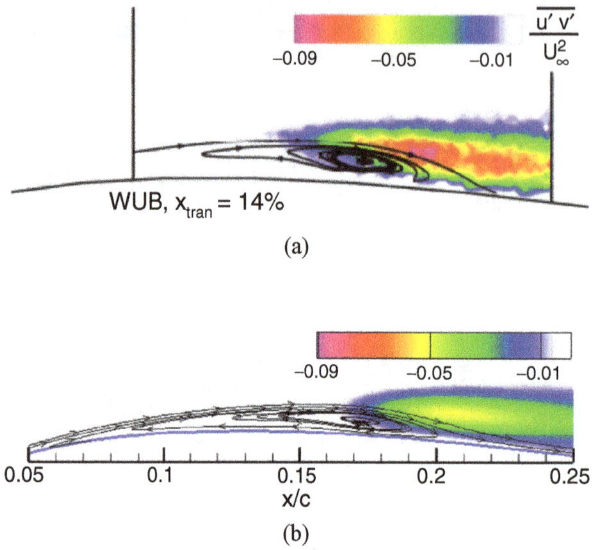

Figure 2.9. Streamlines and turbulent shear stress for the AoA = 8°. (a) Experimental measurement by Radespiel et al. [135] and (b) numerical simulation with $N = 8$ by Lian and Shyy [117].

but does not show considerable change when the Reynolds number increases further. Therefore one does not observe much increase in the lift coefficient even though the LSB length is shorter at higher Reynolds numbers. One can conclude from Figure 2.11d that the enhancement of lift-to-drag ratio is mainly due to the reduction of friction drag at high Reynolds numbers. As the Reynolds number increases, the form drag does not vary as much as does the friction drag.

2.2 Factors Influencing Low Reynolds Number Aerodynamics

Because of the influence of flow separation and laminar-turbulent transition, the preferred airfoil shapes in the low Reynolds number regime are different from those

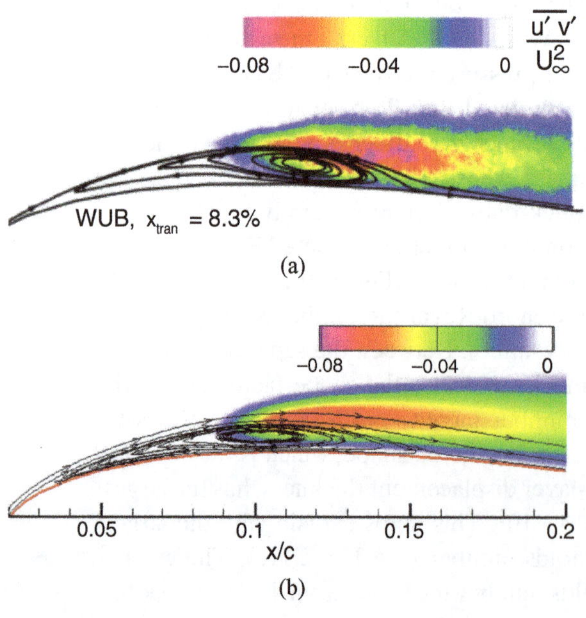

Figure 2.10. Streamlines and turbulent shear stress for the AoA = 11°. (a) experimental measurement by Radespiel et al. [135] and (b) numerical simulation with $N = 8$ by Lian and Shyy [117].

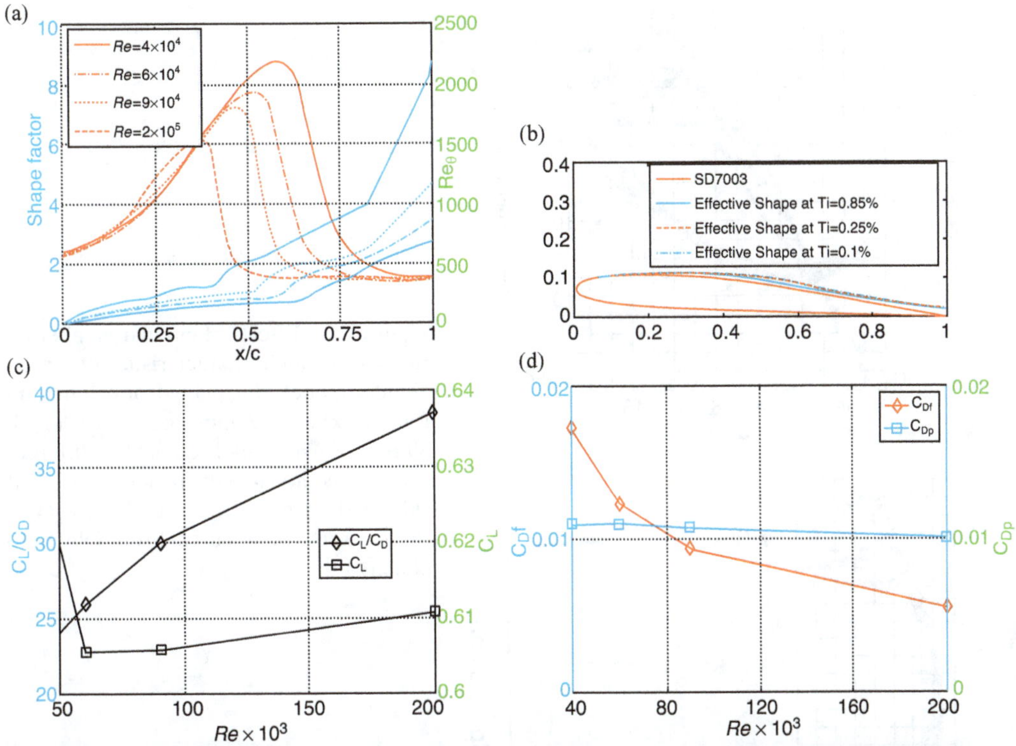

Figure 2.11. Reynolds number effect on the LSB profile and aerodynamic performance at the AoA = 4° for an SD7003 airfoil [117]: (a) shape factor and momentum-thickness-based Reynolds number, (b) effective airfoil shape, (c) lift-to-drag ratio, and (d) drag coefficient.

in the high Reynolds number regime. Furthermore, in addition to the Reynolds number, the airfoil camber, thickness, surface smoothness, free-stream unsteadiness, and aspect ratio all play important roles in determining the aerodynamic performance of a low Reynolds number flyer. These factors are discussed in this section.

2.2.1 *Re* = 10^3–10^4

Okamoto et al. [139] experimentally studied the effects of wing camber on wing performance with Reynolds numbers as low as 10^3 to 10^4. Their experiment used rectangular wings with an *AR* of 6, constructed from aluminum foil or balsa wood. Figure 2.12 illustrates the effects of camber on the aerodynamic characteristics. As the camber increases, the lift coefficient slope and the maximum lift coefficient increase as well. The increase in camber pushes both the maximum lift coefficient and maximum lift-to-drag ratio to a higher AoA. More interestingly, the 3 percent camber airfoil shows a stall-resisting tendency, with the lift just leveling off above an AoA of 10°. Although it has the disadvantage of a high drag coefficient, the low-camber airfoil is less sensitive to the AoA and therefore does not require sophisticated steering.

Sunada et al. [140] compared wing characteristics at the Reynolds number of 4×10^3 using fabricated rectangular wings with an *AR* of 7.25; representative wings are shown in Figure 2.13. After testing 20 wings, they concluded that the wing

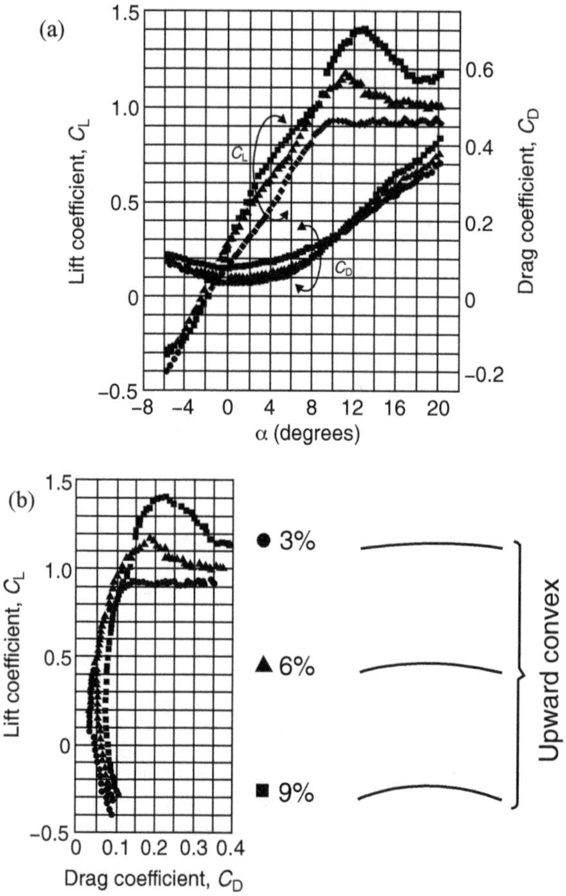

Figure 2.12. Effects of circular camber on the aerodynamic characteristics of a rectangular model wing made from aluminum foil, thickness 0.3 mm and chord length 30 mm. Each symbol refers to a different camber, as shown in the panel on the right side of the figure. (a) C_L and C_D against the AoA; (b) polar curve. From Okamoto et al. [139].

performance can be improved with a modest camber of around 5 percent. Figure 2.14 shows the lift and drag coefficients against the AoA. At the Reynolds number, $Re = 4 \times 10^3$, the effect of camber on aerodynamics found by Sunada and colleagues is similar to that reported by Okamoto et al. [139]. In both experiments, the lift-curve slope increases with the camber; a higher camber wing has a higher stall AoA and generally a larger drag coefficient than a lower camber wing at the same AoA. If we further compare wings of comparable cambers, we notice that they have almost the same stall angle. Sunada et al. [140] also investigated the impact of maximum-camber location, as shown in Figure 2.15. They found that both lift and drag coefficients increase as the position of the maximum camber approaches the trailing edge. In terms of lift-to-drag ratio, the maximum value is obtained when the maximum camber is positioned at 25 percent chord.

Okamoto et al. [139] also studied the effects of airfoil thickness. They found that the wing's aerodynamic characteristics deteriorate as its thickness increases (see Fig. 2.16). In contrast to conventional airfoils, which are smooth and streamlined, insect airfoils exhibit rough surfaces; for example, the cross-sectional corrugations of dragonfly wings (shown in Fig. 217c) or scales on the wing surface (butterfly and moth). Evidence has shown that the corrugated wing configuration confers both structural and aerodynamic benefits to the dragonflies. First, it is of critical

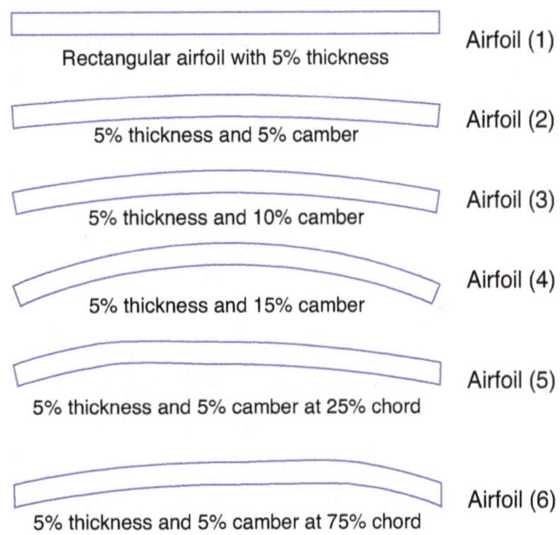

Figure 2.13. Airfoil shapes tested in Sunada et al. [140]. Redrawn from the original reference with permission.

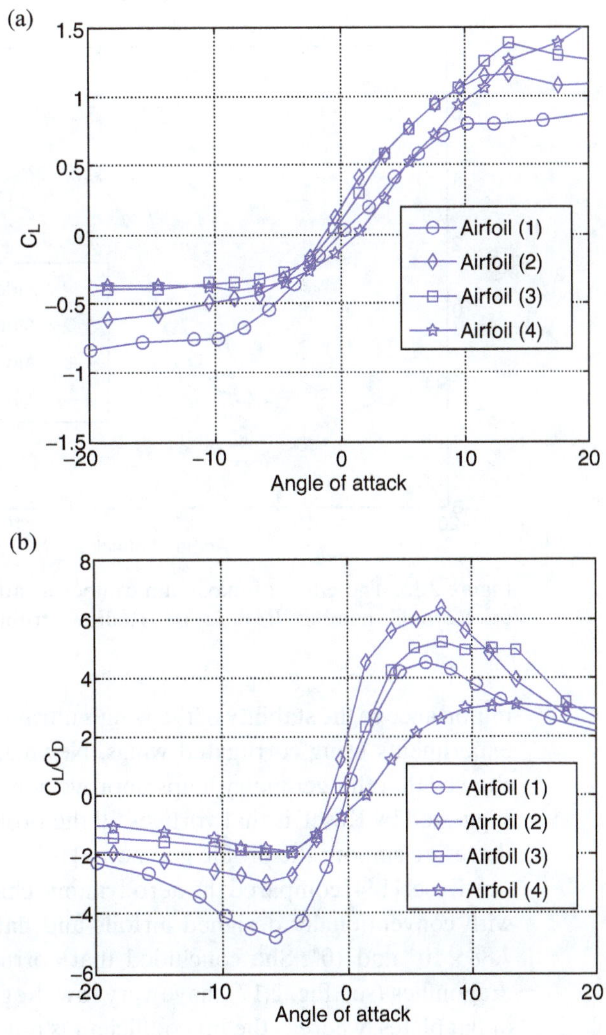

Figure 2.14. The effect of maximum camber location on the aerodynamic characteristics at the Reynolds number, $Re = 4 \times 10^3$. Redrawn from Sunada et al. [140] with permission.

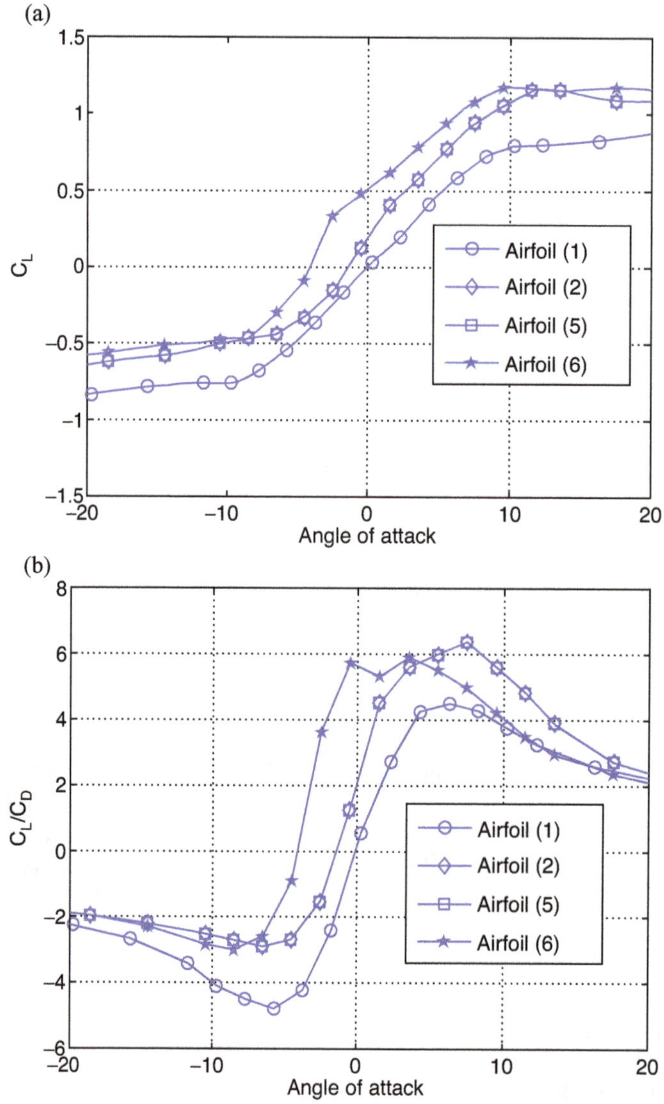

Figure 2.15. The effect of maximum camber location on the aerodynamic characteristics at the Reynolds number, $Re = 4 \times 10^3$. Redrawn from Sunada et al. [140] with permission.

importance to the stability of the wing's ultralight construction. Second, in visualizing experiments using corrugated wings, Newman and Low [141] and Buckholz [142] showed that this geometry helps improve aerodynamic performance. The reason, as suggested by Kesel, is that vortices fill the profile valleys formed by these bends and therefore smooth the profile geometry [143].

Kesel [144] compared the aerodynamic characteristics of dragonfly wing sections with conventionally designed airfoils and flat plates at the Reynolds numbers of 7.88×10^3 and 10^4. She concluded that corrugated airfoils, such as those seen in dragonflies (see Fig. 2.17), have very low drag coefficients closely resembling those of flat plates, whereas the lift coefficient is much higher than those of flat plates. She also investigated the performance of the airfoil by simply filling the valleys with solid

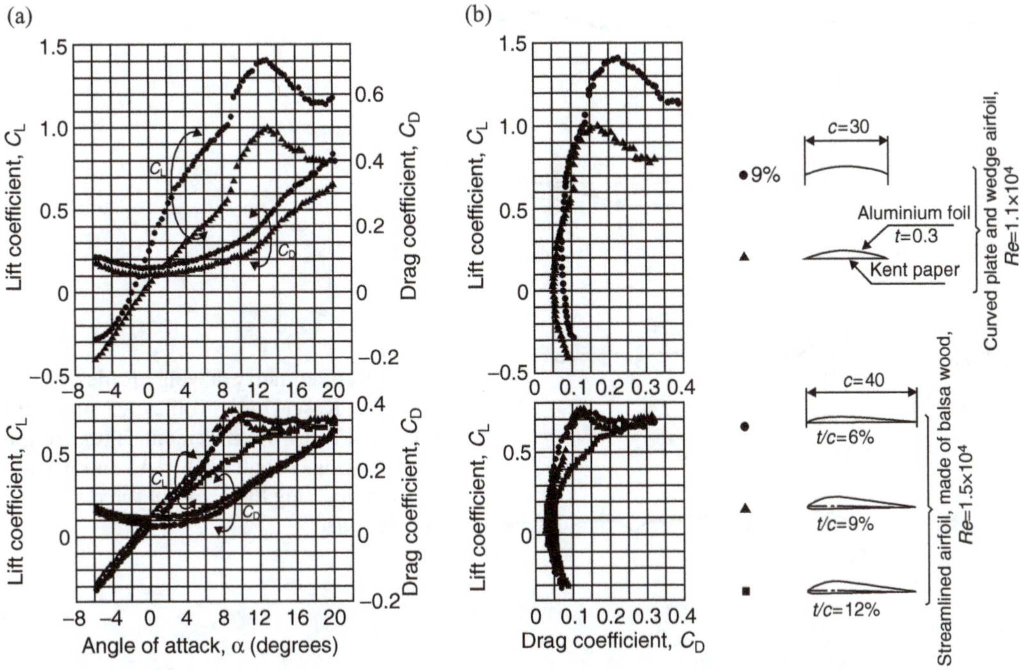

Figure 2.16. Effects of thickness on the aerodynamic characteristics of a curved-section model wing (camber 9%). Each symbol refers to a different airfoil shape as shown in the panel on the right; *c*, chord length; *t*, thickness; *Re*, Reynolds number. All dimensions are given in millimeters. (a) C_L and C_D against the AoA; (b) polar curve. From Okamoto et al. [139].

Figure 2.17. (a) Dragonfly wings exhibit both flexibility and anisotropic, corrugated structures. (b) Like many other natural flyers, the dragonfly's performance is robust, capable of performing adequately even with damaged wings. (c) Close-up view of the dragonfly wing structure; shown on the left is the hindwing and on the right is the forewing. (d) A dragonfly approaches the target.

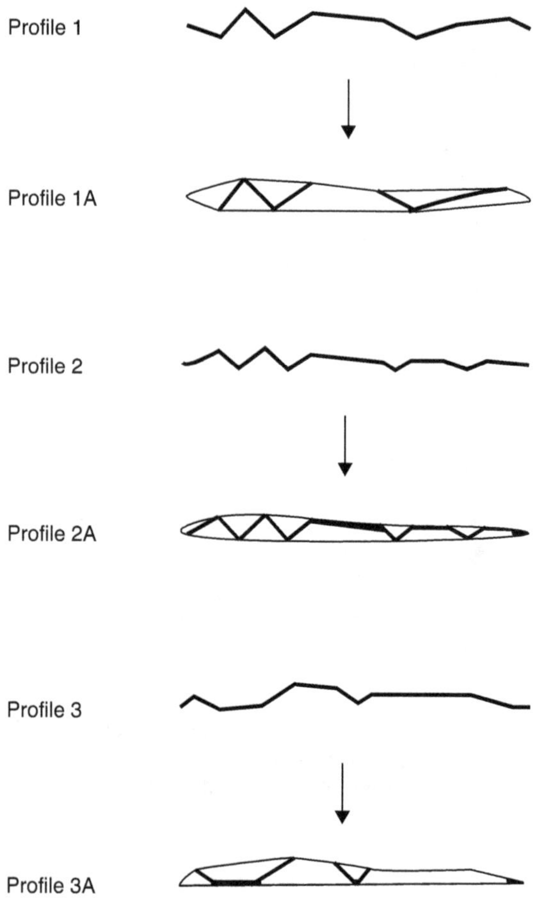

Figure 2.18. Geometry of wing profiles used in the study of Kesel [144]. Profiles 1, 2, and 3 are constructed using measurement taken from a dragonfly wing. Profiles 1A, 2A, and 3A are built by connecting the peaks of the respective cross-sections.

materials (as illustrated in Fig. 2.18). Figure 2.19 highlights the key features of the lift and drag values against AoAs between the three natural and filled airfoil profiles. Figure 2.20 shows the corresponding lift-drag polar. These plots, taken from Kesel [144], show that the filled airfoils have less favorable aerodynamic performances. Therefore it is clear that the performance of such corrugated airfoils is influenced by their "effective" shape, which is characterized by the viscous effects, as previously discussed. In particular, the viscosity and associated vortical structures result in an airfoil with cambered geometry [144].

Vargas et al. [145] numerically investigated the effects of the pleats on the aerodynamic performance at $Re = 10^2$–10^4 with the range of AoA from 0° to 10°. The pleats' effect on the flow is most evident at low AoAs where the flow is basically attached to both surfaces of the wing section. At those AoAs, although the pleated airfoil experiences an increase in the pressure drag, this increase is more than offset by a concomitant decrease in the shear drag. The reduction in the shear drag occurs because there are recirculation zones inside the cavities formed by the pleats, which lead to a negative shear drag contribution (see Fig. 2.21). Also noted by Vargas et al. [145] is that, beyond the Reynolds number of 5.0×10^3 and an AoA of 5°, the pleated airfoil performs better than the flat plate and the profiled airfoil. Kim et al. [146] also numerically investigated the aerodynamic performance of the dragonfly wing at $Re = 1.5 \times 10^2$, 1.4×10^3, and 10^5 with an AoA ranging from 0 to 40°. Their

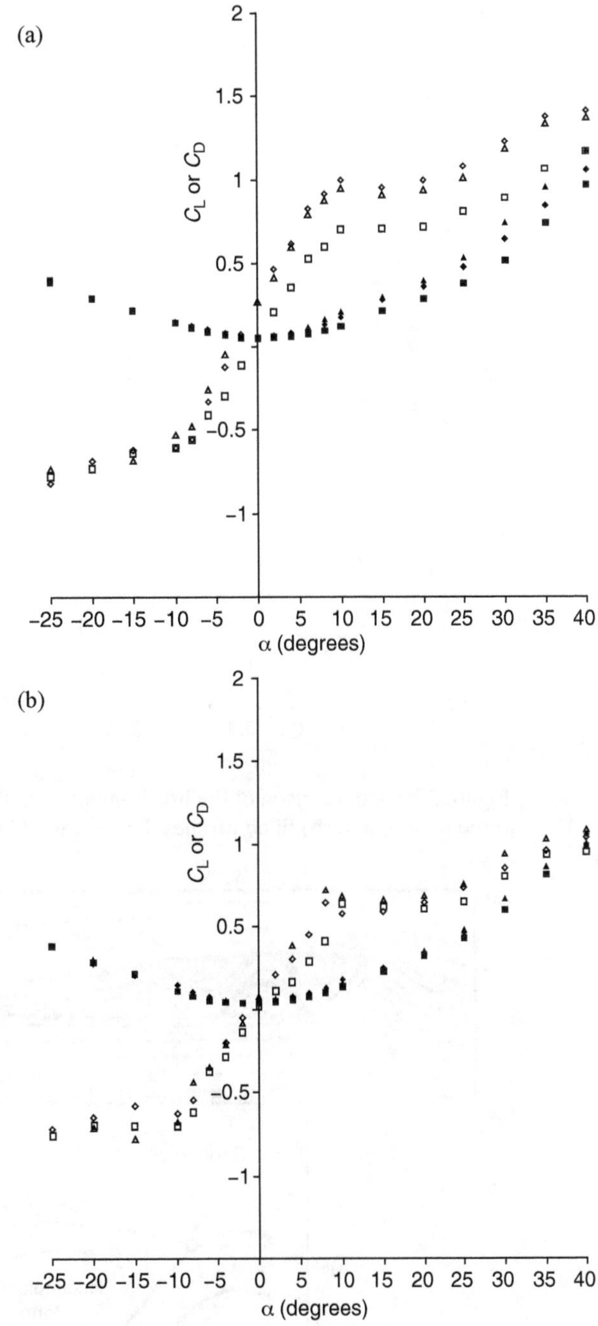

Figure 2.19. Comparison of the lift (open symbols) and drag (filled symbols) coefficients versus the AoA, at the Reynolds number, $Re = 1 \times 10^4$, for (a) three natural and (b) filled profiles. From Kesel [144].

results showed that the wing corrugation enhances lift generation with little influence on drag. The vortex in the valley of corrugation and near the edge of the corrugation is locally different from those of an elliptic wing. However, such local flows have little influence on the time-averaged wing performance. The suction side corrugations of the wing have very little influence on increasing the lift coefficient at a positive angle of attack.

In short, as McMaster [12] has interpreted and as shown in Figure 2.17, the implication of the corrugated wing shape is that, in a low Reynolds number regime

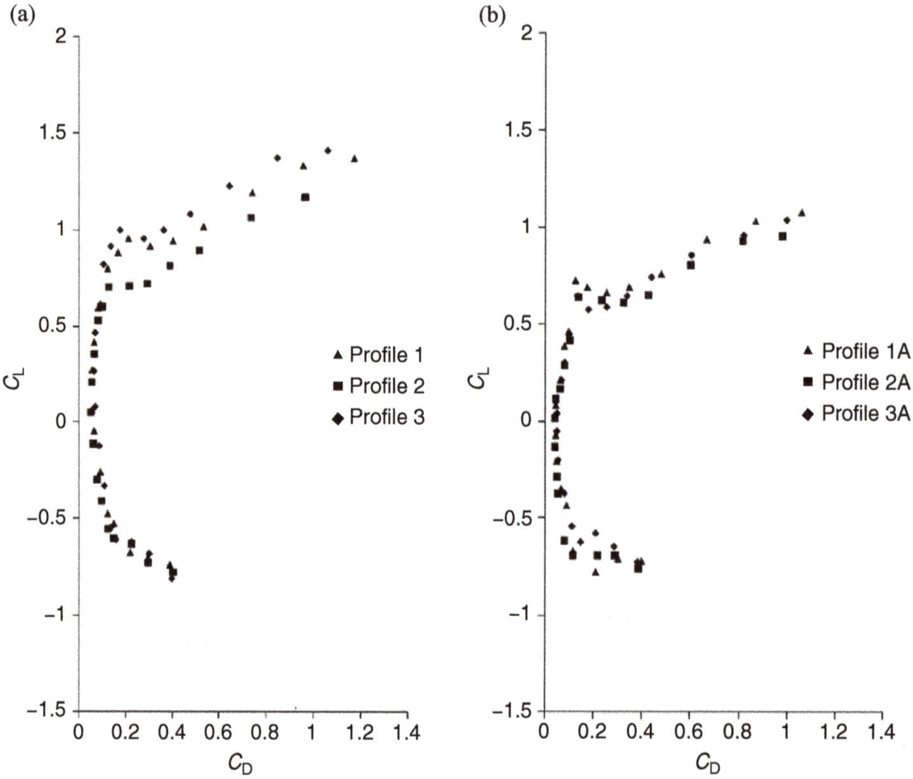

Figure 2.20. Comparison of the lift-drag polars at the Reynolds number, $Re = 1 \times 10^4$, for (a) three natural and (b) filled profiles. From Kesel [144].

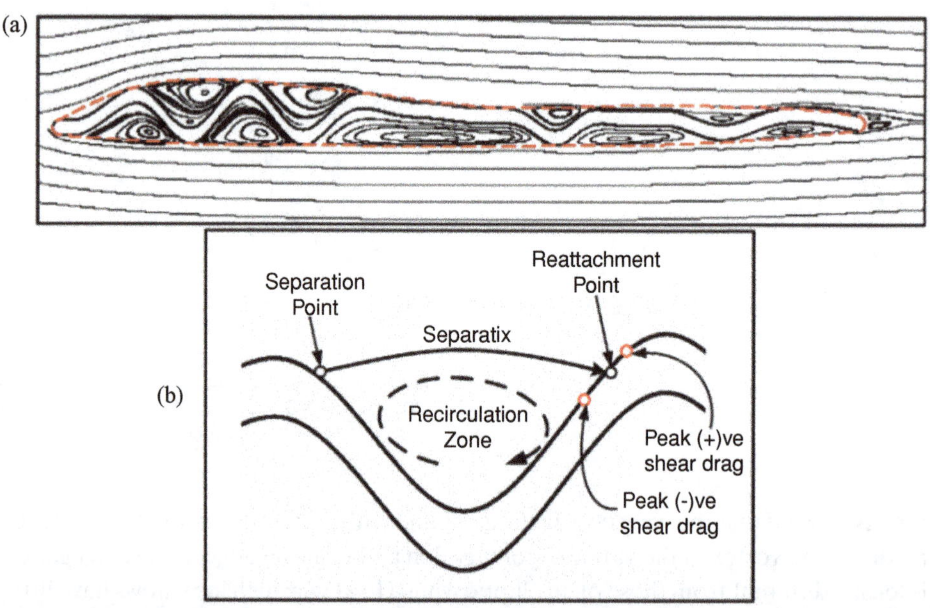

Figure 2.21. (a) Time-averaged streamlines generated by the pleated airfoil at $Re = 10^4$ with the outline of the profile airfoil depicted by dashed lines superimposed to demonstrate that flow looks as if it pertains to this airfoil. (b) Schematic showing the correspondence of surface shear with the flow pattern in typical corrugation. From Vargas et al. [145].

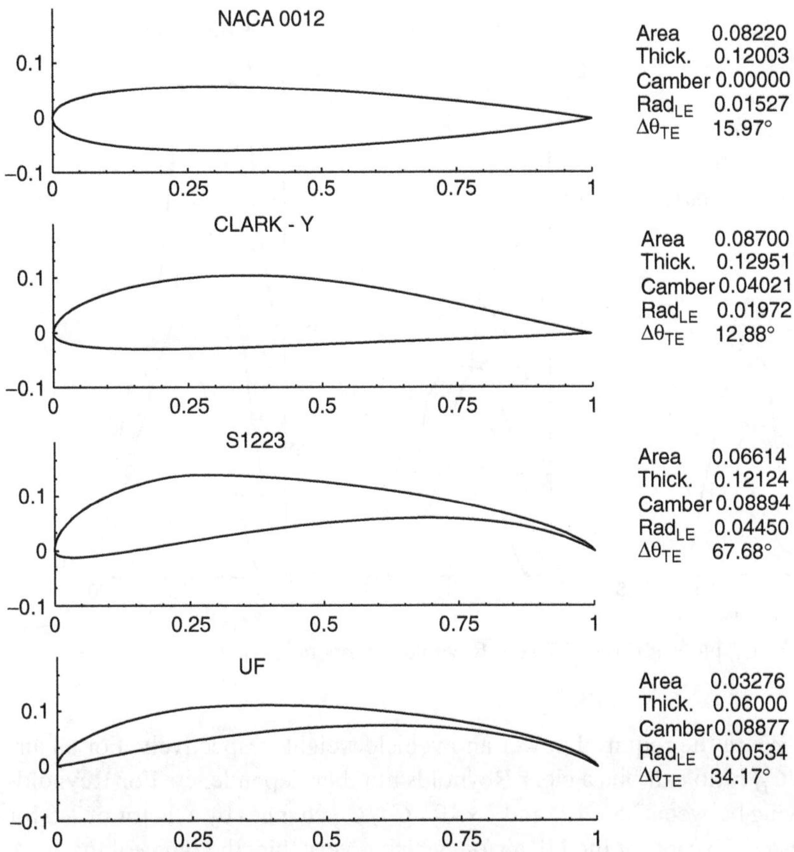

Figure 2.22. Four airfoils chosen for assessment [147].

due to the viscous effects of the flow around the wing, the original airfoil shape behaves effectively like a smooth envelope contour resembling a conventional airplane airfoil. Of course, for dragonfly, this explanation is incomplete because the unsteady fluid physics associated with the flapping motion and deformable wing shapes and the interactions between fore- and hindwings interplay with significant effects.

2.2.2 *Re* = 10^4–10^6

Shyy et al. [147] evaluated the aerodynamics between the chord Reynolds number of 7.5×10^4 and 2×10^6, using the XFOIL code [96], for two conventional airfoils – NACA 0012 and CLARK-Y – and two low Reynolds number airfoils: S1223 [148] and an airfoil modified from S1223, which is called UF (see Fig. 2.22). Figures 2.23 and 2.24 show the power index, $C_L^{3/2}/C_D$, and lift-to-drag ratio, C_L/C_D, plots at three Reynolds numbers: 7.5×10^4, 3×10^5, and 2×10^6. It is noted that for steady-state flight, the power required for maintaining a fixed-wing vehicle in the air is

$$P = W\left(C_D/C_L^{3/2}\right)\sqrt{\frac{2W}{\rho_f S}}, \tag{2–20}$$

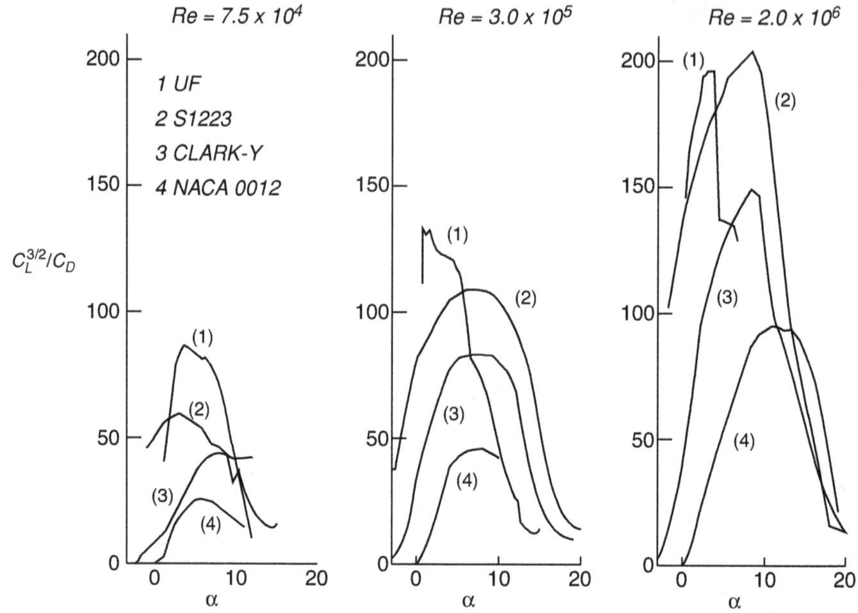

Figure 2.23. $C_L^{3/2}/C_D$ plots at three different Reynolds numbers [147].

where P and W are the required power and vehicle weight, respectively. For all airfoils, the C_L/C_D ratio exhibits a clear Reynolds number dependency. For Reynolds numbers varying between 7.5×10^4 and 2×10^6, C_L/C_D changes by a factor of 2–3 for the airfoils tested. Except for the UF airfoil, which is very thin, the range of the AoA within which aerodynamics is satisfactory becomes narrower as the Reynolds number decreases. Clearly, the camber is important. NACA 0012, with 0 percent camber, and CLARK-Y with 3.5 percent camber, yield a less satisfactory performance under all three Reynolds numbers. S1223 and UF, both with 8.89 percent camber, perform better.

Finally, NACA 0012, CLARK-Y, and S1223 all have maximum thickness of about 0.12c. The UF airfoil, in contrast, is considerably thinner, with a maximum thickness of 0.06c. It is interesting to compare the Reynolds number effect. At the Reynolds number $Re = 2 \times 10^6$, S1223 and UF airfoil have comparable peak performances in terms of $C_L^{3/2}/C_D$ and C_L/C_D; however, S1223 exhibits a wider range of acceptable AoAs. At the Reynolds number $Re = 7.5 \times 10^4$, the situation is quite different. UF, the thinner airfoil with identical camber, exhibits a substantially better aerodynamic performance while maintaining a comparable range of acceptable AoAs. This is consistent with the finding of Okamoto et al. [139] discussed previously.

Murphy and Hu [149] experimentally measured the aerodynamic characteristics of a bio-inspired corrugated airfoil compared with a smooth-surfaced airfoil and a flat plate at the chord Reynolds numbers of 5.8×10^4 and 1.25×10^5. Their measurement result revealed that the corrugated airfoil has better performance in providing higher lift and preventing large-scale flow separation and airfoil stall at low Reynolds numbers ($<10^5$) than the smooth-surfaced airfoil and the flat plate, as shown in Figures 2.25 and 2.26. However, the corrugated airfoil was found to have higher drag coefficients compared to those of the smooth-surfaced airfoil and the flat plate at low

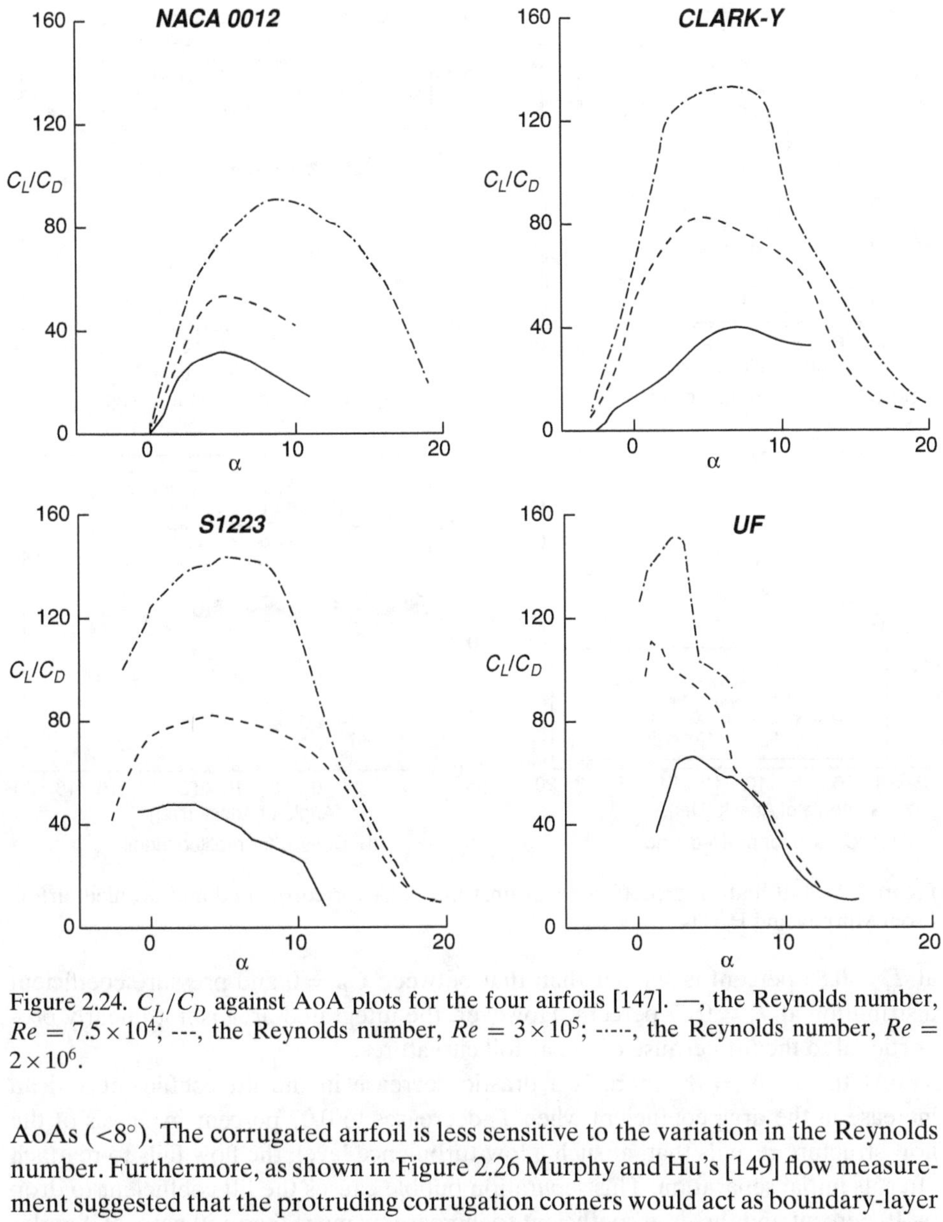

Figure 2.24. C_L/C_D against AoA plots for the four airfoils [147]. —, the Reynolds number, $Re = 7.5 \times 10^4$; ---, the Reynolds number, $Re = 3 \times 10^5$; -·-·-, the Reynolds number, $Re = 2 \times 10^6$.

AoAs (<8°). The corrugated airfoil is less sensitive to the variation in the Reynolds number. Furthermore, as shown in Figure 2.26 Murphy and Hu's [149] flow measurement suggested that the protruding corrugation corners would act as boundary-layer trips to promote the transition of the boundary layer from laminar to turbulent while remaining "attached" to the envelope profile of the high-speed streamlines.

2.2.3 Effect of Free-Stream Turbulence

When both the AoA and chord Reynolds number are fixed, increasing the free-stream turbulence level prompts an earlier transition. The aerodynamic characteristics under different turbulence intensities were investigated by Lian and Shyy [117]. The lift and drag coefficients from their research are shown in Figure 2.27. At the AoA = 4°, there is no noticeable difference in the lift and drag coefficients among the five tested turbulence levels. This seemingly contradicts the pressure coefficient plot in Figure 2.28 because the integrated area between $C_P = 0$ and C_P distribution

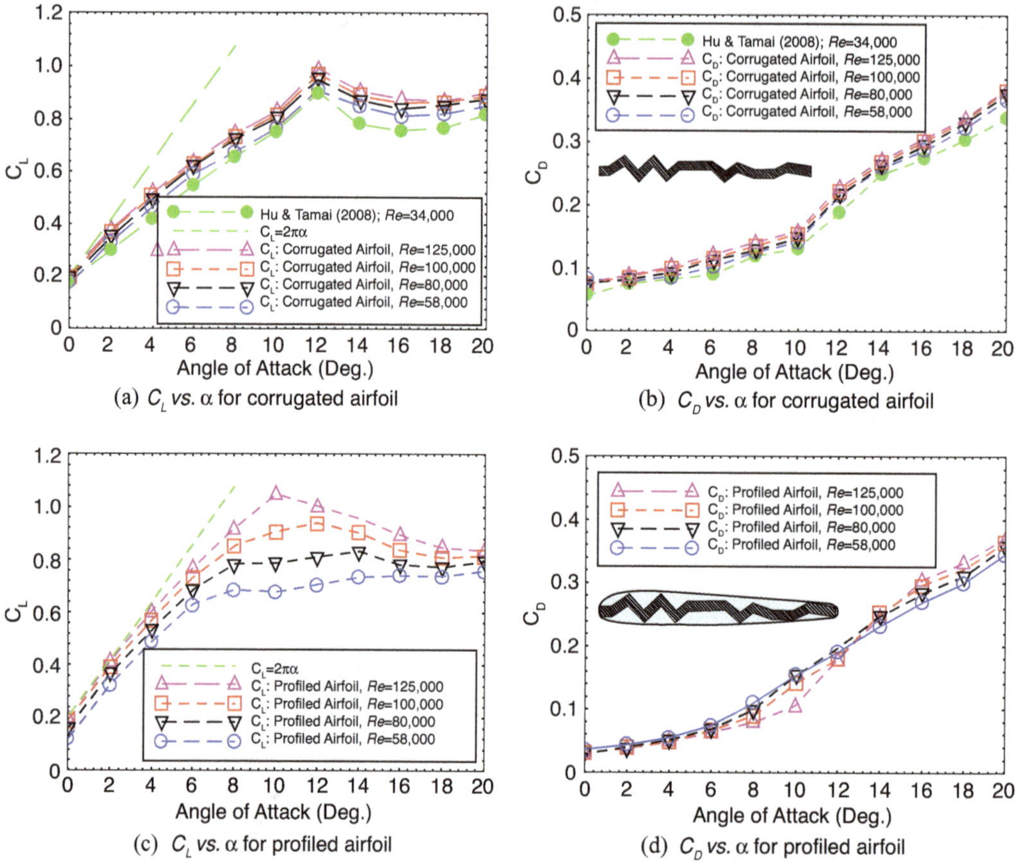

Figure 2.25. Lift and drag coefficients against the AoA for corrugated and profiled airfoil. From Murphy and Hu [149].

at $T_i = 0.85$ percent is smaller than that between $C_P = 0$ and pressure coefficient distribution at $T_i = 0.07$ percent. However, the integrated area is not linearly proportional to the lift because of the airfoil curvatures.

At the AoA $= 8°$, there is a drastic decrease in the lift coefficient and an increase in the drag coefficient when T_i decreases to 0.07 percent. Analysis of the flow structure reveals that, at such a low-turbulence level, the flow fails to reattach after its initial separation. This separation bubble causes the lift coefficient to drop by 10 percent and the drag coefficient to increase by more than 150 percent. Similar conclusions can also be drawn for the case of AoA of 11°.

In general, with the increase in the free-stream turbulence level, the LSB becomes thinner and shorter. This is clearly shown in Figure 2.29. From the same figure it can also be seen that the shear stress decreases with the turbulence level. Because of the viscous effect, the boundary layer and the LSB change the effective shape of the airfoil. As shown in Figure 2.30, the free-stream with a higher turbulence level results in a relatively thinner effective airfoil than that with a lower turbulence level.

O'Meara and Mueller [150] experimentally studied the effects of free-stream turbulence on the separation bubble characteristics of a NACA 663–018 airfoil. They reported that, as the disturbance level is increased, the bubble is reduced in

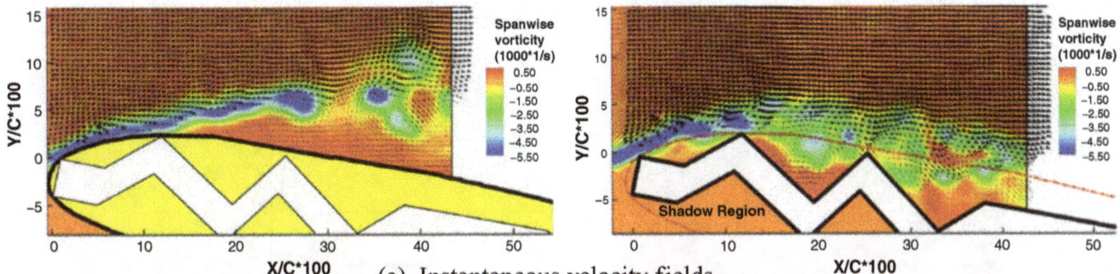

(a) Instantaneous velocity fields

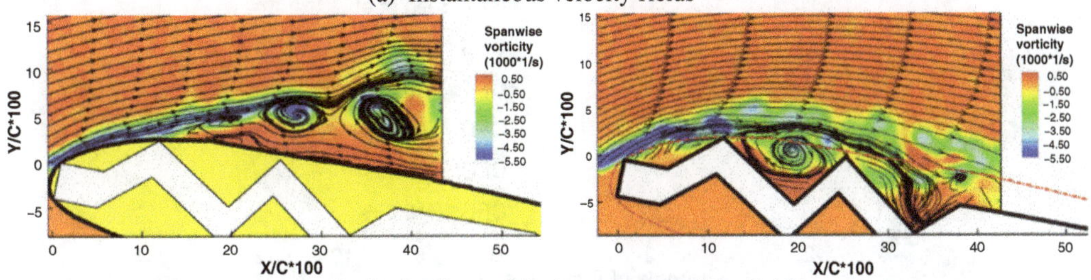

(b) Streamlines of the instantaneous flow fields

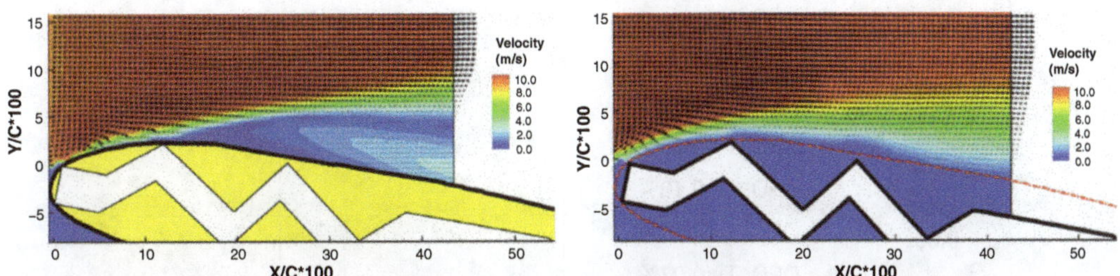

(c) Ensemble-averaged velocity fields

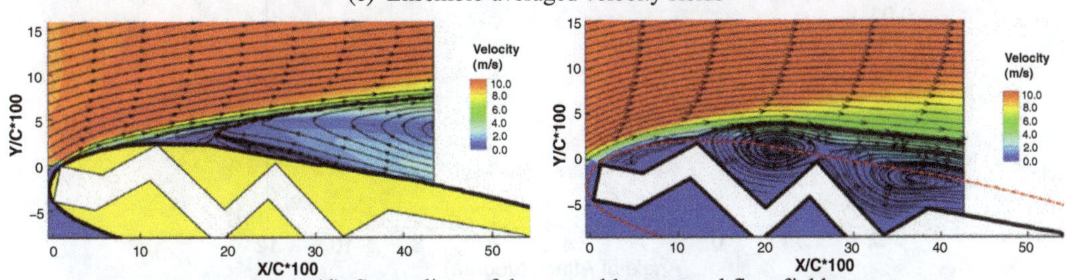

(d) Streamlines of the ensemble-averaged flow fields

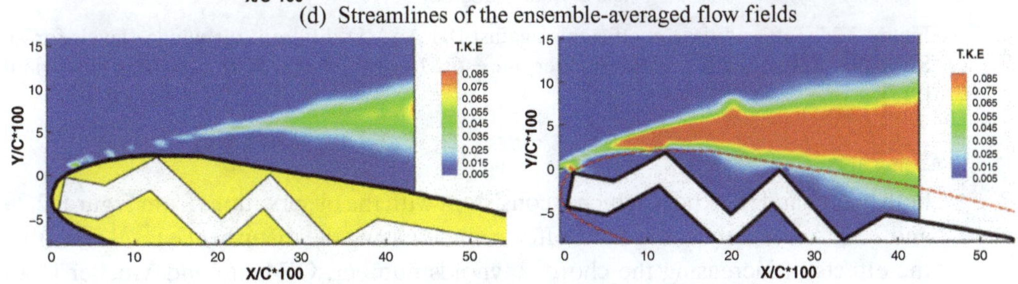

(e) Normalized turbulent kinetic energy distributions

Figure 2.26. Refined PIV measurements with AoA $= 12^\circ$ at the Reynolds number, $Re = 5.8 \times 10^4$. The results are from Murphy and Hu [149].

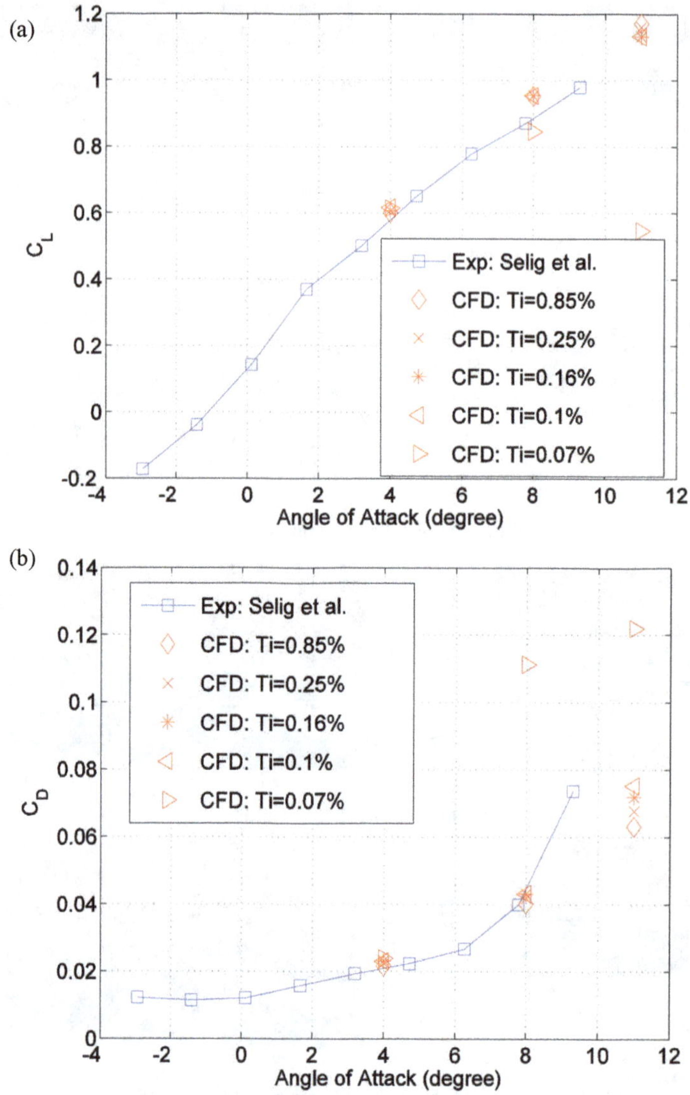

Figure 2.27. Lift and drag coefficients against the AoA at different turbulence levels for an SD7003 airfoil at the Reynolds number, $Re = 6 \times 10^4$: (a) lift coefficient; (b) drag coefficient [117].

both length and thickness, which is consistent with the observations from Figures 2.29 and 2.30. As we discuss later, the effects of increasing the disturbance level resemble the effects of increasing the chord Reynolds number. O'Meara and Mueller [150] also reported that the suction peak grows in absolute magnitude with the disturbance level. However, as shown in Figure 2.28, the pressure peak over the SD7003 airfoil is not sensitive to the disturbance level. These two conclusions are drawn based on different test cases, in which the bubble size and Reynolds number are quite different. The results in Figure 2.31 are obtained at a chord Reynolds number of 1.4×10^5; occupying around 7 percent of chord length, the bubble is short and, as previously discussed, only locally affects the pressure distribution. In contrast, in

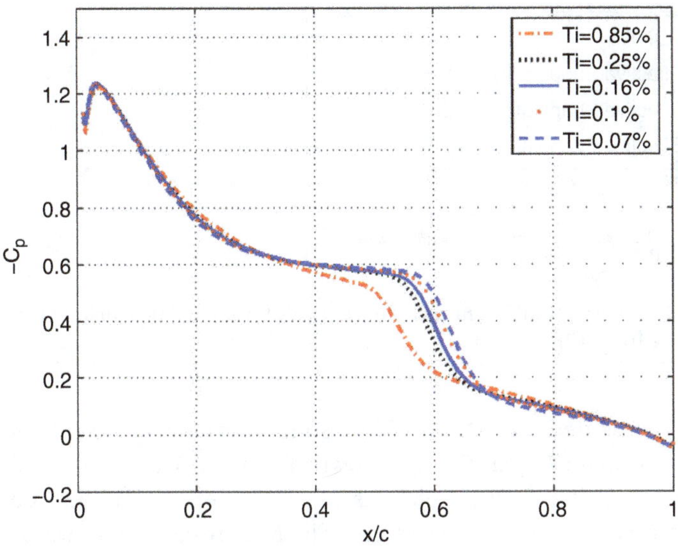

Figure 2.28. Pressure coefficient on the suction surface at the AoA $= 4^\circ$ at different turbulence levels for the SD7003 airfoil at the Reynolds number, $Re = 6 \times 10^4$ [117].

the test of Lian and Shyy [117], the bubble covers more than 30 percent of the upper surface at the Reynolds number of 6×10^4 and the AoA $= 4^\circ$, and the bubble falls into the long-bubble category. This hypothesis is further confirmed by the fact that, at the Reynolds number of 6×10^4 and the AoA $= 8^\circ$, wherein the bubble is 8 percent of the chord, the pressure peak magnitude does increase with the increase in the disturbance level.

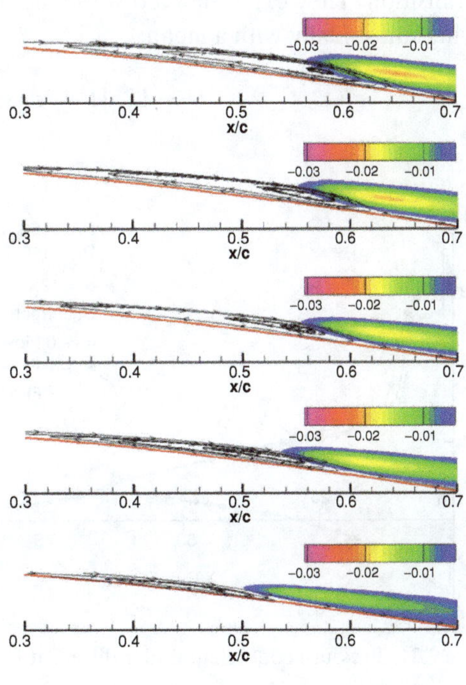

$Ti = 0.07\%$

$Ti = 0.1\%$

$Ti = 0.16\%$

$Ti = 0.25\%$

$Ti = 0.85\%$

Figure 2.29. Streamlines and normalized shear-stress contours at the AoA $= 4^\circ$ for different turbulence levels for an SD7003 airfoil at the Reynolds number, $Re = 6 \times 10^4$ [117].

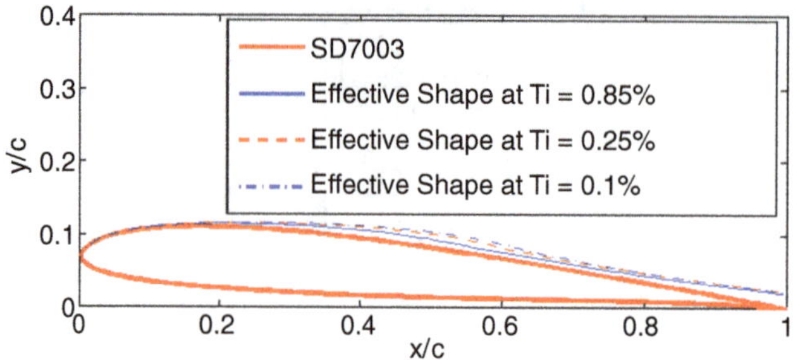

Figure 2.30. Effective airfoil shapes at different turbulence levels for an SD7003 airfoil at the Reynolds number, $Re = 6 \times 10^4$ [117].

Mueller et al. [151] presented the effects of free-stream turbulence on lift and drag performances of a Lissaman 7769 airfoil. As shown in Figure 2.32, the hysteresis characteristics of the lift and the drag coefficients can be observed for the free-stream disturbance intensity of around 0.10 percent. The hysteresis loop, however, disappears as the free-stream turbulence intensity is increased to 0.30 percent. They suggested that the surface roughness can also produce the same result. Furthermore, the disappearance of the hysteresis loop for aerodynamic lift and drag coefficients at high free-stream turbulence intensity seems to be related to the change in flow structure.

2.2.4 Effect of Unsteady Free-Stream

The real operating conditions for MAVs are quite different from the conventionally low-turbulence wind- and water-tunnel setup. In real flight, MAVs often operate in gusty environments. Obremski and Fejer [152] studied the effect of unsteady flow on transition. They experimented with a flat plate subject to a free-stream velocity varying sinusoidally with a mean:

$$U = U_{\text{ref}}[1 + N_A \sin(2\pi f t)], \tag{2–21}$$

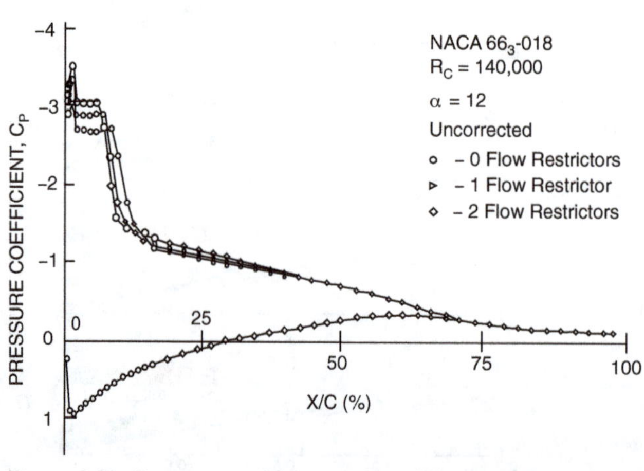

Figure 2.31. Pressure coefficient under different free-stream turbulence levels [150].

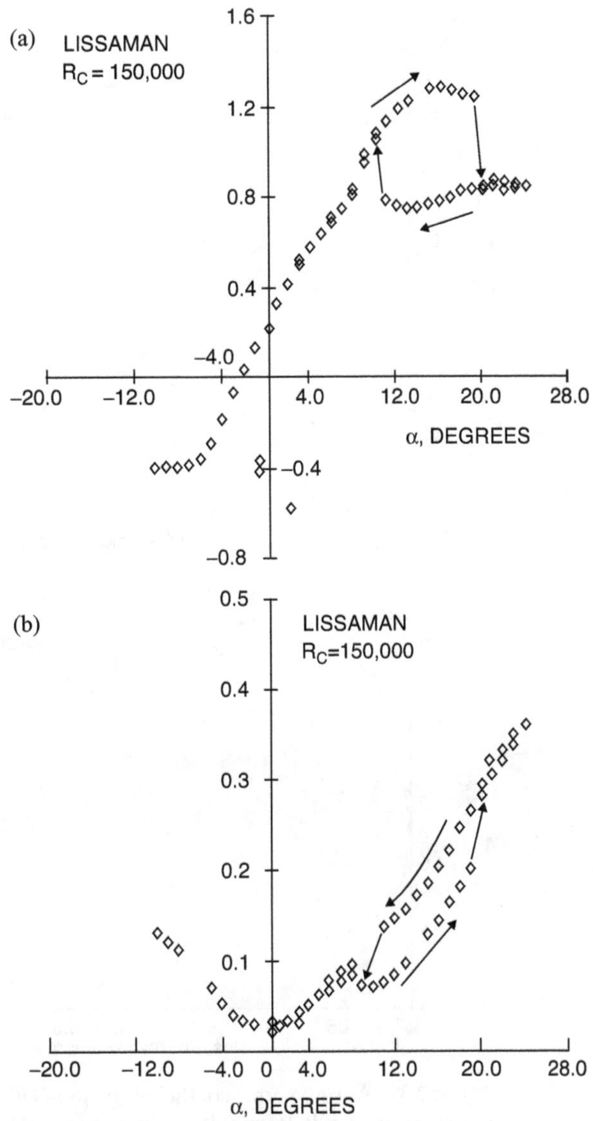

Figure 2.32. Lift and drag coefficients against the AoA for a smooth Lissaman airfoil [151]. (a) Lift coefficient; (b) drag coefficient.

where N_A is the amplitude ratio, f is the frequency, and the reference velocity U_{ref} is the mean free-stream velocity. They found that the transition Reynolds number is affected by the free-stream oscillation when the so-called non-steady Reynolds number, $Re_{ns} = N_A U_{ref}^2 / 2\pi f \nu$, is above a critical point of about 2.6×10^4. Below the critical value, the unsteady free-stream has little impact on the transition process. Obremski and Morkovin [153] observed that, in both high and low Re_{ns} ranges, the initial turbulent bursts are preceded in space and time by a disturbance wave packet. By applying a quasi-steady stability model, they concluded that in the high Re_{ns} range the wave packet is amplified rapidly and bursts into turbulence, whereas in the low range the wave packet bursts into turbulence at a much higher Reynolds number. Guided by their study, Lian and Shyy [117] investigated the influence of free-stream oscillations on the transition for separated flows. In their first test, they set $N_A = 0.33$ and $\omega = 0.3$, resulting in a Strouhal number of 0.0318 and a Re_{ns} of

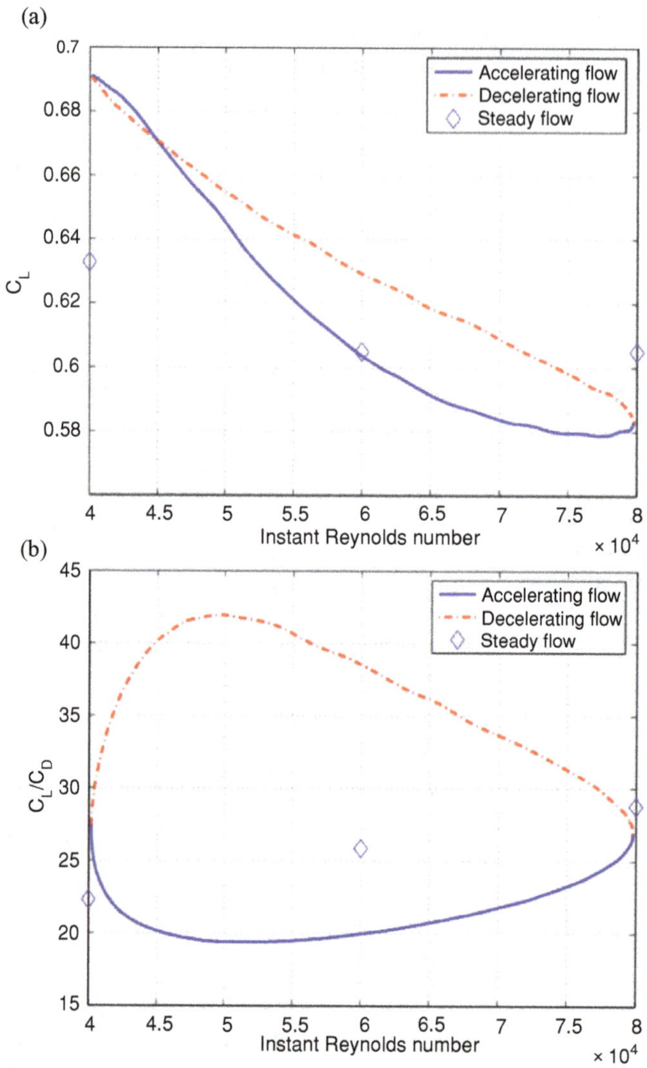

Figure 2.33. Aerodynamic coefficient of an SD7003 airfoil in a gusty environment during one cycle for non-steady Reynolds number, $Re_{ns} = 9.9 \times 10^4$, showing the hysteresis phenomenon: (a) lift coefficient; (b) lift-to-drag ratio [117].

9.9×10^4. They kept the frequency ω well below the range of the expected unstable TS wave frequency, which is around 10 Hz.

Figure 2.33 shows the lift coefficient and lift-to-drag ratio during one selected cycle. Clearly, under a gust situation, the aerodynamic parameters display the hysteresis. For example, when flow accelerates (the Reynolds number increases from 6×10^4 to 8×10^4), the lift coefficient does not immediately reach its corresponding steady-state value. Instead, the steady-state value is reached in the decelerating stage. Compared with a steady incoming flow, the gust leads to a higher lift coefficient at the low-velocity end and a lower lift coefficient at the high-velocity end. The lift-to-drag ratio variation during one cycle is substantial. For example, at the Reynolds number of 6×10^4, the lift-to-drag ratio with a steady-state free-stream is

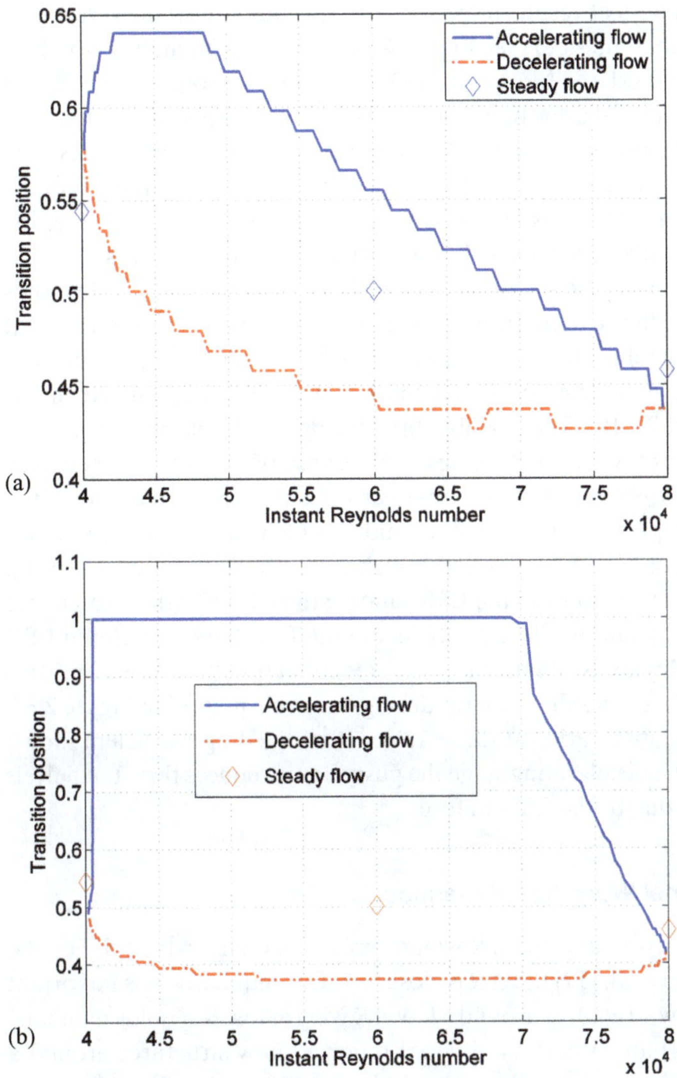

Figure 2.34. Transition position on an SD7003 airfoil during one cycle of the gust at (a) the non-steady Reynolds number, $Re_{ns} = 9.9 \times 10^4$; (b) the non-steady Reynolds number, $Re_{ns} = 1.98 \times 10^4$ [117].

around 26; for gust flow, the instantaneous lift-to-drag ratio reduces to 20 when the flow accelerates, but elevates to 38 when the flow decelerates.

Along with the variations in lift and drag, the transition position is also affected by the gust. As shown in Figure 2.34 the transition position moves toward the leading edge when the flow is accelerating, and it moves toward the trailing edge when flow is decelerating. During the accelerating stage, the instantaneous Reynolds number is increasing. As the Reynolds number increases, the flow experiences an early transition. In the simulation of Lian and Shyy [117], the transition point is simply linked to the computational grid point without further smoothing, resulting in a stair-stepped plot in Figure 2.34.

Lian and Shyy [117] also investigated a higher frequency of $f = 0.24$, five times higher than the previous case, resulting in a Re_{ns} of 1.98×10^4, which is lower than the critical value. Their numerical result shows that the transition position varies with the instantaneous Reynolds number (see Fig. 2.34). This result seemingly contradicts the observation of Obremski and Morkovin [153]. However, it should be noted that Obremski and Morkovin [153] drew their conclusion based on experimentation over a flat plate at a high Reynolds number (10^6), in which the flow is the Blasius type and experiences natural transition. In the test of Lian and Shyy [117], in contrast, the separated flow amplifies the unstable TS wave at such a great rate that it results in a faster transition to turbulence, typical of the bypass-transition process.

Comparison of the transition position at two different non-steady Reynolds numbers reveals that the flow experiences transition for the entire oscillation cycle at a higher non-steady Reynolds number, whereas at the lower value the flow becomes laminar at the early accelerating state and remains laminar until the instant Reynolds number reaches around 7×10^4. It is possible that during the decelerating stage the transition position moves toward the trailing edge because of the lowered Reynolds number. At a higher non-steady Reynolds number (i.e., at a lower frequency), the deceleration has less impact on the transition and the LSB can sustain itself; at a lower non-steady Reynolds number (i.e., at a higher frequency), the deceleration has more impact on the transition and the LSB cannot adjust itself with the high rate of change needed to maintain the closed bubble and the LSB bursts. A closed LSB forms only when the Reynolds number reaches 7×10^4. To better appreciate this phenomenon, see the phase and shape factor during one cycle plotted in Figure 2.35.

Another interesting observation at $Re_n = 1.98 \times 10^4$ is the drag coefficient shown in Figure 2.36. During the decelerating stage the gusty flow produces thrust. Analysis shows that the thrust is due to the friction force.

2.3 Three-Dimensional Wing Aerodynamics

Low Reynolds number flyers use low AR wings, typically no larger than 5. For the MAVs developed by Ifju et al. [17] the AR is close to 1. Consequently, it is important to investigate the 3D flow structures around a low AR wing at low Reynolds numbers.

Lian and Shyy [154] and Viieru et al. [155] reported flow structures around a low AR rigid wing. The geometry follows the design of Ifju et al. [17] as discussed earlier. The wing has a span of 15 cm, a camber of 6 percent, a root chord of 13.3 cm, and a wing area of 160 cm^2.

To confirm the capabilities of the Navier-Stokes solver, the computational results are first compared with wind-tunnel data measured for a MAV rigid wing with a 12.5 cm span, which has a smaller planform area than those used by Lian and Shyy [154] and Viieru et al. [155]. However, the overall shape and AR are similar. The experiment is conducted in a horizontal, open-circuit low-speed wind tunnel. It has a square entrance of a bell-mouth-inlet type, and it has several screens that provide low turbulence levels, less than 0.1 percent, in the test section. The test section is 91.4 cm $\times$ 91.4 cm and has a length of 2 m. The model under test is attached to a six-component strain-gauge sting balance used to measure the aerodynamic forces and moments. The AoA is controlled by computer and can be set in any sequence, steady or variable, in time. The force balance is calibrated from 1 gram to 500 grams, from precisely defined loading points. For more detailed information on the experimental measurements and uncertainty, we refer to Albertani et al. [156].

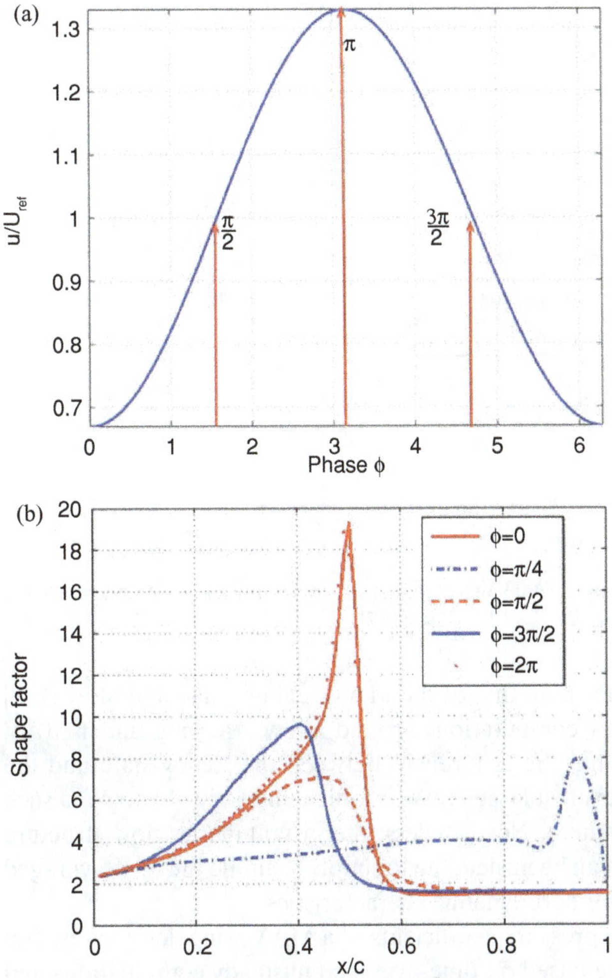

Figure 2.35. Phase and shape factors during one gust cycle on an SD7003 airfoil at the non-steady Reynolds number, $Re_{ns} = 1.98 \times 10^4$: (a) phase; (b) shape factor [117].

The 12.5 cm wing configuration is tested at two different Reynolds numbers (7.1×10^4 and 9.1×10^4) based on the root chord length. The experimental data are obtained by averaging the values from multiple tests for each AoA and Reynolds number. In Figure 2.37a the lift versus drag curves are plotted for the two Reynolds numbers just mentioned. The figure demonstrates agreement between the computational and experimental data. As shown in Figure 2.37b, within the considered Reynolds number range, the lift-to-drag ratio does not vary much. Furthermore, both experimental and computation data show that the best lift-to-drag ratio is reached for an AoA between 4° and 9°.

2.3.1 Unsteady Phenomena at High AoAs

Vortex shedding causes more than just unsteadiness in aerodynamic performance. Cummings et al. [157] reported that, at large AoAs, the unsteady computations predict noticeably lower lift coefficients than do the steady computations. The Reynolds

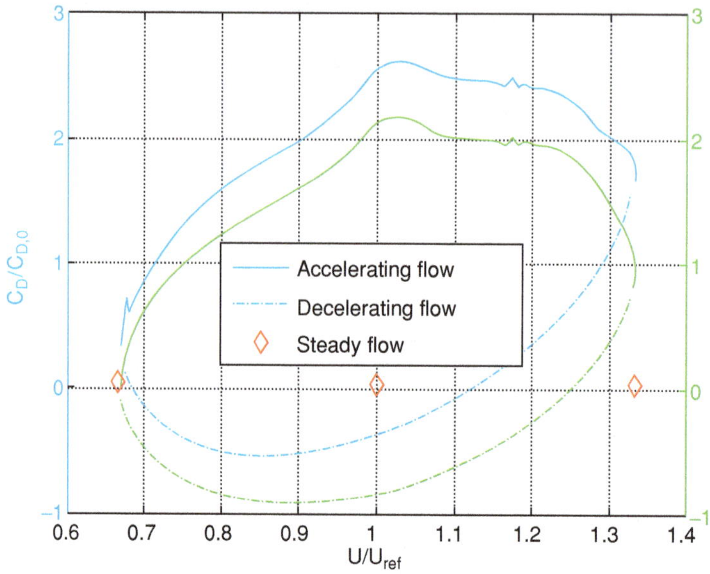

Figure 2.36. Drag coefficient of an SD7003 airfoil in a gusty environment during one cycle for the non-steady Reynolds number, $Re_{ns} = 1.98 \times 10^4$ [117].

number in their study is higher than that of the MAV regime. Lian and Shyy [158] performed Navier-Stokes flow computations around a low AR wing under MAV flight conditions and found that the differences between the steady-state and the time-averaged lift are small, even at large AoAs in which unsteady phenomena such as vortex shedding are prominent. Nevertheless, the instantaneous flow structure varies substantially. Hence it can be misleading to simply examine the time-averaged flow field to estimate the MAV aerodynamic characteristics.

Figure 2.38 compares the pressure coefficients of a MAV wing designed by Ifju and co-workers[17], which are based on time-averaged unsteady computations and steady-state computations. In this design, the camber gradually decreases from the root toward the tip of the wing. Hence, the flow tends to separate first in the root region. At the AoA = 6° the time-averaged pressure coefficient closely matches the steady-state result. The time-averaged value yields a smooth pressure distribution; the steady-state result indicates a small recirculation zone. As the AoA becomes higher, there is little difference in the leading-edge region; on the contrary, clear differences exist in the separated regions.

2.3.2 Aspect Ratio and Tip Vortices

Tip vortices (TiVs) exist on a finite wing because of the pressure difference between the upper and lower wing surface. The TiV establishes a circulatory motion over the wing surface and exerts great influence on the wing aerodynamics. Specifically, it increases the drag force. The total drag coefficient for a finite wing at subsonic speed can be written [44] as

$$C_D = C_{D,P} + C_{D,F} + \frac{C_L^2}{\pi e AR}, \qquad (2\text{–}22)$$

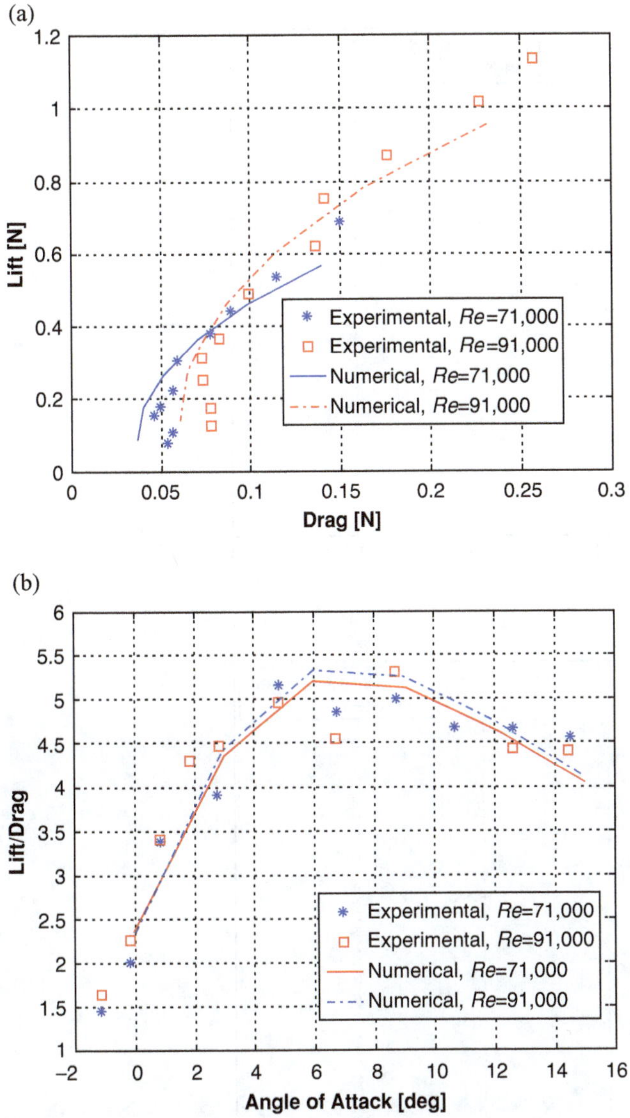

Figure 2.37. Numerical and experimental assessments of lift and drag over a MAV wing for different Reynolds numbers and AoAs [155]: (a) polar curve; (b) lift-to-drag ratio against the AoA.

where $C_{D,P}$ is the drag coefficient due to pressure; $C_{D,F}$ is the drag coefficient due to skin friction; e is the span efficiency factor, which is less than 1; AR is the aspect ratio; and $C_L^2/(\pi eAR) = C_{D,i}$ is the induced drag coefficient due to the existence of TiVs. Equation (2–22) demonstrates that the induced drag varies as the square of the lift coefficient; at a high AoA, the induced drag can be a substantial portion of the total drag. Furthermore, it illustrates that, as AR is decreased, the induced drag increases. The MAV wing presented by Ifju et al. [17] has a low AR of 1.4; therefore, it is important to investigate TiV effects on the wing aerodynamics. In general, TiV effects are twofold: (i) TiV causes downwash that decreases the effective AoA and increases the drag force [44], and (ii) it forms a low-pressure region on the top surface of the wing, which provides additional lift force [159].

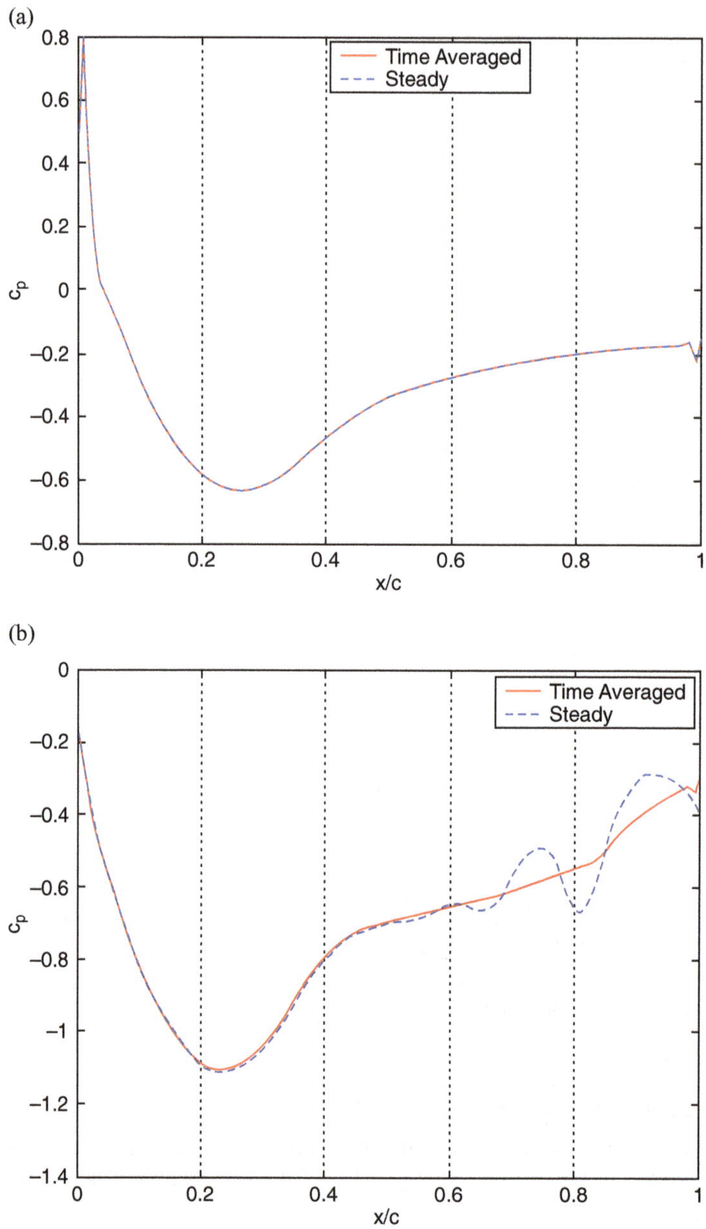

Figure 2.38. Comparisons of C_p on a rigid wing at the root for steady and unsteady computations. (a) AoA $= 6°$; (b) AoA $= 15°$. From Lian and Shyy [158].

Figure 2.39 shows TiVs around the wing surface together with the streamlines at an AoA of 39° [160]. The vortical flow is usually associated with a low-pressure zone as shown in Figure 2.40. The pressure drop further strengthens the swirl by attracting more fluid toward the vortex core; meanwhile, the pressure decreases correspondingly in the vortex core. The low-pressure region created by the vortex generates additional lift. Toward the downstream, the pressure recovers to its ambient value, the swirling weakens, the diameter of the vortex core increases, and the vortex core loses its coherent structure.

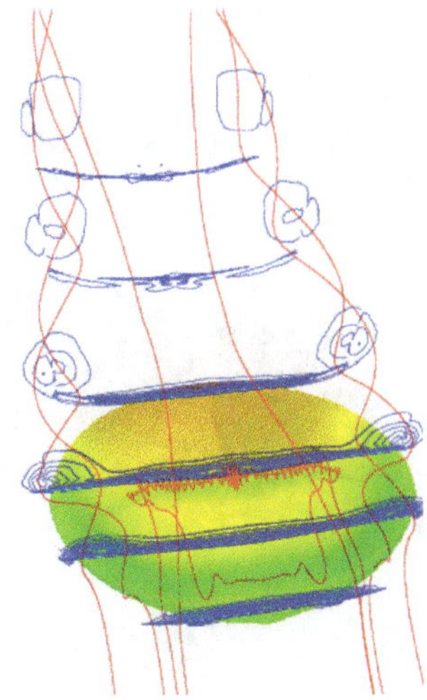

Figure 2.39. Streamlines and vortices for a rigid wing at the AoA = 39°. The vortical structures are shown on selected planes. From Lian et al. [163].

Figure 2.41 visualizes the evolution of the vortical structure with increasing AoA; it also presents the pressure distribution on the upper surface. At the AoA = 6°, TiVs are clearly visible even though they cover a small area and are of modest strength. The flow is attached to the upper surface and follows the chord direction. A low-pressure region is observed near the tip, caused by the vortical structure there.

Even though the flow on the upper surface near the root tends to separate, the flow remains attached in the outer portion of the wing; hence the lift still increases with the AoA until, of course, massive separation occurs on most of the upper surface. For low AR wings, TiVs make considerable contributions to the lift. This

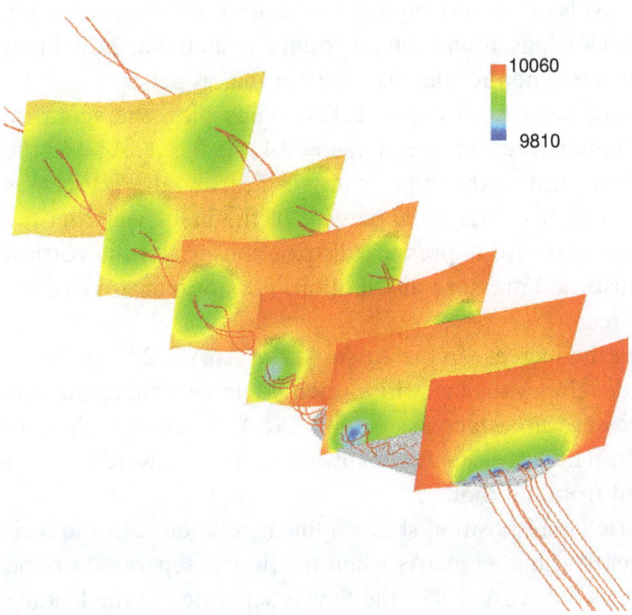

Figure 2.40. Pressure distribution around the rigid wing in the cross-sections with streamlines at the AoA = 39°. From Lian et al. [163].

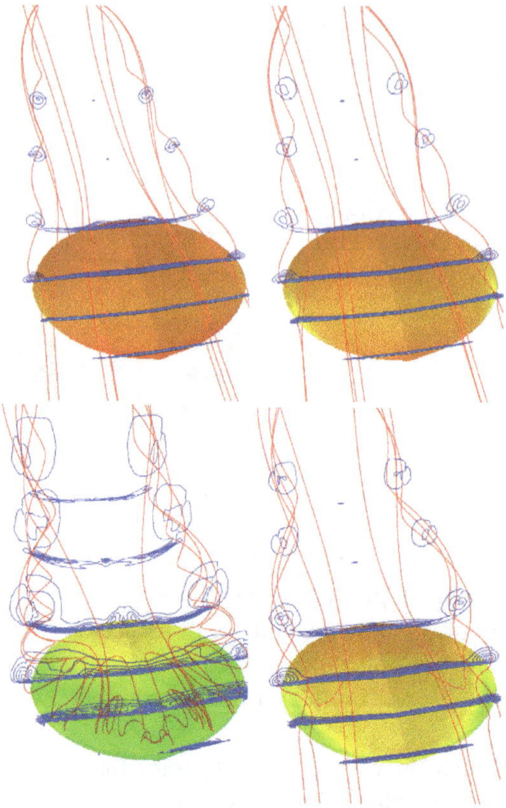

Figure 2.41. Evolution of flow pattern for rigid wing vs. AoAs. From left to right, top to bottom, 6°, 15°, 27°, and 51°. From Lian and Shyy [154]

case is similar to that for delta wings. In his numerical study, Lian [161] observed that the low AR wing suffers less from separation. The wing is not subjected to sudden stall, but the lift coefficient levels off at very high AoAs. Torres and Mueller [162], in their experiments on low AR wings, found similar results. It should be noted that the analysis by Lian [161] included neither the fuselage nor the propeller.

This pressure drop can be seen from Figure 2.42a, which plots the spanwise pressure coefficient on the upper-wing surface at $x/c = 0.4$. At the AoA $= 6°$ the spanwise pressure is almost uniform on the upper-wing surface, and the TiV causes the pressure drop to occur at approximately 90 percent of the half-span from the root. Figure 2.42 is illustrative in regard to pressure distributions versus the vortical structures. Note that the illustrated pressure distributions are not indicative of the total level of the pressure force.

Vortices strengthen with the increase in the AoA. At the AoA $= 27°$, as shown in Figure 2.41, tip vortices develop a strong swirl motion while entraining the surrounding flow. The low-pressure area increases as the AoA becomes higher. In Figure 2.42a, the pressure drop moves along the spanwise direction toward the root and now occurs at 75 percent from the root.

At lower AoAs, the vortex core position shows a linear relation with the incidence. This relation disappears at higher AoAs when the flow is separated on the upper surface. For example, at the AoA $= 45°$, the flow is separated at the leading edge, and the low-pressure zone covers more than 40 percent of the wing surface, which helps maintain the increase in lift force. At the AoA $= 51°$, a considerable

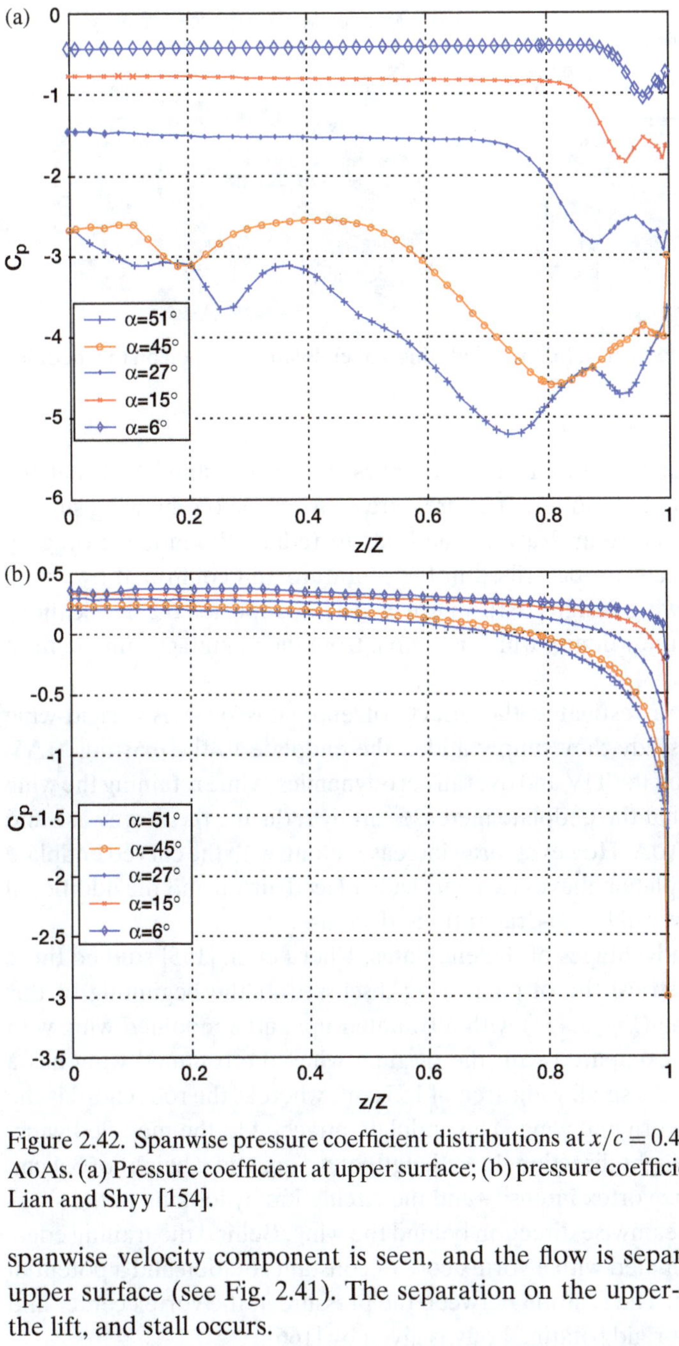

Figure 2.42. Spanwise pressure coefficient distributions at $x/c = 0.4$ for rigid wing at different AoAs. (a) Pressure coefficient at upper surface; (b) pressure coefficient at lower surface. From Lian and Shyy [154].

spanwise velocity component is seen, and the flow is separated from most of the upper surface (see Fig. 2.41). The separation on the upper-wing surface decreases the lift, and stall occurs.

As observed earlier, the TiVs have an important effect on the aerodynamics of low *AR* wings. Again, one major effect is the increase in induced drag for low *AR* wings. Equation (2–22) shows that the smaller the *AR*, the larger the induced drag.

2.3.3 Wingtip Effect

The wing shape chosen in Section 2.3.2 strives to maximize the wing area, and hence the lift, for a given dimension. However, the TiVs associated with the present low

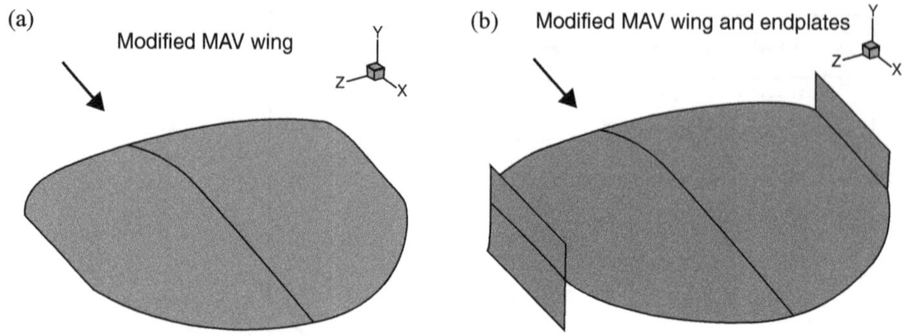

Figure 2.43. Wing-shape geometry: (a) modified wing; (b) endplates' location on the modified wing [155].

AR wing also substantially affect its aerodynamics. It is well established that the TiV causes a downwash that modifies the pressure distribution on the wing surface and increases the induced drag. Various methods to reduce the induced drag by decreasing the TiV effects are described in the literature and confirmed by actual applications to aircraft wing design [164]. Viieru et al. [155] reported the implications of placing endplates at the wingtip, which is simple from the manufacturing point of view.

Viieru et al. [165] investigated the effects of endplates on MAV rigid-wing aerodynamics. In that study they simply added the endplate to the existing MAV wing to probe its effect on the TiV and overall aerodynamics, while retaining the wing shape. They observed that the endplate increases lift by reducing the downwash and increases the effective AoA. However, drag increases along with the curved endplate in part because the endplate behaves as a vertically placed airfoil and the additional form drag causes the overall lift-to-drag ratio to decrease.

To remedy the disadvantages of the endplates, Viieru et al. [155] studied three alternative wing geometries: the original wing discussed at the beginning of this chapter, a modified wing (Fig. 2.43a) with a trimmed tip, and a modified wing with endplates (Fig. 2.43b). Compared with the original wing, the trimmed wing has a shorter span of 14 cm and a small wing area of 155 cm^2, whereas the root chord is the same length as with the original wing. The endplate attached to the modified wing, which is parallel to the flight direction, has a length of 4.4 cm and a height of 3.4 cm.

One can observe the vortex intensity and the circulation by looking at the slices perpendicular to the streamwise direction behind the wing. Behind the trailing edge, the flow can be approximated with a vortex core of constant rotation and a potential motion outside the core. The relation between the pressure at the vortex center and the circulation around a rigid rotating body is given by [166]:

$$\Gamma^2 = 4\pi^2 r_1^2 p_{\text{center}}/\rho, \tag{2–23}$$

where r_1 is the rigid body radius and P_{center} is the pressure at the rigid body center. Equation (2–23) shows that the vortex strength, measured by its circulation, is proportional with the pressure drop in the vortex core and its radius. In Figure 2.44, the pressure coefficient is plotted along the vortex core diameter at $x/c = 3$ behind the wing and $x/c = 5$. The amount of pressure drop inside the vortex core indicates

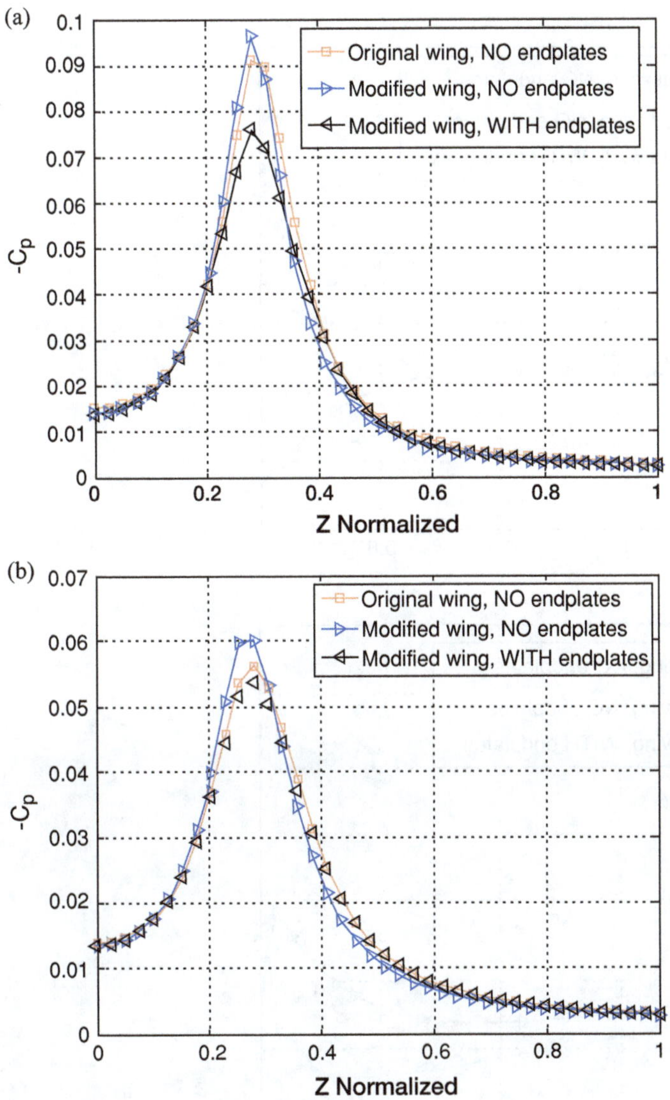

Figure 2.44. Pressure coefficient along the vortex core behind the wing at AoA = 6°: (a) $x/c = 3$; (b) $x/c = 5$ [155].

that the endplates reduce the vortex strength. Also, the modified wing without the endplates shows the strongest vortex.

From the pressure contours and horizontal velocity contours, one observes that the endplate affects the flow field over the wing. The endplate slows down the flow near the wingtip. This decrease in velocity reduces the pressure drop on the upper-wing surface corresponding to the vortex core (Fig. 2.45a). In contrast, a lower velocity slightly below the wing increases the high-pressure area there, because more momentum is transferred to the wing as pressure instead of being shed as vorticity at the wingtip. The increase in the high-pressure zone on the lower wing surface in the presence of the endplate can be clearly seen from the spanwise pressure coefficient on the lower-wing surface plot (Fig. 2.45b).

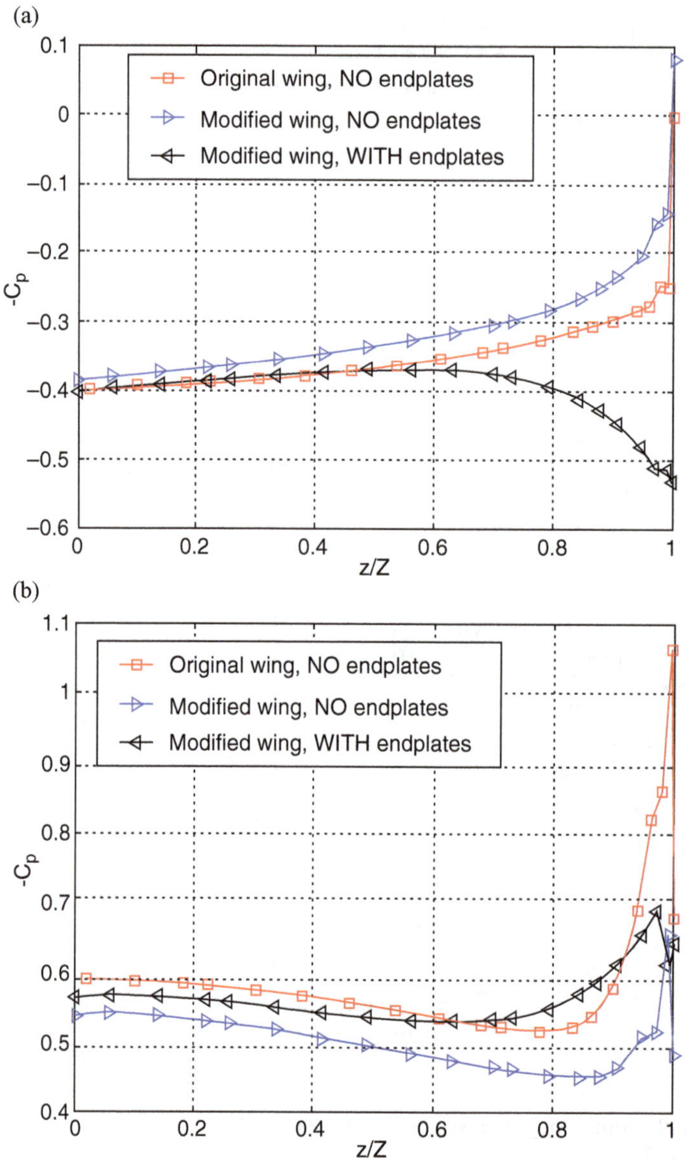

Figure 2.45. Pressure coefficient on the wing surface at $x/c = 0.34$ and AoA $= 6^{\circ}$: (a) lower-wing surface; (b) upper-wing surface [155].

Figure 2.46 plots the spanwise lift distribution obtained by integrating the pressure difference along the local chord at a specified spanwise location. It clearly shows that when the endplates are attached the lift on each cross-section is higher compared to the wing without the endplate. With a smaller overall wing area, the modified wing with the endplates produces almost the same lift as the original wing. Furthermore, the modified wings (with and without an endplate) experience lower drag over almost 75 percent of the wingspan starting from the root.

In Table 2.1 the overall aerodynamic performance parameters are presented for the AoA $= 6^{\circ}$. The modified wing configuration with endplates has a better lift-to-drag ratio than the baseline configuration (10 percent improvement). This

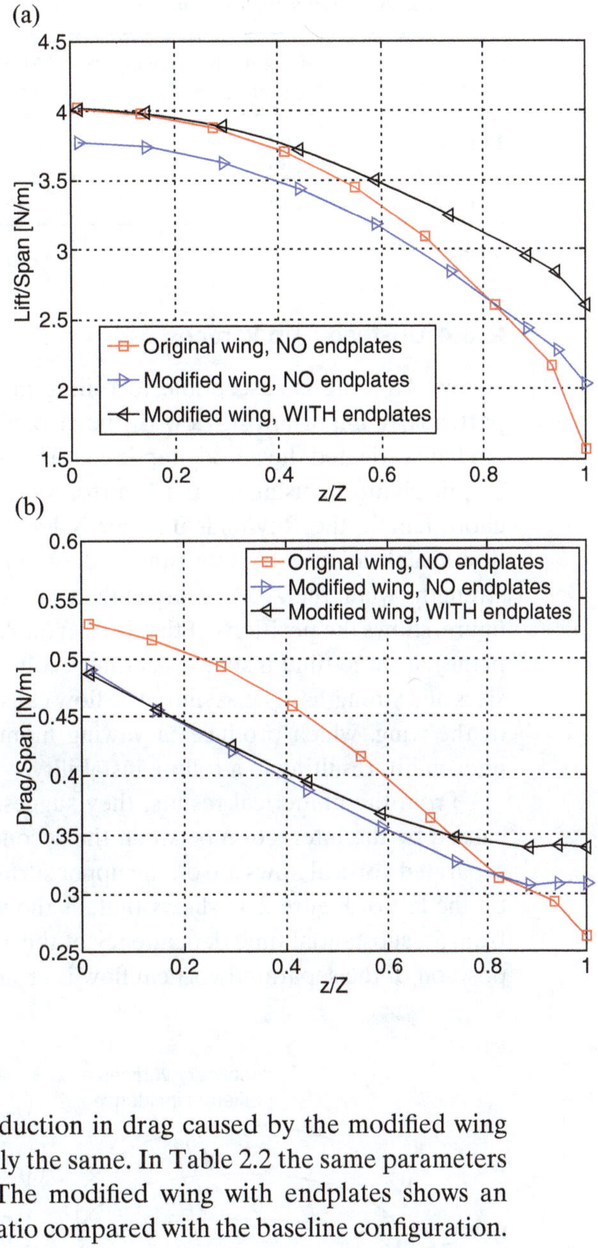

Figure 2.46. Spanwise lift and drag distribution at AoA = 6°: (a) lift and (b) drag [155].

improvement is mainly due to the reduction in drag caused by the modified wing shape because the total lift is essentially the same. In Table 2.2 the same parameters are presented for the AoA = 15°. The modified wing with endplates shows an increase of 1.4 percent in lift-to-drag ratio compared with the baseline configuration.

Table 2.1. *Aerodynamic forces at a 6° AoA*

AoA = 6°	Original MAV wing, no endplates	Modified MAV wing, no endplates	Modified MAV wing, with endplates
Lift (N)	0.49	0.44	0.49
Drag (N)	0.074	0.065	0.067
Lift/drag (-)	6.64	6.85	7.39

Source: [155].

Table 2.2. *Aerodynamic forces at a 15° AoA*

AoA = 15°	Original MAV wing, no endplates	Modified MAV wing, no endplates	Modified MAV wing, with endplates
Lift (N)	0.92	0.86	0.87
Drag (N)	0.22	0.21	0.21
Lift/drag (-)	4.16	4.15	4.22

Source: [155].

2.3.4 Unsteady Tip Vortices

A low AR wing is susceptible to rolling instabilities (wobbling). This problem is particularly important in view of the strong gust effect on MAVs. Tang and Zhu [167] investigated the aerodynamic characteristic of a low AR wing. The wing has an elliptic planform, using the E-174 airfoil with an AR of 1.33. Based on the maximum chord length, the Reynolds number is 1×10^4. Through numerical simulation and flow visualization in a water tunnel, they found that TiVs are unsteady in sizes and strengths when the AoA is larger than 11°. Figure 2.47a-c, on the right side of the figure, shows the positions of the TiVs at an AoA of 25° in the vertical plane (Trefftz plane) at three time instants. As time evolves, the left and right TiVs change their sizes and strengths. The asymmetric flow causes unequal drag between the two sides of the wing, which produces a yawing moment; the asymmetric flow also causes uneven lift, resulting in a rolling instability.

From the numerical results, they suggested that this unstable phenomenon is caused by the interaction between the secondary vortical flows and the TiVs. The separated vortical flows are on the upper surface of the wing. The schematic diagram on the left of Figure 2.47 shows that, as the wing incidence progressively increases from 5°, substantial time dependency of the TiVs is observed. At the AoA = 5°, the position of the separated vortical flow is around the trailing edge. As the incidence

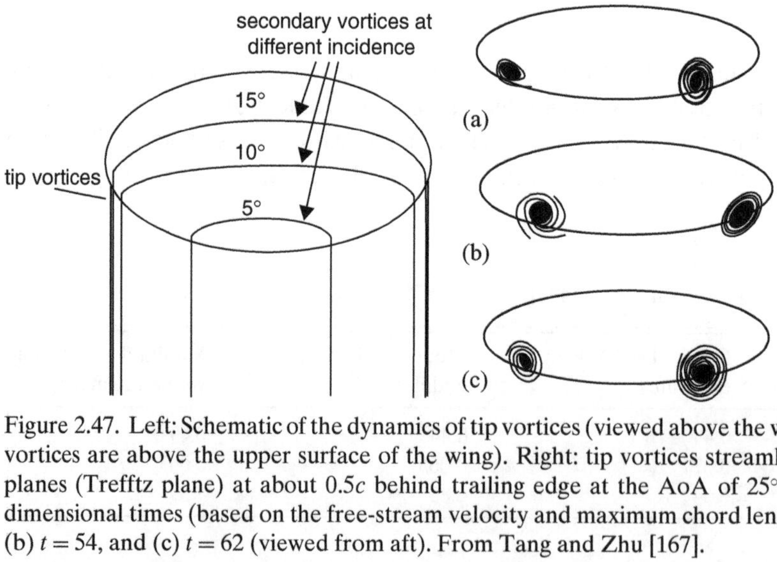

Figure 2.47. Left: Schematic of the dynamics of tip vortices (viewed above the wing, secondary vortices are above the upper surface of the wing). Right: tip vortices streamlines in vertical planes (Trefftz plane) at about $0.5c$ behind trailing edge at the AoA of 25°, at three non-dimensional times (based on the free-stream velocity and maximum chord length): (a) $t = 42$, (b) $t = 54$, and (c) $t = 62$ (viewed from aft). From Tang and Zhu [167].

increases, the separating flow moves toward the leading edge. When the incidence reaches 15° or higher, the separating flows above the wing interact with the TiVs, causing them to become substantially unsteady. To date, MAV flight tests have not reported such rolling instabilities as a major barrier. This is apparently because the airfoil shapes used for MAV flyers are much thinner and do not induce as many separating flows above the wing surface. Nevertheless, the issue of unsteady TiVs needs to be investigated in the MAV design and flight test process.

2.4 Concluding Remarks

This chapter highlighted the rigid fixed-wing aerodynamics at the low Reynolds number range between 10^3 and 10^6. As the Reynolds number drops from 10^6 to 10^4 or lower, the lift-to-drag ratio of an airfoil substantially decreases. A thinner airfoil with modest camber is preferable for low Reynolds number flyers because it generates better lift-to-drag ratio and power efficiency compared to conventional airfoils.

At the Reynolds number around 10^4, the laminar-to-turbulent transition and LSB play important roles in determining airfoil performance. In this flow regime, the lift-to-drag polar exhibits zigzag characteristics that are due to the formation and burst of the LSB. Because of the effect of transition, the wing performance is expected to be sensitive to the free-stream turbulence intensity and wind gust. For low Reynolds number flight, drastically unconventional wing shapes may be beneficial. For example, a corrugated wing is less sensitive to the variation in Reynolds number, and it can provide a more favorable lift than a non-corrugated wing because the viscous effect substantially modifies the effective airfoil shape. In essence, protruding corrugation corners act as boundary-layer trips to promote the transition of the boundary layer from laminar to turbulent while remaining "attached" to the envelope profile of the high-speed streamlines; however, a corrugated wing can also experience higher drag coefficients compared with those of the smooth-surfaced airfoil.

A downwash movement induced by a TiV reduces the effective AoA of a wing. For a low AR and low Reynolds number wing, the induced drag by the TiV substantially affects its aerodynamic performance. The TiVs affect not only lift and drag generation but also potentially flight stability. Moreover, wind gust is a prominent factor in low Reynolds number flyers. The low Reynolds number aerodynamics often exhibits hysteresis in a gusty environment. The transition position varies with the local Reynolds number, and depending on the flow parameters, either drag or thrust can be generated from the unsteady aerodynamics.

3 Rigid Flapping-Wing Aerodynamics

As we have explored in the previous chapters, flying animals flap wings to create lift and thrust, as well as to perform remarkable maneuvers with rapid accelerations and decelerations. Insects, bats, and birds provide illuminating examples of using unsteady aerodynamics that can guide the design of MAVs.

The first experimental work confirming the possibility of thrust generation on the unsteadily moving wing was conducted by Katzmayr in 1922 [168]. He investigated a fixed wing placed into an oscillating flow field. His studies validated the Knoller-Betz hypothesis [169] [170]. Both Knoller and Betz observed that the vertical motion of a flapping wing creates an effective AoA, generating an aerodynamic force with both lift and thrust components. Polonskiy [171] and Bratt [172] performed detailed visualizations of the large-scale vortex structures shed from harmonically plunging foils in a uniform flow and observed the characteristics of the vortex structures behind the airfoils. These experimental observations confirmed the Karman-Burgers thrust-generation hypothesis (i.e., the formation of a reverse Karman vortex street). In their experiments Polonskiy [171] and Jones et al. [173] showed the existence of different types of large-eddy structures generated by oscillating wings, as well as the vortex structures that are shed at an angle to the free-stream. Other researchers [174]–[178] studied the 2D flow structure behind oscillating foils and thrust generation, confirming that, depending on the parametric conditions, the wake structure can change from simple sinusoidal perturbations to two or four large-scale eddies. These typical flow structures were experimentally captured in flow visualization by Lai and Platzer [179]. Figure 3.1 shows the typical Karman vortex street behind a stationary NACA 0012, in which clockwise rotating vortices are shed from the upper surface and counterclockwise rotating vortices are shed from the lower surface; Figure 3.2a shows two pairs of vortices shed from the trailing edge of a NACA 0012 per plunge cycle, whereas Figures 3.2b and c show a single pair with a reverse Karman vortex street pattern. When an airfoil plunges at zero angle of incidence without an incoming flow, as shown in Figure 3.3a, a jet is produced by the flapping airfoil, and the streamwise velocity downstream of the airfoil is greater than the peak plunge velocity [180]. The jet appears to be biased toward the half-plane above the airfoil. This phenomenon was also observed by Jones et al. [181]. It occurs because, as soon as St exceeds approximately 0.8, the vortices shed from the trailing edge come too close together and start to interact with each other. In another experiment with a circular cylinder,

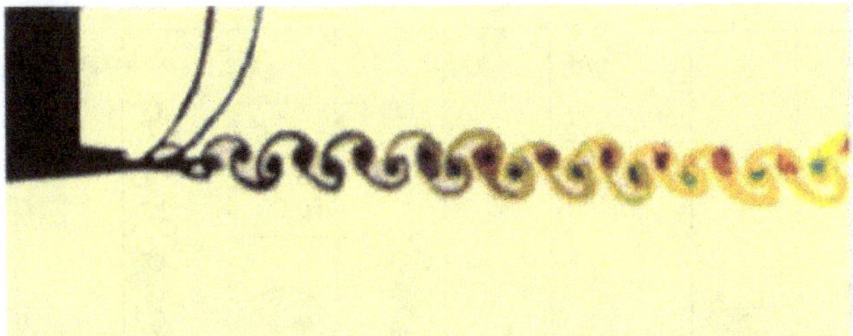

Figure 3.1. Vortical structure behind a stationary NACA 0012 airfoil for a free-stream velocity of 0.2 m/s [179].

the jet flow is not observed (see Fig. 3.3b). Therefore, it seems that the jet flow is caused by the detailed geometry such as curvature and asymmetry of the solid object. In another study, Taneda investigated the influence of traveling-wave characteristics associated with a flexible plate [182]. He revealed that the turbulence in the boundary layer is suppressed when the speed of propagation of the traveling wave exceeds that of the uniform incoming flow.

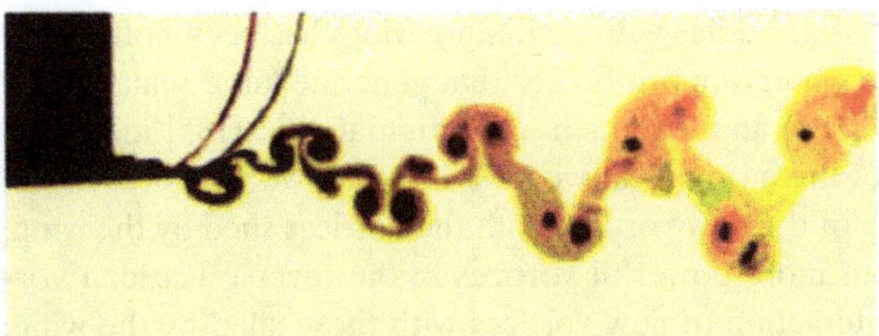

(a) $h_a = 0.0125$ ($St = 0.098$)

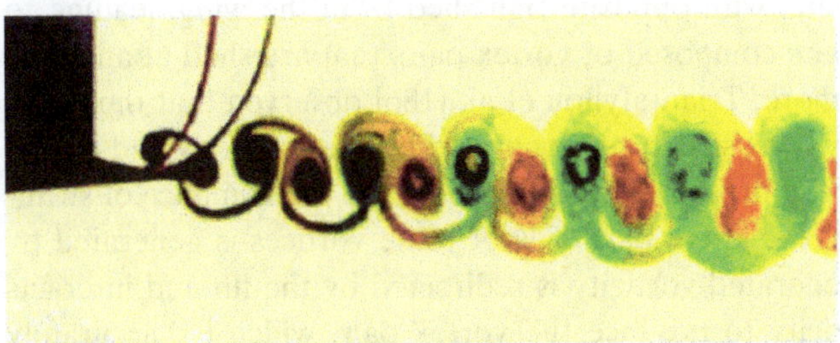

(b) $h_a = 0.025$ ($St = 0.196$)

(c) $h_a = 0.05$ ($St = 0.392$)

Figure 3.2. Vortex patterns for a NACA 0012 airfoil oscillated in plunge for a free-stream velocity of approximately 0.2 m/s, a frequency of $f = 2.5$ Hz, $Re = 2.1 \times 10^4$, and various amplitudes of oscillation [179].

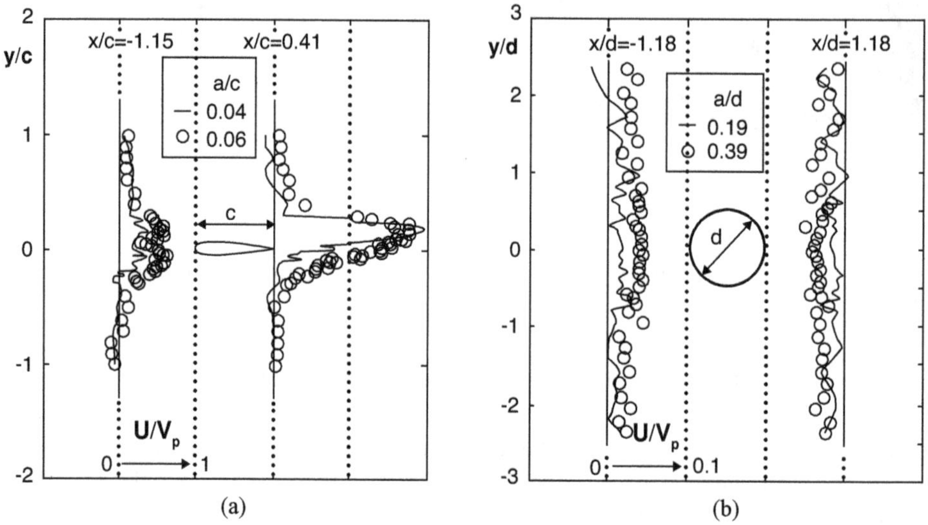

Figure 3.3. Non-dimensional mean streamwise velocity profiles generated by a (a) plunging airfoil, and (b) plunging cylinder at $f = 5$ Hz [180].

The interaction of large-scale eddies with oscillating wings has been noticed for many years in the context of observing fish bodies that generate large-scale eddies [183]. Gopalkrishnan et al. [184] and Streitlien and Triantafyllou [185] identified three types of interactions of the harmonically oscillating wing with vortices in the wake: (i) optimal interaction of the new vortices with the vortices shed by the wing, resulting in the generation of more powerful vortices in the reverse Karman vortex street; (ii) destructive interaction of new vortices with those shed by the wing, resulting in the generation of weaker vortices in the reverse Karman street; and (iii) interaction of vortex pairs with opposite sign shed from the wing, leading to the generation of a wide wake composed of vortex pairs that are shed at an angle to the free-stream. Furthermore, Triantafyllou et al. [186] observed that the interaction between large-scale vortex structures generated by the fish body and vortex structures of the fin are important factors in determining the performance of swimming. In the "normal" case, the initial pair of large-scale vortices is generated by the body. Then the body-generated vorticity is redirected by the fin and interacts with the fin-generated vorticity to produce the vortex pair, which is "accurately controlled" by the fish. The timing of vortex formation, propagation, and its instantaneous position are critical for efficient maneuvering and acceleration. Thus the control of vortex generation plays an important role in achieving high locomotion efficiency. In summary, depending on the features of the interactions between airfoil movement and the associated flow structure, either thrust or drag generation can be observed.

Another experiment discovered the delay of the leading-edge flow separation in unsteady motion. Devin et al. [187] indicated that, for a rigid wing with an AR of from 1 to 4 and a NACA 0012 section, no separation is observed up to the instantaneous AoA of 45°; in contrast, separation normally occurs at AoA $= 15°$ ($Re = 10^5$) in the steady case. Moreover, investigations [188]–[190] have found that the stall does not come instantly when a wing is rapidly pitched beyond the static stall angle. Figure 3.4

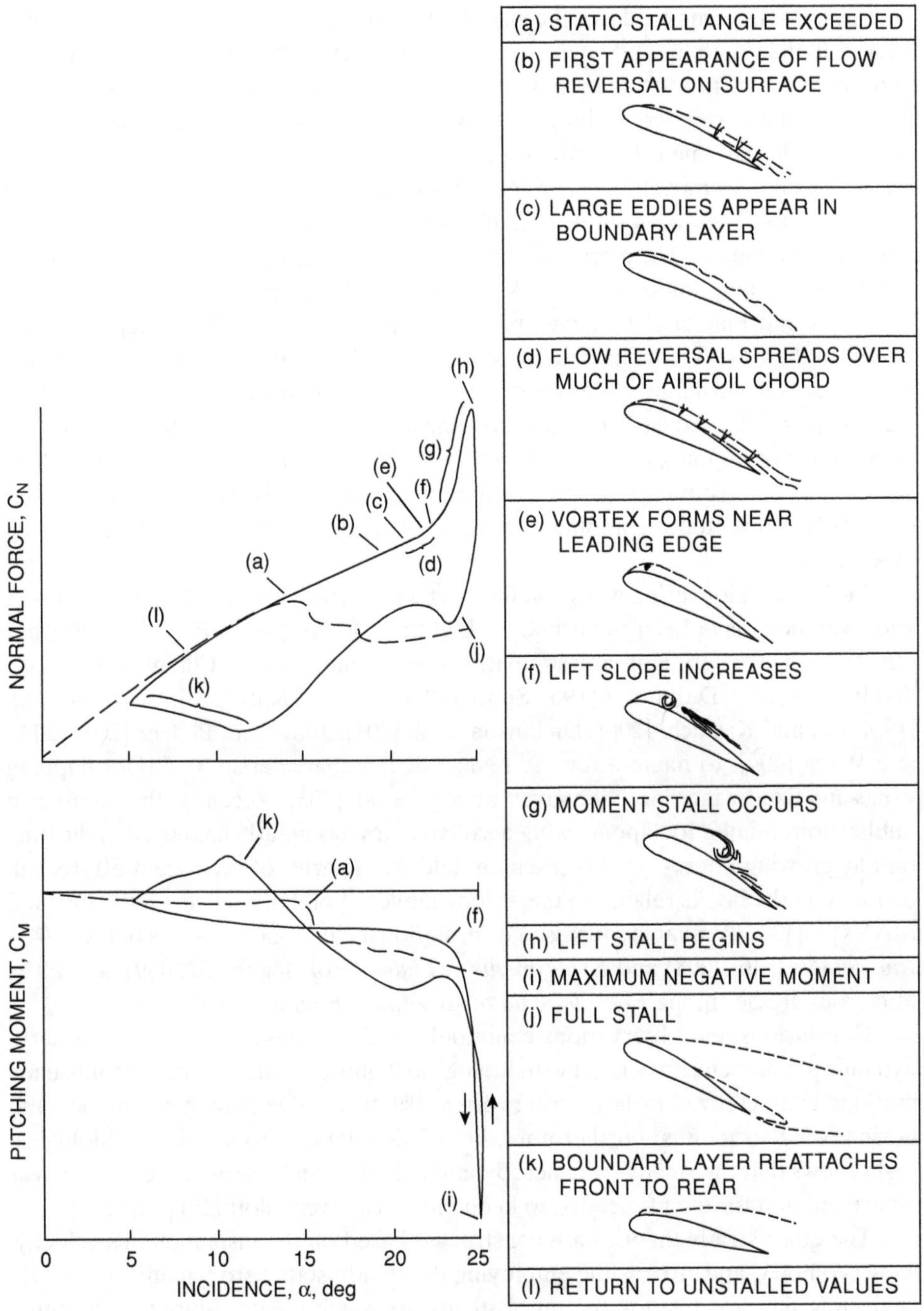

Figure 3.4. Sequence of events charactering dynamic stall of an NACA 0012 airfoil. From McCroskey and Fisher [191].

depicts the evolution of the flow structures in dynamic stall for a rapidly pitching NACA 0012 airfoil [191]. The reverse flow affects the pressure distribution (point b in Fig. 3.4) after the wing rapidly exceeds the static stall angle (point a in Fig. 3.4). This reversal progresses up on the airfoil upper surface and forms a vortex. This vortex initially appears near the leading edge of the airfoil (point c in Fig. 3.4), enlarges, and then moves down the airfoil. The pitching moment reaches its negative peak, and then both lift and pitching momentum start to drop dramatically (points d and f in Fig. 3.4), producing the phenomenon known as dynamic stall. As the AoA decreases, the vortex moves into the wake and a fully separated flow develops on the airfoil. At the time instant when the AoA reaches its minimum AoA, lift has not reached its minimum value, which indicates that the dynamic stall process forms a hysteresis loop. Figure 3.4 shows such characteristics for the development of lift and pitching momentum. The amplitude and the shape of the hysteresis loop depend on the oscillation amplitude, mean AoA, and reduced frequency.

Jones and Platzer [192] developed and experimentally tested a flapping wing MAV. In their MAV configuration, lift is generated by a fixed forewing while thrust is produced by two flapping hindwings (see Fig. 1.15c) based on the thrust generation mechanisms discussed earlier. It is interesting to note that they also showed different possibilities of flapping airfoil usage, such as the reduction or suppression of flow separation behind blunt or cusped airfoil trailing edges. Willis et al. [193] and Persson et al. [194] presented a computational framework to design and analyze flapping MAV flight.

Moreover, pioneering works on flapping wing aerodynamics of biological flyers and swimmers have been published by Lighthill [40] and Weis-Fogh [68]. Further efforts, both in experiments and simulations, are documented by Ellington [65], Katz and Plotkin [195], DeLaurier [196], Smith [197], Vest and Katz [198], Ellington et al. [199], Liu and Kawachi [200], Dickinson et al. [201], Jones and Platzer [202] [203], and Wang [204], to name a few. A review of the characteristics of both flapping wings and fixed-wings has been given by Shyy et al. [205]. Recently, the number of publications related to flapping wing aerodynamics has greatly increased, indicating rapidly growing interest in this research field. Summaries of the recent efforts can be found in the books related to the aerodynamics of biological flyers [28] [206] and MAVs [36] [207].Valuable sources of information include special issues of the *AIAA Journal* (Vol. 46, 2008) and *Experiments in Fluids* (Vol. 46, No. 5, 2009), as well as numerous articles in the *Encyclopedia of Aerospace Engineering* (2010).

Conclusions and observations commonly made in these studies are that aerodynamic phenomena associated with biological flights prominently feature unsteady motions, characterized by large-scale vortex structures, 3D flapping kinematics, and flexible wing structures. Furthermore, knowledge gained from studying biological flight shows that the steady-state aerodynamic theory can be seriously challenged as an explanation for the lift needed to keep biological flyers aloft [26] [65] [199].

The quasi-steady theories are constructed based on the instantaneous velocity, wing geometry, and AoA while employing the steady-state aerodynamic model. By neglecting the flow history, the quasi-steady approach greatly simplifies the time-dependent problem by converting it to a sequence of independent, steady-state problems, and so it has been frequently used in interpreting biological flight characteristics [4] [43] [60] [68] [80] [208]–[210]. For example, this approach has been used to estimate the mechanical power requirements of hummingbirds [211] and bumblebees [212]. However, based on the theoretical analyses [213] and experimental measurements of tethered insects [214] [215], it has been found that the quasi-steady model is insufficient to predict the lift needed to support the weight of the insect body. In contrast, two studies [216] [217] involving a dynamically scaled, rigid-winged, flapping robotic flyer flapping in mineral oil suggest that the quasi-steady 2D blade element models can yield satisfactory agreement with the experimental measurement of aerodynamic forces. Further discussions and assessments regarding the quasi-steady aerodynamic model are presented in Section 3.6.

In this chapter we present various issues related to the aerodynamics of flapping flight of rigid wings. First, we discuss the scaling of flapping wing flight in terms of reduced frequency, Reynolds number, and Strouhal number. These non-dimensional parameters are important for investigating fluid physics of both rigid and flexible wings; discussion of flexible wing structures and aeroelasticity is presented in Chapter 4. Then, we discuss the main unsteady lift-enhancing aerodynamic mechanisms associated with flapping wings, including delayed stall of leading-edge vortex (LEV); rapid pitch-up; wake capturing; interactions between LEV, trailing-edge vortex (TEV), and tip vortex (TiV); and clap-and-fling mechanisms. Subsequently, we look at the fluid physics in two different Reynolds number regimes in more detail. For both Reynolds number regimes, we investigate the effects of wing kinematics on the resulting flow field and aerodynamics.

For the Reynolds number regime of $O(10^2)$, we compare the aerodynamics of the 3D hovering wing to its 2D counterpart and discuss the effects of free-stream fluctuations on aerodynamic performance. In this Reynolds number regime, intriguing fluid dynamics phenomena are observed for a free-to-move vertically plunging rigid wing [218]. The vertical motion of the wing is imposed with sinusoidal kinematics, whereas the wing is free to move in its horizontal directions. When the plunging frequencies are below a threshold value (i.e., $Re\ (= \rho f h_a c/\mu) < 3.9 \times 10^2$), the wing remains stationary in the horizontal plane, and the wakes shed in the flow form a symmetric structure. For the frequencies above this threshold value, such that $Re > 3.9 \times 10^2$, the symmetry of the wake breaks, resulting in an inverted von Karman vortex street that is indicative of propulsion: the flapper moves forward. When the rigid wing is replaced with a flexible plate with the same geometry but with lower elastic modulus, the resulting forward speed is significantly greater than that of the rigid wing [219]. For more details we refer the reader to the literature (e.g., [218] [220] [221] for rigid wings and [219] [222] [223] for passively pitching wings).

For the Reynolds number regime of $O(10^4)$ we focus on the effects of airfoil shapes on pitching and plunging wings in forward flight. Furthermore, we review approximate analysis for non-stationary airfoil aerodynamics and discuss several quasi-steady models. Finally, we highlight fluid physics associated with biological flapping flyer-like models, including the flyer's scale effects on resulting flow structures such as LEVs and spanwise flow.

3.1 Flapping Wing and Body Kinematics

Figure 3.5 illustrates the wing and body movement of a hummingbird, as well as definitions of some key terms used in this section. The body kinematics is represented by the body angle χ (inclination of the body), which is relative to the horizontal plane, and by the stroke plane angle β (indicated by the solid lines), which is defined

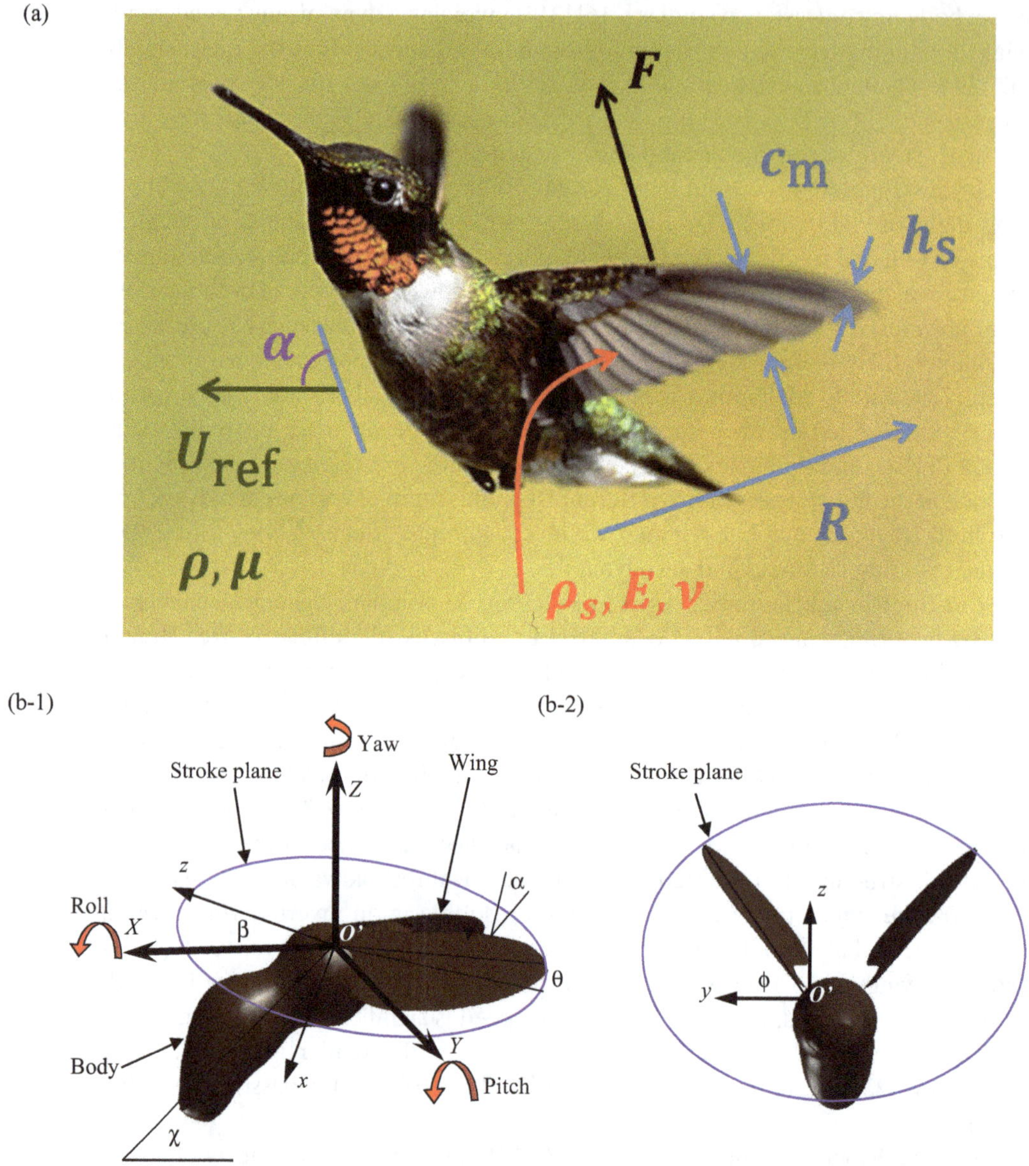

Figure 3.5. (a) Physical variables shown in the case of a hummingbird. (b) Schematic diagram of coordinate systems and wing kinematics. (b-1) The local wing-base-fixed and the global space-fixed coordinate systems. The local wing-base-fixed coordinate system (x, y, z) is fixed on the center of the stroke plane (origin O' at the wing base) with the x-direction normal to the stroke plane, the y-direction vertical to the body axis, and the z-direction parallel to the stroke plane; (b-2) definition of the positional angle ϕ, the feathering angle (AoA) α, and elevation angle ϕ of the flapping wing, the body angle χ, and the stroke plane angle β.

as the plane that includes the wing base and the wingtip's positions at the maximum and the minimum sweep. The wing-beat kinematics is described by three angles relative to the stroke plane: (i) flapping about the x-axis in the wing-fixed coordinate system described by the positional (flapping) angle ϕ, (ii) rotation of the wing about the z-axis described by the elevation (deviation) angle θ; and (iii) rotation (feathering) of the wing about the y-axis described by the AoA (feathering angle) α. The time history of AoA includes high-frequency modes that are characteristic of some insects [224]. The body angle and the stroke plane angle vary in accordance with the flight speed and flapping wing kinematics of biological flyers [70] [225]. For a general 3D case, a definition of the positional angle, the elevation angle, and the AoA/feathering, all in radians, can be given as follows, for the first four modes:

$$\phi(t) = \sum_{n=0}^{3} [\phi_{cn} \cos(2n\pi ft) + \phi_{sn} \sin(2n\pi ft)], \tag{3–1}$$

$$\theta(t) = \sum_{n=0}^{3} [\theta_{cn} \cos(2n\pi ft) + \theta_{sn} \sin(2n\pi ft)], \tag{3–2}$$

$$\alpha(t) = \sum_{n=0}^{3} [\alpha_{cn} \cos(2n\pi ft) + \alpha_{sn} \sin(2n\pi ft)], \tag{3–3}$$

Note that n indicates the order of the Fourier series. The Fourier coefficients $\phi_{cn}, \phi_{sn}, \theta_{cn}, \theta_{sn}, \alpha_{cn},$ and α_{sn} can be determined from the empirical and measured kinematics data. Wing and body kinematics of biological flyers can be measured by high-speed cameras [226]–[234], laser techniques (a scanning projected line method [235], a reflection beam method [236], a fringe shadow method [237], and a projected comb fringe method [238]), and a combination of high-speed cameras and a projected comb-fringe technique with the Landmarks procedure [239]. Advancement in measurement techniques also enables quantification of flapping wing and body kinematics along with the 3D deformation of the flapping wing. Recently, data on the instantaneous wing kinematics involving camber along the span, twisting, and flapping motion have been reported (e.g., a hovering honeybee [240], a hovering hoverfly [241], a free-flying hawkmoth [242], and a bat [243]). Clearly, these efforts have helped establish more complex and useful computational models as well as to develop bio-inspired MAVs. Examples of the kinematics of a hovering hawkmoth [226] and free-flying hoverflies are plotted in Figure 3.6.

Although 3D effects are important for predicting low Reynolds number flapping wing aerodynamics, 2D experiments and computations do provide valuable insight into the unsteady fluid physics related to flapping wings. Tang, Viieru, and Shyy [244] discussed two hovering modes that are observed in nature: the "water treading" mode [175] and the "normal hovering" mode [217]. The plunging and pitching of the airfoil are described by symmetric, periodic functions:

$$h(t) = h_a \sin(2\pi ft + \varphi), \tag{3–4}$$

$$\alpha(t) = \alpha_0 + \alpha_a \sin(2\pi ft), \tag{3–5}$$

where h_a is the plunging amplitude, f is the plunging frequency, α_0 is the initial pitching angle, α_a is the pitching amplitude, and ϕ is the phase difference between plunging and pitching motion. The schematics of the two hovering modes are presented in Figure 3.7. The initial pitching angle for the water treading mode is zero ($\alpha_0 = 0°$), whereas for the normal-flapping/hovering mode $\alpha_0 = 90°$. Figure 3.8 shows

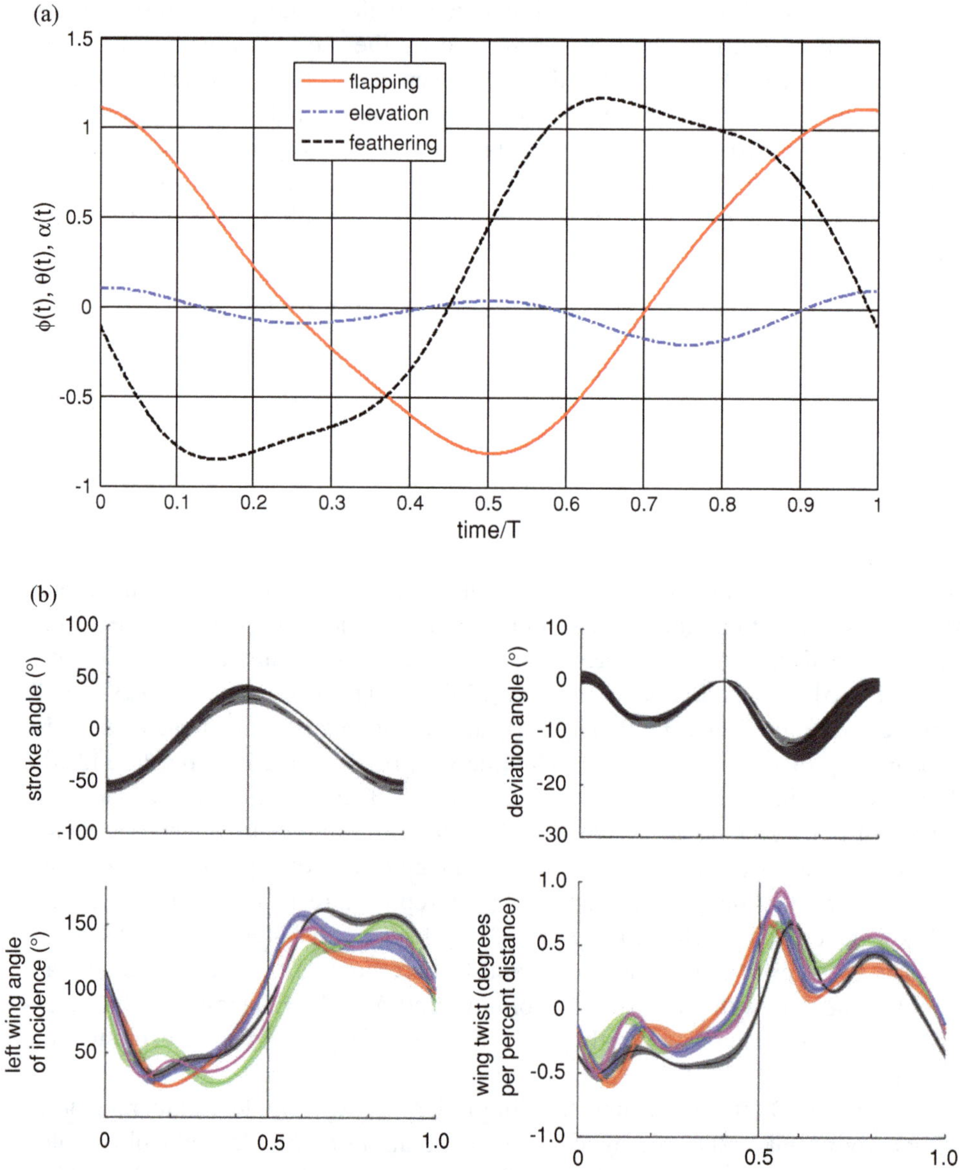

Figure 3.6. Variation of (a) positional ($\phi(t)$), elevation ($\theta(t)$) and feathering (α) angles (in radians) based on a hovering hawkmoth during one flapping period. From Liu and Aono [225]. (b) Stroke (positional), deviation, angle of incidence (feathering), and twisting angle variation for one period for free-flying hoverflies. From Walker et al. [241].

other 2D flapping motions – the pitching and plunging motions – that are more commonly observed for forward flight. The fluid dynamics based on these kinematics are highlighted in Section 3.5.

For normal hovering, based on the phase relationship between the translation and rotation of the wing, Dickinson et al. [201] categorized the wing motion into advanced ($\varphi > 90^\circ$), synchronized ($\varphi = 90^\circ$), and delayed modes ($\varphi < 90^\circ$). The advanced mode (Fig. 3.9a) is the pattern in which the wing rotates before it reverses direction at the end of each stroke. The synchronized mode (Fig. 3.9b) is the pattern

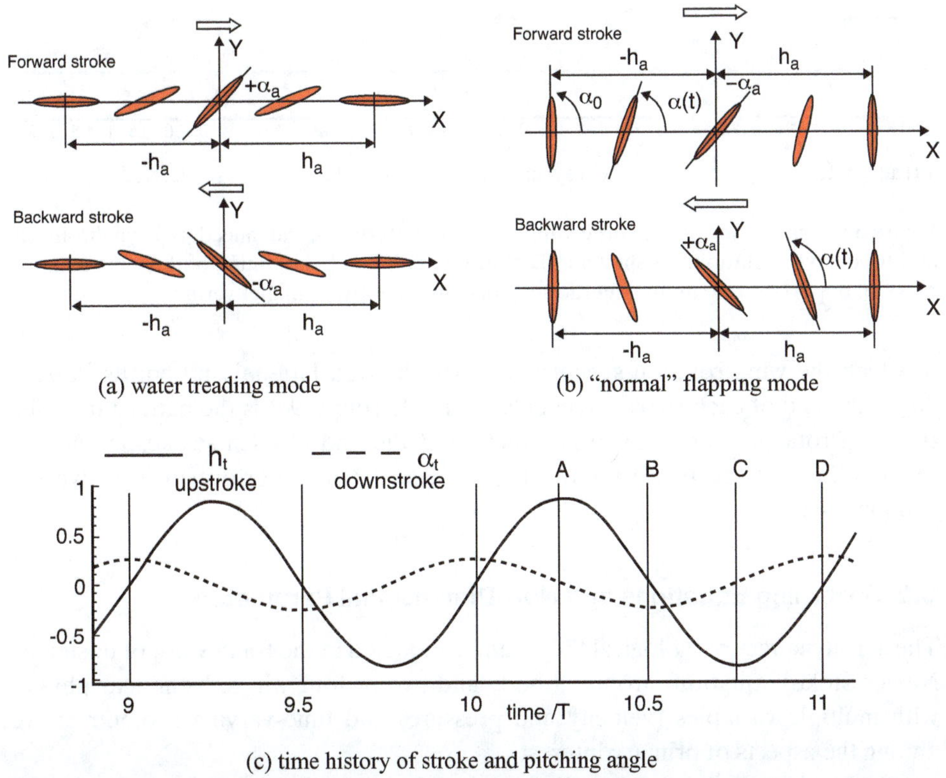

Figure 3.7. Schematic of 2D flapping motion [245]. (a) Water treading mode; (b) in the normal flapping mode, an airfoil moves from left to right in the forward stroke and vice versa in the backstroke; (c) time histories of airfoil stroke (solid line: $h(t)$ shown in Eq. (3–4)) and pitch angle (dash line: $\alpha(t)$ shown in Eq. (3–5)) employed for both modes.

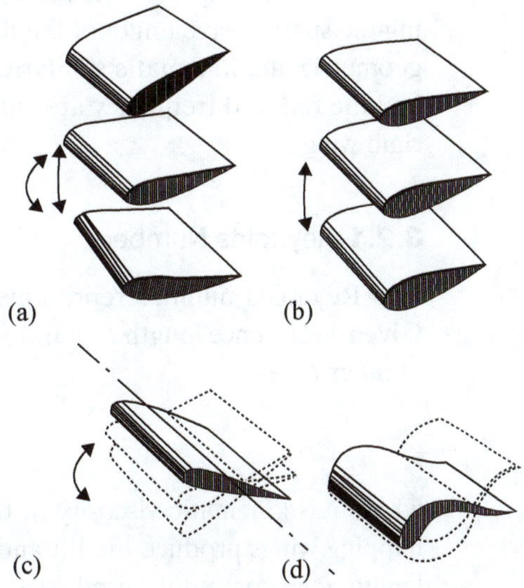

Figure 3.8. Four typical unsteady motions in plunging and pitching motion. (a) Combined translational-rotational oscillations, (b) purely translational oscillations, (c) purely rotational oscillations, and (d) advancing wave-type deformations. From Rozhdestvensky and Ryzhov [246].

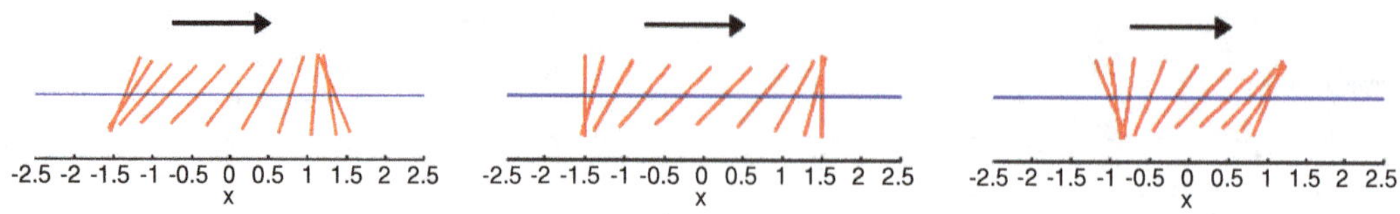

(a) advanced (b) synchronized (c) delayed

Figure 3.9. Schematics of the three wing rotation patterns: (a) advanced, (b) synchronized, and (c) delayed rotation. As shown in Dickinson et al. [201], the timing of the wing rotation plays an important role in lift generation and, consequently, in maneuvering.

in which the wing rotation synchronizes with its translational motion: its AoA is 90° at the end of each stroke. The delayed mode (Fig. 3.9c) is the pattern in which the wing rotates after it reverses direction at the end of each stroke. Discussion of the effects of the phase relationship on aerodynamic performance is given in Section 3.4.1.

3.2 Governing Equations and Non-Dimensional Parameters

The aerodynamics of biological flight can be modeled in the framework of unsteady, Navier-Stokes equations around a body and two or four wings. Nonlinear physics with multiple variables (velocity and pressure) and time-varying geometries are among the aspects of primary interest.

Scaling laws are useful to reduce the number of parameters, to clearly identify the physical flow regimes, and to offer guidelines to establish suitable models for predicting the aerodynamics of biological flight. Three main dimensionless parameters in the flapping flight scaling are (i) the Reynolds number (Re), which represents the ratio of inertial and viscous forces; (ii) the Strouhal number (St), which describes the relative influence of forward versus flapping speeds in forward flight; and (iii) the reduced frequency (k). The reduced frequency measures unsteadiness by comparing the spatial wavelength of the flow disturbance to the chord [186]. Together with geometric and kinematic similarities, the Reynolds number, the Strouhal number, and the reduced frequency are sufficient to define the aerodynamic similarity for a rigid wing.

3.2.1 Reynolds Number

The Reynolds number represents the ratio between inertial and viscous forces. Given a reference length L_{ref} and a reference velocity U_{ref}, one defines the Reynolds number Re as

$$Re = \frac{U_{\mathrm{ref}} L_{\mathrm{ref}}}{\nu}, \tag{3–6}$$

where ν is kinematic viscosity of the fluid. In flapping wing flight, considering that flapping wings produce the lift and thrust, either a mean chord length c_m or a wing length R is commonly used as the reference length, whereas the body length is typically used in swimming animals. Note that the definition of mean chord length is an averaged chord length in the spanwise direction. The reference velocity U_{ref} is the free-stream velocity in forward flight, but it is defined differently in hovering flight. In *hovering*, the mean wingtip velocity may be used as the reference velocity, which is also written as $U_{\text{ref}} = 2\Phi fR$, where Φ is the wing-beat amplitude (measured in radians). Therefore, the Reynolds number for a 3D flapping wing, Re_{f3}, in hovering flights can be cast as

$$Re_{f3} = \frac{U_{\text{ref}} L_{\text{ref}}}{\nu} = \frac{2\Phi fRc_m}{\nu} = \frac{2\Phi fR^2}{\nu}\left(\frac{4}{AR}\right), \tag{3–7}$$

where the aspect ratio AR as described in Chapter 1 has been introduced in a form of $AR = 4R^2/S$, with the wing area being the product of the wingspan ($2R$) and c_m. Note that the Reynolds number here is proportional to the wing-beat amplitude, the flapping frequency, and the square of the wing length, R^2, but is inversely proportional to the AR of the wing. In insect flights, the wing-beat amplitude and the aspect ratio of the wing do not vary significantly, whereas the flapping frequency increases as the insect size is reduced. In general these characteristics result in Re ranging from $O(10^1)$ to $O(10^4)$. In addition, given a geometrically similar wing model that undergoes flapping hovering with the same wing-beat amplitude, the product of fR^2 can preserve the same Reynolds number. This implies that a scaled-up but low-flapping-frequency wing model can be built mechanically to mimic insect flapping flight based on the aerodynamic similarity. In fact, robotic model-based studies [199] [201] have been established on such a basis provided that the second parameter, the reduced frequency, can be satisfied simultaneously.

The Reynolds number can also be defined with an alternative reference length and/or reference velocity. For example, using the wing length R as the reference length and the wing velocity, $U_{\text{ref}} = 2\pi fr_2R$, where r_2 is the radius of the second moment of wing area (approximately 0.52 for the hawkmoth, *Manduca sexta* [247] [248]), the Reynolds number Re_{f3} is proportional with $(\Phi fR^2)/\nu$ and is not dependent on the aspect ratio of the wing. Note that the reference velocity here is almost half of that at the wingtip.

For a 2D hovering wing, the Reynolds number can be defined by the maximum plunging velocity given by Eq. (3–8a) or by the mean plunging velocity, Eq. (3–8b), with a factor of $\pi/2$ difference. Both definitions have their benefits, but in this chapter, we use Eq. (3–8b) as the Reynolds number for a 2D hovering motion unless stated otherwise.

$$Re_{f2} = \frac{U_{\text{ref}} L_{\text{ref}}}{\nu} = \frac{2\pi fh_a c_m}{\nu}, \tag{3–8a}$$

$$Re_{f2} = \frac{U_{\text{ref}} L_{\text{ref}}}{\nu} = \frac{4fh_a c_m}{\nu}. \tag{3–8b}$$

3.2.2 Strouhal Number and Reduced Frequency

The Strouhal number (*St*) is well known for characterizing the vortex dynamics and shedding behavior for flows around a stationary cylindrical object, such as the von Karman vortex street behind a cylindrical object, and for characterizing induced unsteady flows about 2D airfoils undergoing pitching and plunging motions. At certain Strouhal numbers, the pitching and plunging airfoils produce forward thrust, and the vortices in the wake have a flow structure that is similar to the von Karman vortex street, but with reversed direction of vorticity. Such vortex structures are named reverse von Karman vortices [179]. In general, for flapping flight, the dimensionless parameter *St* describes the dynamic similarity between the wing velocity and the characteristic velocity; it is usually defined as

$$St = \frac{fL_{\text{ref}}}{U_{\text{ref}}} = \frac{2fh_a}{U}. \tag{3–9}$$

This definition offers a measure of propulsive efficiency in flying and swimming animals. For natural flyers and swimmers in cruising conditions, it was found that the Strouhal number is within a narrow region of $0.2 < St < 0.4$ [186] [249]. For hovering motions, the Strouhal number has no specific meaning because the reference velocity is also based on the flapping velocity.

In addition to the Strouhal number, an important dimensionless parameter that characterizes the unsteady aerodynamics of pitching and plunging airfoils is the reduced frequency, defined in Eq. (1–19). In hovering flight, for which there is no forward speed, the reference speed U_{ref} is defined as the mean wingtip velocity $2\Phi fR$, and the reduced frequency can be reformed as

$$k = \frac{2\pi fL_{\text{ref}}}{2U_{\text{ref}}} = \frac{\pi}{\Phi AR}, \tag{3–10}$$

where the aspect ratio AR is introduced here again as in Eq. (1–7). For the special case of 2D hovering airfoils, here the reference velocity U_{ref} is the maximum translational flapping velocity, and the reduced frequency is defined as

$$k = \frac{2\pi fL_{\text{ref}}}{2U_{\text{ref}}} = \frac{c_m}{2h_a}, \tag{3–11}$$

which is the inverse of the normalized stroke amplitude. Based on the definition of the reference velocity and reduced frequency, the airfoil kinematics Eqs. (3–4) and (3–5) can be rewritten as

$$h(t/T) = h_a \sin(2kt/T + \varphi) \tag{3–12}$$

$$\alpha(t/T) = \alpha_0 + \alpha_a \sin(2kt/T) \tag{3–13}$$

where t/T is a dimensionless time, which is non-dimensionalized by a reference time T. Another interpretation of the reduced frequency is that it gives the ratio between the fluid convection time scale, T, and the motion time scale, $1/f$.

In the case of forward flight, another dimensionless parameter is the advance ratio, J. In a 2D framework, J is defined as

$$J = \frac{U_{\text{ref}}}{2\pi fh_a}, \tag{3–14}$$

which is related to *St* because $J = 1/(\pi St)$. In Eq. (3–14), the reference velocity (U_{ref}) is the forward flight velocity (U).

With the reduced frequency the 3D wing kinematics as illustrated in Eqs. (3–1), (3–2), and (3–3) can be further reformed as

$$\phi(t/T) = \sum_{n=0}^{3}[\phi_{cn}\cos(2nkt/T) + \phi_{sn}\sin(2nkt/T)], \tag{3–15}$$

$$\theta(t/T) = \sum_{n=0}^{3}[\theta_{cn}\cos(2nkt/T) + \theta_{sn}\sin(2nkt/T)], \tag{3–16}$$

$$\alpha(t/T) = \sum_{n=0}^{3}[\alpha_{cn}\cos(2nkt/T) + \alpha_{sn}\sin(2nkt/T)], \tag{3–17}$$

where t/T is a dimensionless time, which is non-dimensionalized by a reference time T, resulting in a dimensionless period of π/k.

If we choose c_m, U_{ref}, and $1/f$ as the length, velocity, and time scales, respectively, for non-dimensionalization, then the corresponding momentum equation for constant density fluid yields

$$\frac{k}{\pi}\frac{\partial}{\partial t^*}(u_i^*) + \frac{\partial}{\partial x_j^*}\left(u_j^* u_i^*\right) = -\frac{\partial p^*}{\partial x_i^*} + \frac{1}{Re}\frac{\partial^2}{\partial x_j^{*2}}(u_i^*), \tag{3–18}$$

where * designates a dimensionless variable. The reduced frequency and Reynolds number appear in the momentum equation, whereas the Strouhal number comes up in the kinematics.

Morphological and flight parameters for the fruit fly (*Drosophila melanogaster*), bumblebee (*Bombus terrestris*), hawkmoth (*Manduca sexta*), and hummingbird (*Lampornis clemenciae*) are summarized in Table 3.1. For all these flyers, the flapping frequency is around 20–200 Hz, and the flight speed is about several m/s, yielding a Reynolds number from 10^2 to 10^4 based on the mean chord and the forward flight speed. In this flight regime, the unsteady, inertia, pressure, and viscous forces are all important.

3.3 Unsteady Aerodynamic Mechanisms in Flapping Wings

As previously discussed with geometric and kinematic similarities, the dynamic similarity can be maintained by scaling up the wing dimension while appropriately lowering the flapping frequency, thereby rendering the Reynolds number and the reduced frequency unchanged. For example, Ellington et al. [199] used a robotic wing model to investigate flow over the wings of a hovering hawkmoth and discovered that the LEV spiraled out toward the wingtip. Their findings provided a qualitative explanation of one particular high-lift mechanism. Dickinson et al. [201] also used a robotic wing model representing a fruit fly to directly measure the forces and visualize the flow patterns around the flapping wing. They demonstrated two force peaks during the rotation phase: (i) the rotational mechanism associated with fast pitch-up and (ii) the wake-capture mechanism resulting from the airfoil and wake interactions. Although different explanations of the two force-generation mechanisms have been offered, as described in the following section, it is clear that a robotic model provides an insightful experimental framework in studying flapping wing flight. Furthermore, Birch and Dickinson [254] observed substantially different

Table 3.1. *Morphological and flight parameters for selected species*

		Fruit fly (*Drosophila melanogaster*) [250]	Bumblebee (*Bombus terrestris*) [251]	Hawkmoth (*Manduca sexta*) [226] [227]	Hummingbird (*Lampornis clemenciae*) [252] [253]
Morphological parameters					
Total mass (body) (kg)	m_b	2.0×10^{-6}	1.75×10^{-4}	1.579×10^{-3}	8.400×10^{-3}
Wing mass(both wings) (kg)	m_w	9.6×10^{-9}	9.0×10^{-7}	9.4×10^{-5}	5.88×10^{-4}
Wing length (m)	R	3.0×10^{-3}	1.3×10^{-2}	4.9×10^{-2}	8.5×10^{-2}
Wing area (both wings) (m^2)	S	2.9×10^{-6}	1.06×10^{-4}	1.782×10^{-3}	3.524×10^{-3}
Wing loading (N/m^2)	W/S	7	16	9	23.5
Aspect ratio (-)	AR	2.4	6.6	5.3	8.2
Kinematics parameters					
Flapping frequency (Hz)	f	200	150	25	23
Stroke amplitude (rad)	Φ	2.6	2.1	2.0	2.6
Flight speed (m/s)	U_{ref}	2.0	4.5	5.0	8.0
Reynolds number (-)	Re	1.3×10^2–2.1×10^2	1.2×10^3–3.0×10^3	4.2×10^3–5.3×10^3	1.1×10^4
Advance ratio (-)	J	-	0.66	0.91	0.34

flow patterns around the same model used in their previous work [201] at different Reynolds numbers (moths and flies) and investigated the impact of the scaling parameters on the aerodynamic outcome. Further refining the experimental techniques, Fry et al. [228] recorded the wing and body kinematics of free-flying fruit flies performing rapid flight maneuvers with a 3D infrared high-speed video and "replayed" them on their robotic model to measure the aerodynamic forces produced by the wings. They reported that the fruit fly generates sufficient torque for rapid turn with subtle modifications in wing motion and suggested that inertia, not viscous force, dominates the flight dynamics of flies. In addition, numerous studies using flow visualization around biological flyers have been conducted, including smoke visualizations [255]–[260] and particle image velocimetry (PIV) measurements [14] [30] [260]–[269] to understand their flight mechanisms. Furthermore, the advance of such experimental technologies has enabled researchers to obtain not only 2D but also 3D flow structures around biological flyers [262] [266]–[269] and/or scaled models [270]–[272] with reasonable resolution in space. For example, Srygley and Thomas

[256] reported a study on the force-generation mechanisms of free-flying butterflies using high-speed, smoke-wire flow visualizations to obtain qualitative images of the airflow around flapping wings. They observed clear evidences of LEV structures. In comparison, in moth and fly flight, the helical structure and the spanwise, axial flow patterns appear to be much weaker. They suggested that free-flying butterflies use a variety of aerodynamic mechanisms to generate force, such as wake capture, LEVs, active and inactive upstrokes, rapid rotation, and clap-and-fling mechanisms. These different mechanisms, discussed in subsequent sections, are often used in successive strokes as seen during takeoff, maneuvering, maintaining steady flight, and landing.

Warrick et al. [14] used PIV to observe the wake around hovering hummingbirds (see Fig. 1.11). They estimated the aerodynamic force based on the circulation computed by integrating the measured vorticity field and observed an asymmetry in the force between upstroke and downstroke. Specifically, hummingbirds generate 75 percent of the lift during the downstroke and 25 percent during the upstroke. They reported an inversion of the cambered wings during the upstroke, as well as evidence of the formation of LEVs, created during the downstroke. Videler et al. [30] measured the flow around a single swift wing in fast gliding in a water-tunnel experiment using the digital particle image velocimetry (DPIV) technique. Their results show that the gliding swifts can generate stable LEVs at small AoAs (5°–10°). Whereas the swept-back hand wings generate lift using the LEVs, the flow around the arm wings seems to remain attached.

This discussion presents a sample of experimental and modeling investigations. Clearly, depending on the size and flow parameters of individual species, various lift-enhancement mechanisms are observed. In the following discussion, we address the flapping wing aerodynamics by focusing on the unsteady flow mechanisms, as well as related scaling, geometric, and kinematic parameters. Overall, several unsteady aerodynamic mechanisms associated with flapping wing aerodynamics for force generation have been reported in the literature:

(i) delayed stall of LEVs
(ii) lift peak due to pitch-up rotation
(iii) wake capture due to vortical flow and airfoil interactions
(iv) tip vortex
(v) a persistent downward jet in the wake region
(vi) clap-and-fling mechanism

Figure 3.10 displays these mechanisms. In the upper left corner, the wake-capture mechanism, which occurs near the beginning of the translation, is illustrated. This mechanism produces an increase in aerodynamic lift by transferring fluid momentum associated with large-scale vortical flow shed from the previous stroke to the wing at the beginning of each half-stroke. The LEV and its delayed stall are shown in the upper right corner, whereas in the lower right corner, the downward jet is depicted in the wake of a hovering wing. The delayed-stall phenomenon has been investigated from both the dynamic-stall [201] [273] and upper-wing LEV [43] [199] viewpoints. Finally, the tip vortices are shown in the lower left corner via the instantaneous streamlines colored by their vertical velocity components. The impact

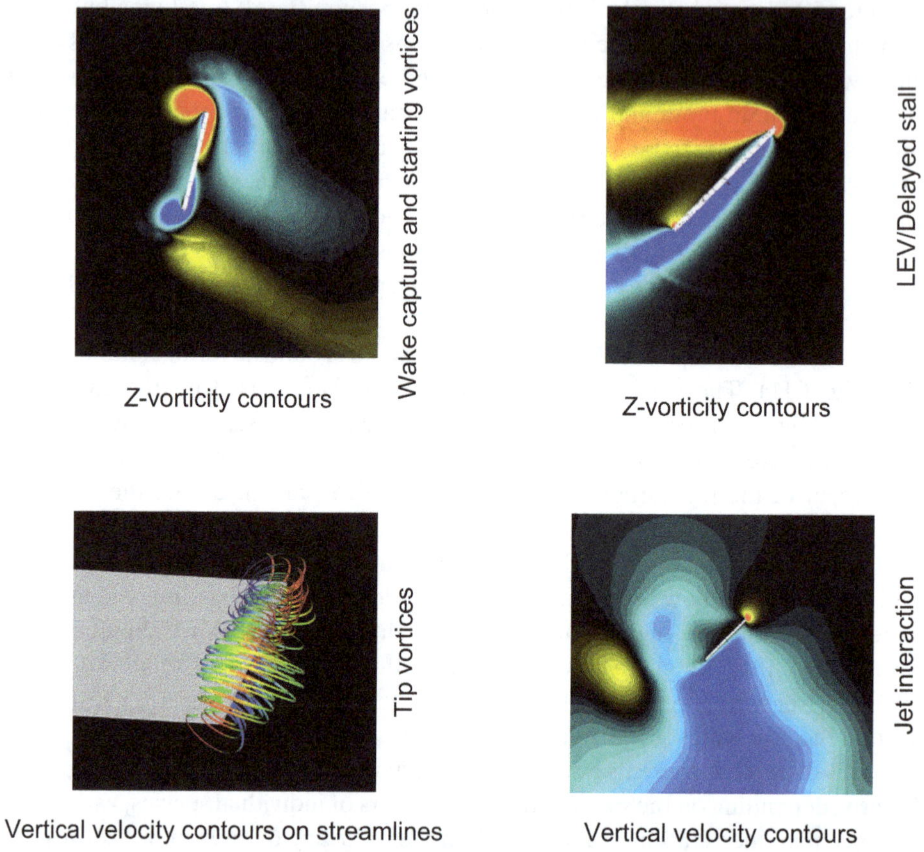

Figure 3.10. Illustration of the time-dependent flow structures affecting the aerodynamics of flapping airfoil during the stroke cycle. Upper left: starting vortices and wake capture. Lower left: tip vortices. Upper right: delayed stall and leading edge vortex. Lower right: jet interaction.

of these mechanisms on the lift and thrust associated with a flapping wing is discussed later.

3.3.1 Leading-Edge Vortices (LEVs)

When an airfoil is accelerated impulsively to a constant velocity, the bound vortex needs time to develop to its final, steady-state strength. Depending on the pace of acceleration, it may take up to six chord lengths of travel for the circulation and lift to reach 90 percent of the final values [213]. This lift enhancement can be explained in part by the so-called Wagner effect [274], which describes the unsteady aerodynamics associated with an accelerating airfoil. Specifically, an impulsively started airfoil only develops a fraction of its steady-state circulation immediately; the steady-state value can be attained only after the airfoil moves through several chord lengths. However, if the airfoil is started at an AoA above its stalling angle, then a large transient vortex forms above the leading edge, which can dramatically increase the lift [51]. Dickinson and Götz [275] measured the aerodynamic forces of an airfoil impulsively started at a high AoA in the Reynolds number range of a fruit fly (Re = 75–225). They observed that, at AoAs above 13.5°, impulsive movement resulted in the production of an LEV that stayed attached to the wing for the first two chord lengths of travel, resulting in an 80 percent increase in lift compared with the performance measured five chord lengths later. For a Reynolds number of 6.0×10^4, Beckwith and Babinsky [276] experimentally investigated the relative importance of delayed stall and the Wagner effect for an impulsively started flat plate ($AR = 4$). They showed that the prestall plate shows a gradual build in lift force similar to Wagner's prediction [274] and that the lift on the poststall plate rises more quickly to levels above steady state, which is a clear example of delayed stall. Force curves presented by Beckwith and Babinsky [276] are similar to those seen by Dickinson and Götz [275] at $Re < 1.0 \times 10^3$, which implies that the Wagner's effect on lift-force generation is less sensitive to the Reynolds number variation.

Most of the research on the dynamic-stall phenomenon has been performed on pitching airfoils. This 2D motion has been useful in highlighting the characteristics of dynamic stall on helicopter blades, fish swimming, and flapping flight. The viscous effects play an important role in these cases. This has led to a more careful investigation of the dynamic-stall process, including evaluation of the type of motion [246] involved (see Fig. 3.8). McCroskey et al. [277] showed the sensitivity to history effects in dynamic stall. They observed that the high-angle part of the oscillating airfoil in a dynamic-stall cycle depends significantly on the rate of change of the AoA near the stall angle; the same lift- and pitching-moment behavior can be attained by matching the rate of change of the AoA at stall limit with different amplitudes.

The potential benefit of trapped or wing-attached vortices in flapping-wing lift enhancement has long been recognized [43] [275] [278]–[280]. In particular, the high-lift mechanism generated by the LEV in a flying insect, which has received substantial attention, was first discovered by Ellington et al. [199]. It appears that the LEV can enhance lift by attaching the bounded vortex core to the upper leading edge during wing translation [199] [281]. The LEVs generate a lower pressure area in the suction side of the wing, which results in a large suction on the upper surface. It seems that the lift enhancement can be sustained for three or four chord lengths of travel before vortex breakdown or complete de-attachment occurs.

As previously discussed, Ellington and co-workers [199] [248] designed a 10:1 scaled-up robotic model for studying the aerodynamics of the hawkmoth *Manduca sexta* (see Fig. 3.11). To maintain both the Reynolds number and the reduced frequency similarity in hovering, as introduced in Section 3.2, they preserve fR^2 between the real insect and the mechanical model. The use of geometrically similar hawkmoth wing models that undergo hovering with the same flapping kinematics can therefore satisfy the aerodynamic similarity. By using smoke streams to visualize the flow around a flapping wing, Ellington et al. [199] demonstrated the presence of a vortex close to the leading edge of the hawkmoth wing model. They observed a small but strong LEV that persists through each half-stroke (downstroke). From these observations, they proposed that the LEV is responsible for enhancing lift-force generation. Furthermore, the LEV has a high axial flow velocity in the core and is stable, separating somewhat from the wing at approximately 75 percent of the wing length spanwise and then connecting to a large, tangled tip vortex. The overall vortical structures are qualitatively similar to those of low AR delta wings [199] that

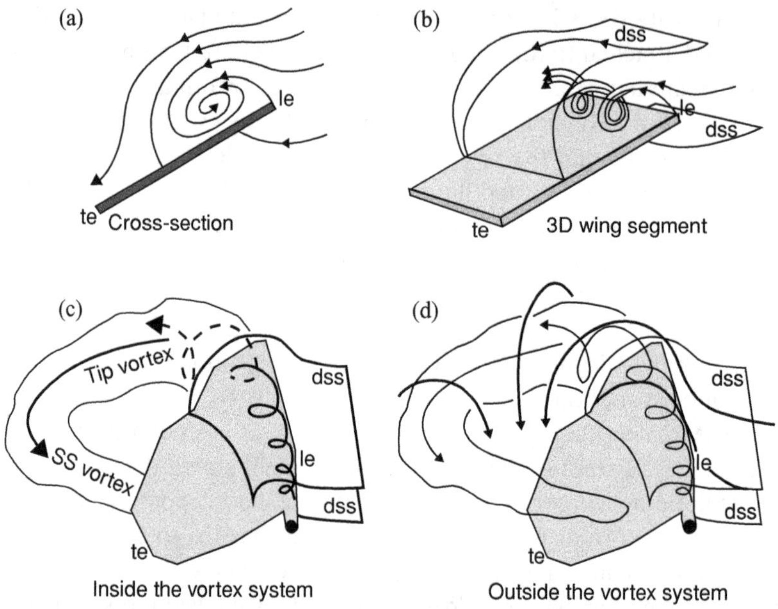

Figure 3.11. Spatial flow structure of LEVs, where le designates leading edge, te designates trailing edge, dss designates dividing stream surface, and SS vortex designates combined starting/stopping vortex. From van den Berg and Ellington [248].

stabilize the LEV due to the spanwise pressure gradient, increasing lift well above the critical AoA. They further suggested that the LEV stability in flapping wings is maintained by a spanwise axial flow along the vortex core (see [199]) that creates "delayed stall," thereby enhancing lift during the translational phase of the flapping motion.

Using the same wing model and kinematics considered in Ellington et al [199], Liu and Kawachi [200] and Liu et al. [247] conducted unsteady Navier-Stokes simulations of the flow around a wing of hawkmoth *Manduca sexta* to probe the unsteady aerodynamics of hovering flight. They demonstrated the salient features of the LEV as well as the spiral axial flow during translational motions. Their results are consistent with those observed by Ellington et al. [199]. Figure 3.12 shows that the LEV created during previous translational motion and the vortical flows established during the rotational motions of pronation and supination together form a complex flow structure around the wing. The computed lift history shows that the lift is produced mainly during the downstroke and in the latter half of the upstroke.

Birch and Dickinson [254] investigated the LEV features of the flapping model fruit fly wing at the Reynolds number of 1.6×10^2. They reported that, in contrast to the LEV on the model hawkmoth wing, which detaches from the wing surface at approximately 75 percent of the wing length with the presence of a strong axial flow in the core, the LEV of the fruit fly exhibits a stable vortex structure without de-attachment during most of the translational phases. Furthermore, little axial flow is observed in the vortex core, amounting to only 2–5 percent of the averaged tip velocity [254]. However, strong spanwise flow is seen at the rear two-thirds of the chord, at about 40 percent of the wingtip velocity. For a fruit fly, the LEV is stably

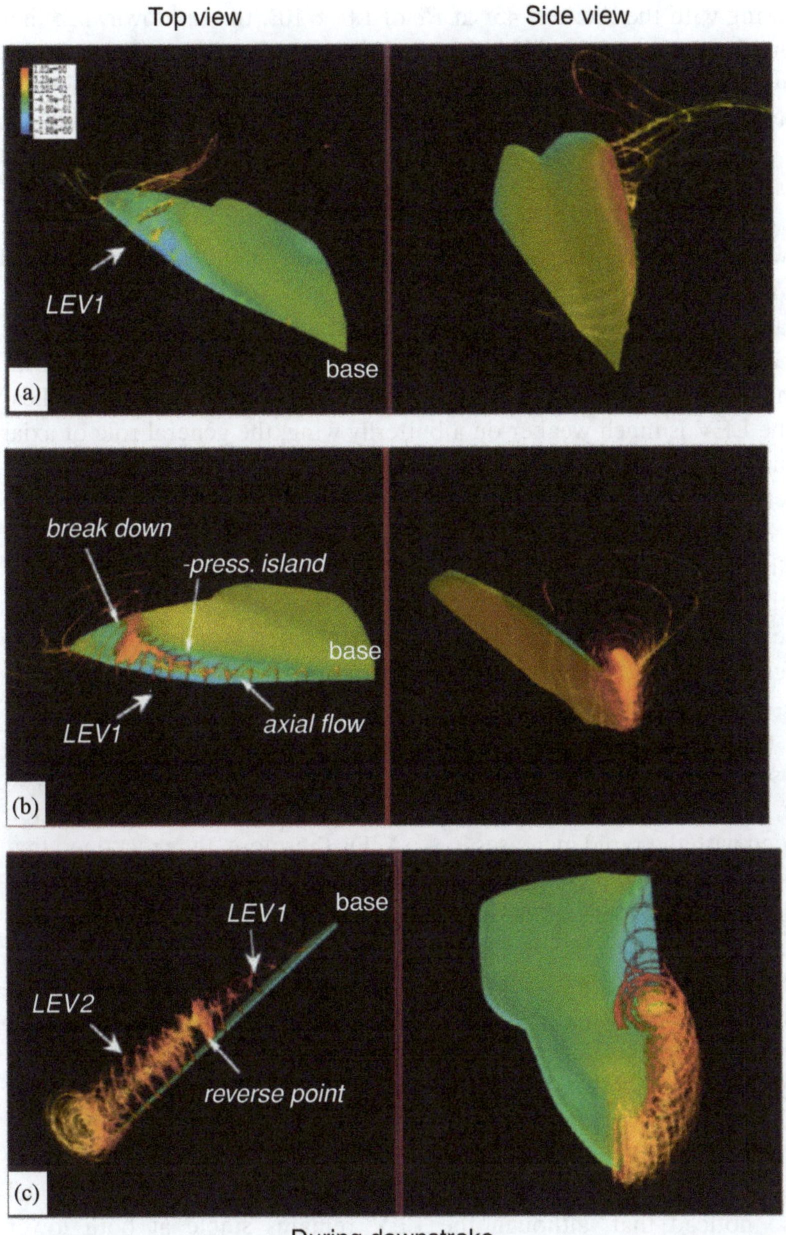

Figure 3.12. Wing surface pressure and streamlines revealing the vortical structures from a 3D numerical simulation of a hovering hawkmoth [247]. (a) Positional angle $\phi = 30°$; (b) $\phi = 0°$; (c) $\phi = -36°$. Reynolds number is approximately 4.0×10^3 and the reduced frequency is 0.37.

attached throughout the half-stroke without breaking up. Based on these considerable differences between fruit fly and hawkmoth models, Birch and Dickinson [254] hypothesized that the attenuating effect of the downwash induced by the tip vortex and wake vorticity limits the growth of the LEV by lowering the effective AoA and prolonging the attachment of the LEV. Recently, using the time-resolved scanning tomography PIV technique David et al. [282] reported that, in the case of a translating

NACA 0012 wing with the AoA of 45° at *Re* of 1.0×10^3, the tip downwash that originates from the formation of TiV indeed reduces the effective AoA and hence also the spanwise vorticity production at the leading edge. However, it should be noted that a NACA 0012 wing is relatively thicker than the model hawkmoth and fruit fly wing.

Another study [256] on large red admiral butterflies, *Vanessa Atlanta*, questioned the existence of axial flow even at the level of the Reynolds numbers comparable to that of hawkmoths. Using smoke trails to visualize the wake behind free-flying butterflies in a wind tunnel, the investigators showed that the LEV spreads from the wing surface to the body of the animal. In contrast to the conical LEV observed in the hawkmoth, the butterfly LEV exhibits a more cylindrical-shaped vortex with constant diameter, and at the end it connects with the tip vortex. Because the helical structure of the LEV is much weaker on a butterfly wing, the general role of axial flow for stabilizing the LEV was again questioned.

Thomas et al. [258] showed that dragonflies attain lift by generating high-lift LEVs using free- and tethered-flight flow visualization. Specifically, in normal free flight, dragonflies use counter stroke kinematics with an LEV on the forewing downstroke and with attached flow on the forewing upstroke and on the hindwing throughout. When the dragonflies accelerate, they switch their kinematics to in-phase wing beats with highly separated downstroke flows, with a single LEV attached across both the fore- and hindwings. Based on their flow visualizations, Thomas and coworkers also suggested that the spanwise flow is not a dominant feature of the flow field. They observed that the spanwise flow sometimes runs from the wingtip only to the centerline, or vice versa, depending on the degree of sideslip. The LEV formation always coincided with the rapid increases in AoA. Furthermore, they reckoned that the flow fields produced by dragonflies differ qualitatively from those published for mechanical models of dragonflies, fruit flies, and hawkmoths, which preclude natural wing interactions. However, parametric assessment showed that, provided the Strouhal number is appropriate and the natural interaction between left and right wings can occur, even a simple plunging plate can reproduce the detailed features of the flow seen in dragonflies. Thomas et al. [258] suggested that stability of the LEV is achieved by a general mechanism whereby the flapping kinematics is configured such that an LEV would be expected to form naturally over the wing and remain attached for the duration of the stroke.

Birch et al. [283] conducted flow visualization around a robotic fruit fly model wing and also noticed that, although the LEV remains stable at both lower ($Re_{f3} = 1.2 \times 10^2$) and higher ($Re_{f3} = 1.4 \times 10^3$) Reynolds numbers, the flow changes from a relatively simple pattern at lower Reynolds numbers to spiral flow at higher Reynolds numbers. Vorticity measurements taken at mid-stroke, in a plane located at 0.65 of the wing length and perpendicular to the spanwise direction, show stronger and larger LEVs for the higher Reynolds numbers associated with intense axial (spanwise) velocity within the LEV core, with magnitudes significantly larger than those of the tip velocity [254]. At a lower Reynolds number, they observed no peak in axial flow in the area of the LEV core, likely because of the stronger viscous effect.

Kim and Gharib [284] experimentally studied spanwise flow generated by a flat plate by rotating and translating motions with a constant AoA of 45° at a *Re* of 1.1×10^3. They observed that, for the translating plate, an LEV develops non-uniformly along the span because of the influence of the wingtip. This deformed LEV induces the spanwise flow over the flat plate and subsequent vorticity transport. However, the spanwise flow is not strong enough to suppress the growth of the LEV near the central region of the plate. For the rotation mode, the vorticity of the LEV is tilted because the size of the LEV increases from the base to the tip. The tilted vorticity induces the spanwise flow over the rotating plate. Contrary to the translation mode, the spanwise flow is also found in the wake; this spanwise flow is due to the streamwise component of vorticity, which is distributed inside a shear layer and a starting vortex in the wake.

The LEV of a flapping wing resembles that of a fixed delta wing. The delta wing owes much of the lift that it is able to generate to the fact that the vortex flow initiates at the leading edge of the wing and rolls into a large vortex over the leeward side, containing a substantial axial velocity component. This high-flow velocity in the core of the vortex is a region of low pressure that generates a suction (i.e., lift). For a delta wing placed at high AoAs, vortex breakdown occurs, causing the destruction of the tight and coherent vortex. The diameter of the core increases, and the axial velocity component is no longer unidirectional. With the loss of axial velocity the pressure increases, and consequently, the wing loses lift. The literature on the subject is immense, and for a more general presentation of the various aspects of vortex breakdown, we refer the reader to several review articles [164] [285] [286]. For a fixed wing, an important trend is that, at a fixed AoA, if the swirl is strengthened, then vortex breakdown occurs at lower Reynolds numbers. In contrast, a weaker swirling flow tends to break down at a higher Reynolds number. Since the fruit fly exhibits a weaker LEV, from this viewpoint, it tends to better maintain the vortex structure than a hawkmoth, which creates a stronger LEV.

Of course, the link between the vortex breakdown and a fixed or flapping wing, if any, is not established. It should be noted that, although helicopter blade models have been used to help explain flapping-wing aerodynamics, spanwise axial flows are generally thought to play a minor role in influencing helicopter aerodynamics [287]. In particular, helicopter blades operate at substantially higher Reynolds numbers and lower AoA. The much higher AR of a blade also makes the LEV harder to anchor.

3.3.2 Rapid Pitch-Up

The LEV-based lift-enhancement mechanism seems to be a main feature during the translational motion of the stroke. In addition, the flapping wings experience rapid wing rotation at the ends of the down- and upstroke, which can enhance lift force in flying insects.

Kramer [288] first demonstrated that a wing can experience lift coefficients above the steady stall value when that wing is rotating from low to high AoAs, which is now termed the Kramer effect. The unsteady aerodynamic characteristics associated with the time-dependent AoA, including hysteresis, are shown in Figure 3.4. Dickinson et al. [201] used their *Robofly* along with a varied rotational pattern, illustrated in Figure 3.9, to investigate the interplay between kinematics and lift generation. They identified two aerodynamic force peaks at the end and the beginning of each

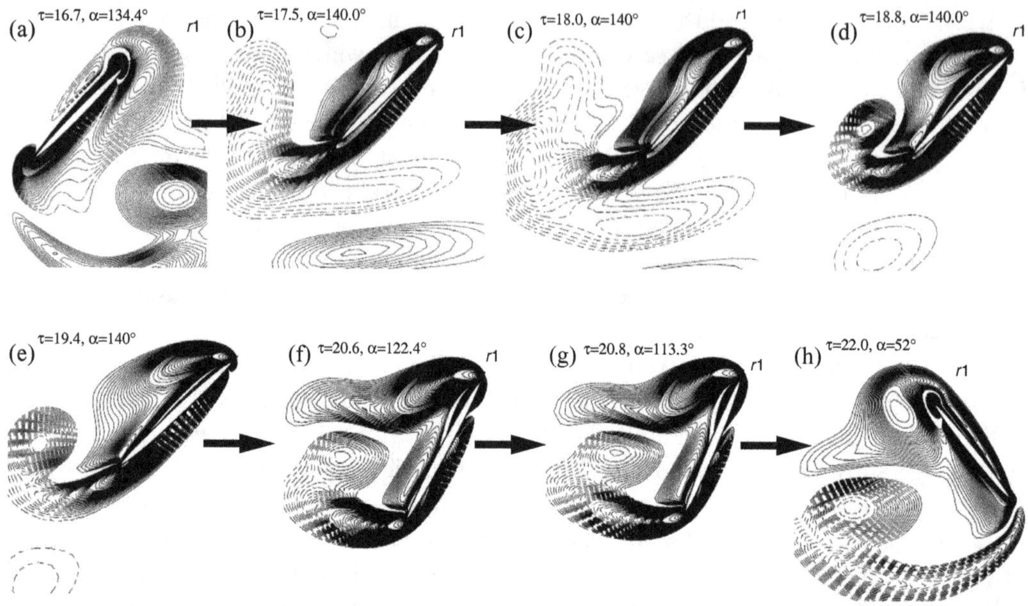

Figure 3.13. Vorticity plot at 75% wingspan. At the beginning of the upstroke the wing accelerates with small changes in the angle of attack (from $\alpha = 46°$ at (a) to $\alpha = 40°$ at (c)). The fast acceleration increases the rate of change of fluid momentum, and an increase in lift is observed. From (c) to (e) the wing translates with constant speed at $\alpha \sim 40°$. The dynamic stall vortex does not shed and a large lift can be maintained. From (e) to (f) the wing pitches up rapidly while translating with constant speed, resulting in a sharp increase in lift and drag; (h) the wing decelerates and the lift is decreasing fast. The Reynolds number is 136. From Sun and Tang [289].

stroke (pronation and supination). The first force peak can be explained based on the rotational circulation. The resulting force enhancement is influenced by the timing of the wing rotation while translating. They found that an advanced rotation produces a mean lift coefficient $C_L = 1.74$, which is almost 1.7 times higher than that of a delayed rotation ($C_L = 1.01$), whereas a symmetric rotation can attain a value of $C_L = 1.67$. These peaks were confirmed by the numerical simulations of Sun and Tang [289] and Ramamurti and Sandberg [290]. In addition, Sun and Tang [291] further investigated three mechanisms responsible for lift enhancement via unsteady aerodynamics: (i) rapid acceleration of the wing at the beginning of a stroke, (ii) delayed stall, and (iii) fast pitch-up rotation of the wing near the end of the stroke. In advanced rotation, the wing flips before reversing its translational direction as illustrated in Figure 3.9, and the leading edge rotates backward relative to the translation. Based on their computational analysis, Sun and Tang [289] suggested that the first peak is due to the increase in rapid vorticity that occurs when the wing experiences fast pitch-up rotation. The pitch-up rotation and the associated vorticity increase are plotted in Figure 3.13f and g. Sane and Dickinson [292] attributed this first force peak to the additional circulation generated to reestablish the Kutta condition during the rotation. Overall, the findings reported by Sun and Tang [289] and Sane and Dickinson [292] are in agreement. The second peak, termed wake capture, is related to the wing-wake interaction and is discussed next. Together, these two peaks contribute to lift enhancement. Because both pitch-up and wake

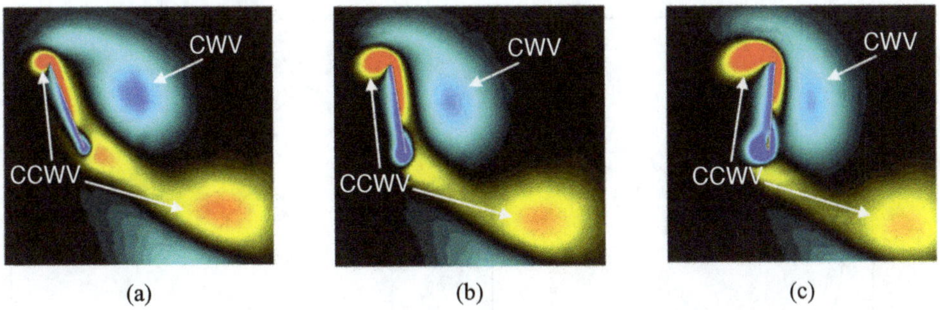

Figure 3.14. Illustrations of the wake-capture mechanism [201] [296]. (a) Supination, (b) beginning of upstroke, and (c) early of upstroke. At the end of the stroke, (a), the wake shed in the previous stroke denoted by CWV is en route of the flat plate. As the flat plate moves into the wake (b–c) the effective flow velocity increases and additional aerodynamic force is generated. The color of the contour indicates the spanwise component of vorticity. CWV and CCWV indicate clockwise and counterclockwise vortex.

capture are strongly influenced by flapping kinematics, more discussion is provided later to help elucidate the parametric variations of these factors.

3.3.3 Wake Capture

As discussed by Dickinson et al. [201] the wing-wake interaction can significantly contribute to lift production in hovering insects. They found that the second peak is generated at the beginning of each stroke of hovering flight when the wings reverse the direction of moving while rotating about the spanwise direction. This physical mechanism, termed *wake capture*, produces aerodynamic lift by transferring fluid momentum to the wing at the beginning of each half-stroke. A wing meets the wake created during the previous stroke after reversing its direction, thus increasing the effective flow speed surrounding the airfoil, which generates the second force peak. The wake-capture mechanism is illustrated in Figure 3.14.

As with rapid pitch-up, the effectiveness of wake capture is a function of wing kinematics and flow structure. This second force peak (occurring right after stroke reversal) is apparently distinct from rotational lift because its timing is independent of the phase of wing rotation. Dickinson et al. [201] showed that the second peak persists even by halting the wing after it rotates, indicating that the wake produced by the wing motion in the previous half-stroke serves as an energy source for lift production.

The sinusoidal motion along a horizontal stroke plane is similar to that shown by Wang et al. [217], who conducted 2D simulation of a hovering elliptic airfoil with the stroke amplitude h_a between $1.4c$ and $2.4c$, leading to a reduced frequency (Eq. (3–11)) k between 0.36 and 0.21. The Reynolds number considered is between 75 and 115. The computational results of Wang et al. [217] and Tang et al. [244] both identify a lift peak after the stroke reversal for the normal hovering mode (see Fig. 3.7). However, for the "water treading" mode, the results of Tang et al. [244] show a continuous increase in lift as the airfoil pitch angle increases to its maximum value without a noticeable peak. More detailed discussion of the lift generation for both normal and water treading hovering modes can be found in Tang et al. [244].

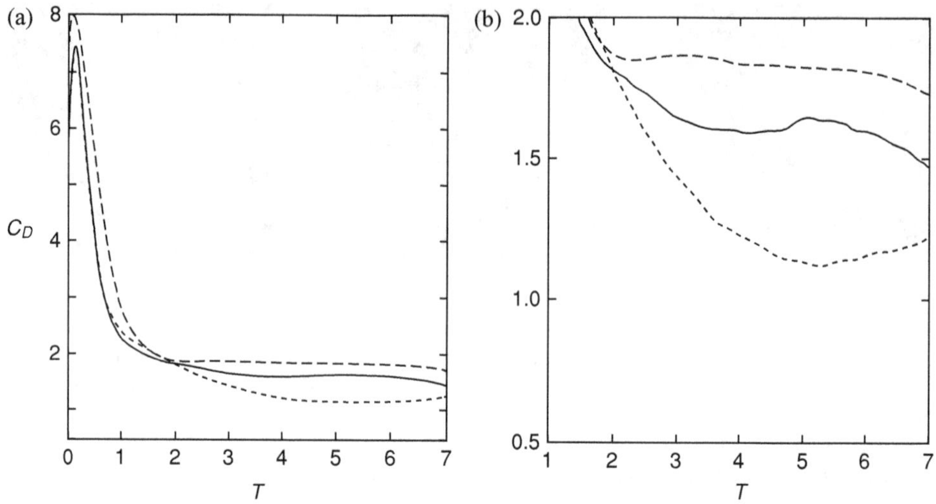

Figure 3.15. Drag coefficients of AR 6 and 2 flat plates during a starting-up translation as a function of T (the number of chord lengths the plate has traveled, see definition of [302]), from Ringuette et al. [299]. (a) Overall view; (b) detailed view of (a). Both (a) and (b) show drag coefficients for the free tip AR of 6 plate (continuous line) and for the same plate with the tip grazing a raised bottom (dash-dotted line); the dashed line is drag coefficient for the AR of 2 plate with the tip free. Low AR of the wing reduces the drag coefficient significantly due to interactions between a TiV and a LEV.

This interpretation of wake-capture force generation has been questioned recently based on the claim that the rotation-independent lift peak is due to a reaction caused by accelerating the added mass of fluid [293]. In general, the inertia of the flapping wing is increased by the mass of the accelerated fluid – termed *added mass* [45] – which can play a significant role in the aerodynamics of insect flight [294]. However, evaluating the added mass, and thus estimating inertial forces, is, not easy. Although the mass of a wing itself may be tiny, the mass of the accelerated fluid need not be [65] [295].

3.3.4 Tip Vortices (TiVs)

Tip vortices (TiVs) associated with fixed finite wings are seen to decrease lift and induce drag [297]. However, in unsteady flows, TiVs can influence the total force exerted on the wing in three ways: (i) by creating a low-pressure area near the wingtip [296], (ii) through an interaction between the LEV and the TiV [296], and (iii) constructing a wake structure by downward and radial movement of the root vortex and TiV [298]. The first two mechanisms (see Fig. 3.15) also were observed for impulsively started flat plates normal to the motion with low aspect ratios: Ringuette et al. [299] presented experimentally that the TiVs contributed substantially to the overall plate force by interacting with the LEVs at $Re = 3.0 \times 10^3$. Taira and Colonius [300] used the immersed boundary method to highlight the 3D separated flow and vortex dynamics for a number of low aspect ratio flat plates at different AoAs. At Reynolds numbers of 3.0×10^2 to 5.0×10^2, they showed that the TiVs could stabilize the flow and exhibited non-linear interaction with the shed vortices. Stronger

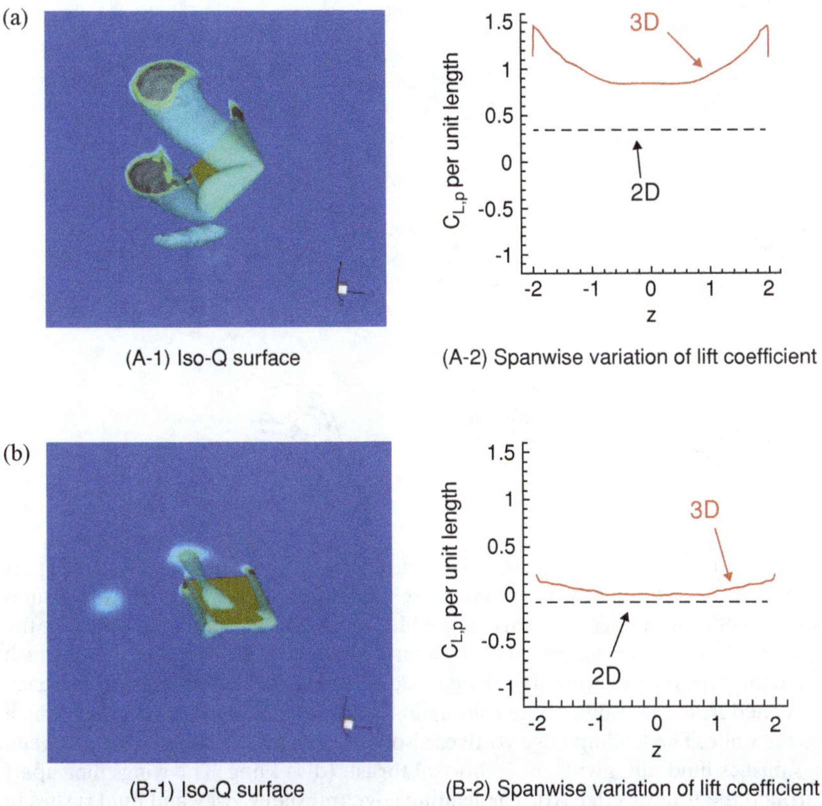

Figure 3.16. Effect of TiVs on lift generation associated with a pitching/plunging flat plate ($AR = 4$) at the middle of stroke; from Shyy et al. [296]. (a) Delayed rotation case ($2h_a/c_m = 2.0$, $\alpha_a = 45°$, and $\varphi = 60°$); (b) synchronized rotation case ($2h_a/c_m = 3.0$, $\alpha_a = 80°$, and $\varphi = 90°$).

influence of the downwash from the TiVs resulted in reduced lift for lower aspect ratio plates.

For flapping motion in hover, however, depending on the specific kinematics, the TiVs could either promote or make little impact on the aerodynamics of a low aspect ratio flapping wing. Shyy et al. [296] demonstrated that for a flat plate with $AR = 4$ at $Re = 64$ with two different wing motions (such as delayed rotation and synchronized rotation; see Fig. 3.16), the TiV anchored the vortex shed from the leading edge, thereby increasing the lift compared to a 2D computation under the same kinematics. In contrast, under different kinematics with a small AoA and synchronized rotation, the generation of TiVs was small, and the aerodynamic loading was well approximated by the analogous 2D computation. They concluded that the TiVs could either promote or make little impact on the aerodynamics of a low aspect ratio flapping wing by varying the kinematic motions [296]. Furthermore, at $Re = 64$, Trizila et al. [301] found that four prominent competing effects were in play due to TiVs:

1. Enhancement of lift due to the proximity of the associated low- pressure region of the tip vortex next to the airfoil

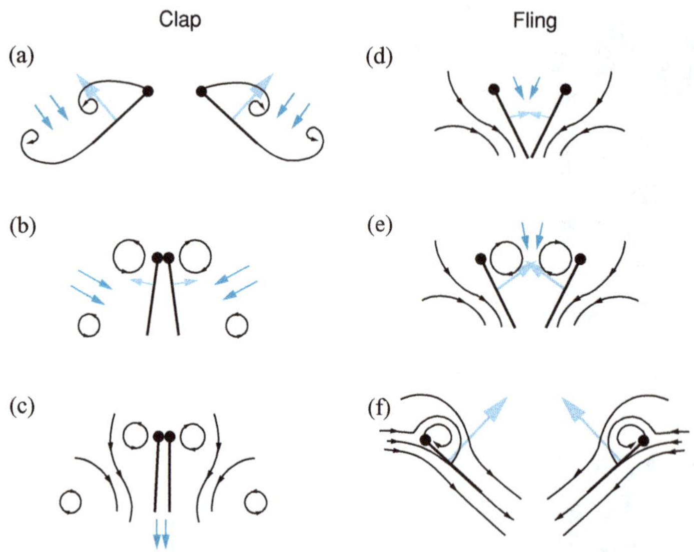

Figure 3.17. Sectional schematic of wings approaching each other to clap (a–c) and fling apart (d–f), from Sane [312] and originally described in Weis-Fogh [68]. Black lines show flow lines and dark blue arrows show induced velocity. Light blue arrows show net forces acting on the airfoil. (a–c) Clap. As the wings approach each other dorsally (a), their leading edges touch initially (b), and the wing rotates around the leading edge. As the trailing edges approach each other, vorticity shed from the trailing edge rolls up in the form of stopping vortices (c), which dissipate into the wake. The leading-edge vortices also lose strength. The closing gap between the two wings pushes fluid out, giving an additional thrust. (d–f) Fling. The wings fling apart by rotating around the trailing edge (d). The leading edge translates away and fluid rushes in to fill the gap between the two wing sections, giving an initial boost in circulation around the wing system (e). (f) A leading edge vortex forms anew, but the trailing-edge starting vortices are mutually annihilated because they are of opposite circulation.

2. Induced downwash acting to reduce the effective AoA along the span and weakening the LEV, hence reducing the instantaneous lift
3. Interaction with the vortices shed from the LEs and TEs and anchoring them from shedding near the wingtips, thereby enhancing the lift
4. Because TiVs pull fluid from the underside of the wing to the upper side, the wing leaves behind a weaker pocket of downward momentum in the flow. On interaction with this downward momentum, a loss in lift is seen, and so a weaker wake valley means higher lift.

3.3.5 Clap-and-Fling Mechanism

One of the most complex kinematic maneuvers in flying animals is the wing-wing interaction of the left and right wings during the dorsal stroke reversal, termed the clap-and-fling mechanism (see Fig. 3.17). This unique procedure enhances lift generation and is particularly observed in the flight of tiny insects [68]. It has been further observed in other studies [70] [303] [304]. A modified kinematics termed "clap-and-peel" was found in tethered flying *Drosophila* [305] and larger insects such as butterflies [26], the bush cricket, mantis [306], and locust [307]. It seems that the clap-and-fling mechanism is not used continuously during flight and is more often

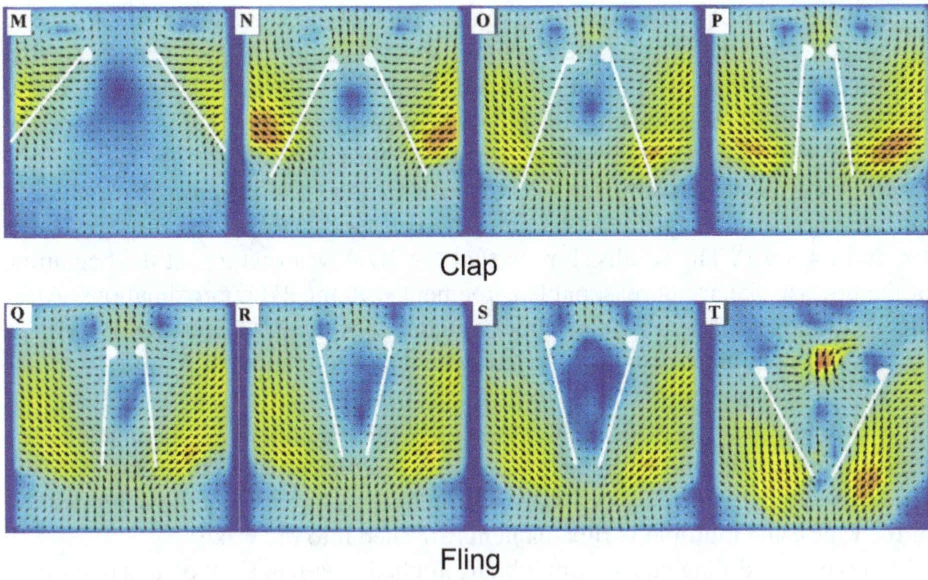

Figure 3.18. Experimental visualization of clap-and-fling mechanism by two wings (M-T) using robotic wing models; from Lehmann et al. [273]. Vorticity is plotted according to the pseudo-color code, and arrows indicate the magnitude of fluid velocity, with longer arrows signifying larger velocities.

observed in insects during maximum flying performance while carrying loads [308] or performing power-demanding flight turns [307]. Marden's experiments [308] on various insect species found that insects with the clap-and-fling wing-beat produce about 25 percent more lift per unit flight muscle (79.2 N kg^{-1}mean value) than insects using conventional wing kinematics (such as flies, bugs, the mantis, dragonflies, bees, wasps, beetles, and sphinx moths; 59.4 N kg^{-1}mean value).

The clap-and-fling mechanism is a close apposition of two wings at the dorsal stroke that reverses preceding pronation. By strengthening the circulation during the downstroke it can generate considerably large lift on the wings. The fling phase preceding the downstroke is thought to enhance circulation due to fluid inhalation in the cleft formed by the moving wings, which causes strong vortex generation at the leading edge. Lighthill [209] showed that a circulation proportional to the angular velocity of the fling was generated. Maxworthy [43], by a flow visualization experiment on a pair of wings, reported that during the fling process, an LEV is generated on each wing and its circulation is substantially larger than that calculated by Lighthill [209].

Lehmann et al. [273] used a dynamically scaled mechanical model fruit fly *Drosophila melanogaster* wing to investigate force enhancement due to contralateral wing interactions during stroke reversal (clap-and-fling; see Fig. 3.18). Their results suggest that lift enhancement during clap-and-fling requires an angular separation between the two wings of no more than 10°–12°. Within the limitations of the robotic apparatus, the clap-and-fling augmented the total lift production by up to 17 percent, but the actual performance depended strongly on stroke kinematics. They measured two transient peaks of both lift and drag enhancement during the fling phase: a prominent peak during the initial phase of the fling motion, which accounts for most of the benefit in lift production, and a smaller peak of force enhancement at the end of the fling when the wings started to move apart. Their investigation indicates that the effect of clap-and-fling is not restricted to the dorsal part of the stroke cycle but that it extends to the beginning of the upstroke. This suggests that the presence of the image wing distorts the gross wake structure throughout the stroke cycle.

Recently, Kolomenskiu et al. [309] investigated the clap-fling-sweep mechanism of hovering insects using 2D and 3D simulations at specific Reynolds numbers: 1.28×10^2 and 1.4×10^3. The results showed that the 3D flow structures at the beginning of the downstroke are in reasonable agreement with the 2D approximations. After the wings move farther than one chord length apart, the 3D effects can be seen in force history and flow structures for both Reynolds numbers. At $Re = 1.28 \times 10^2$, the spanwise flow from the wing roots to the wingtips is driven by the centrifugal forces acting on the mass of the fluid trapped in the recirculation bubble behind the wings. The spanwise flow removes the excess of vorticity and delays the periodic vortex shedding. At $Re = 1.4 \times 10^3$, vortex breakdown occurs past the outer portion of the wings, and multiple vortex filaments are shed into the wake.

This clap-and-fling mechanism is being applied to enhance lift of actual flapping wing MAVs (de Croon et al. [310] and Nakata et al. [311]; see Figure 3.19).

3.4 Fluid Physics in $O(10^2$ to $10^3)$ Reynolds Number Regime

3.4.1 Effects of Kinematics on Hovering Airfoil Performance

Trizila et al. [301] used surrogate modeling techniques [314] to investigate the time-averaged lift and power input subject to various combinations of kinematic parameters and reported numerous interesting findings. They considered a 2 percent thick wing with $AR = 4$. The Reynolds number, based on the average tip velocity, is equal to 64, which is similar to that of a fruit fly, *Drosophila melanogaster*. They used simplified wing kinematics (see Eqs. (3–4) and (3–5)) and selected the mid-chord of the rigid airfoil as the pitching axis. Because the free-stream is absent, the tip velocity is taken as the reference velocity, so that the reduced frequency contains the same information as the inversed normalized stroke amplitude [315]. The rotational (pitching) motion is similarly governed by the flapping frequency, and the angular amplitude, α_a, is a measure of how far the airfoil deviates from the yz-plane (see Fig. 3.20). The time average of the pitching motion is $\alpha_0 = 90°$. Higher angular amplitudes yield lower angles of attack and vice versa. The phase lag between the pitching and plunging motions is denoted as φ.

The three kinematic parameters used as the design variables in surrogate modeling – stroke amplitude (h_a), pitching angular amplitude (α_a), and phase lag (φ) – are varied independently. The range of the design variables is as follows. The normalized stroke amplitude is representative of a range of flapping wing flyers: $2.0 < 2h_a/c_m < 4.0$. Details on pitch amplitude and phase lag are not as plentiful in the literature, so cases are chosen with low AoAs ($\alpha_{min} = 10°$; high pitching amplitude) and high AoAs ($\alpha_{min} = 45°$; low pitching amplitude): $45° < \alpha_a < 80°$. The bounds on phase lag are chosen to be symmetric about the synchronized hovering: $60° < \varphi < 120°$. Although delayed rotation is not a focus of many studies found in the

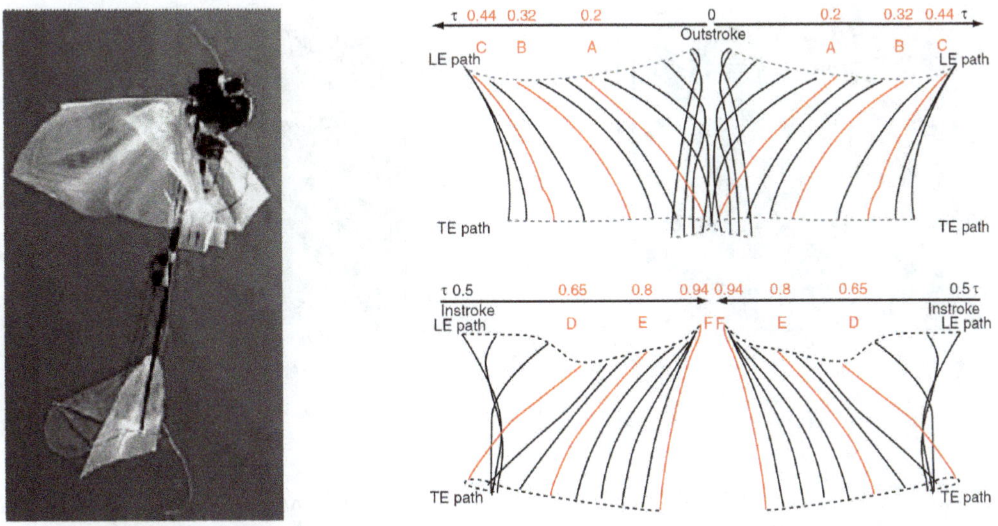

(a) Delfly [314]

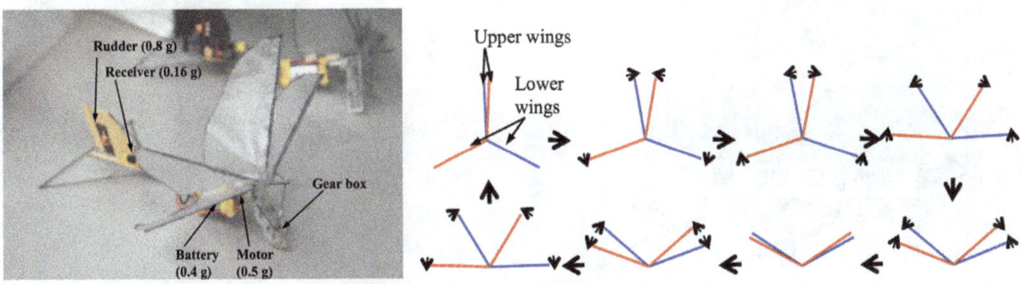

(b) Chiba University flapping MAV [312]

Figure 3.19. Flapping wing MAVs use clap-and-fling mechanism: (a) Delfly and (b) Chiba University flapping MAV.

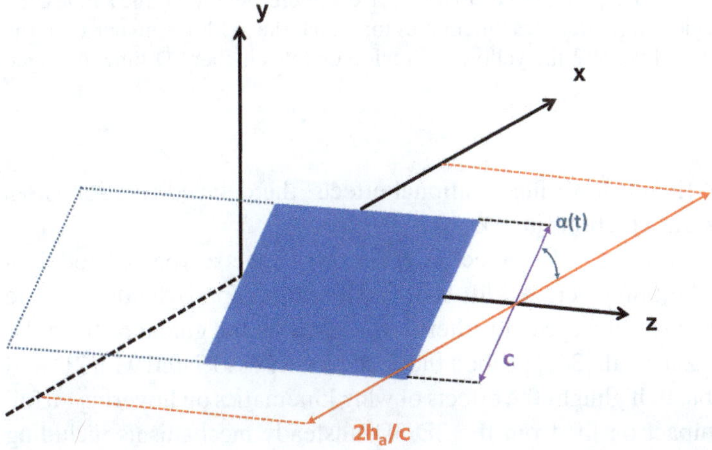

Figure 3.20. Illustration of the finite aspect ratio ($AR = 4$, only half shown) wing plunging in the x-direction and pitching about the z-axis.

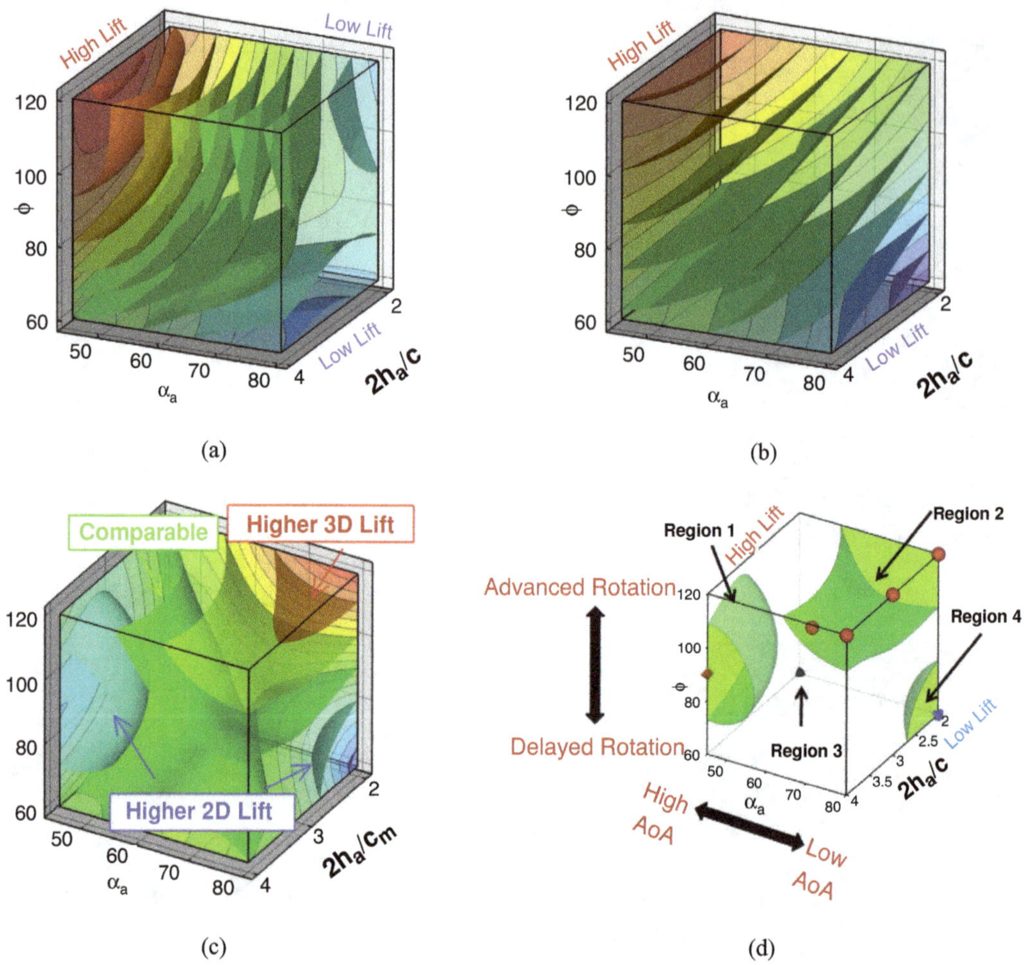

Figure 3.21. Iso-sufaces of 2D lift (a), 3D lift (b), 3D minus 2D lift (c), and (d) where the absolute difference between the 2D and 3D lift equals 0.10. The symbols denote training points in those regions for which detailed force histories and flow field quantities are available; brown octahedra (region 1), circles (region 2), black quarter sphere (region 3), and a blue cube (region 4). The blue region in (c) denotes kinematics for which the 2D lift is higher than the equivalent 3D kinematics. Likewise the yellow/red region denotes higher 3D time-averaged lift.

literature because of the lift-lowering rotational effects, the delayed rotation cases offer interesting wake interactions in 3D.

The quantities of interest – the objective functions in the surrogate modeling nomenclature – are the time-averaged lift coefficient and an approximation of the power required over the stroke cycle. Furthermore, based on the guidance from the surrogate models, Trizila et al. [301] probed the fluid physics associated with 2D and 3D cases and were able to highlight the effects of wing kinematics on hovering airfoil performance. The impact on lift from the 2D/3D unsteady mechanisms including interactions between the LEV and TiVs is detailed in subsequent sections.

Figure 3.21 shows the iso-surfaces of the time-averaged lift coefficient where each axis corresponds to one of the kinematic parameters (i.e., h_a, α_a, or φ). Figure 3.21a-c correspond to the 2D lift, 3D lift, and the difference between the two, respectively. Observations that are immediately apparent are that kinematic combinations with a low α_a (high AoA) and advanced rotation (high φ) have the highest mean lift in 2D and 3D. This result qualitatively agrees with the results of Wang et al. [217] and Sane and Dickinson [292]. Due to the non-monotonic response in phase lag (φ) found in the 2D lift response at high pitching amplitudes (see Fig. 3.21a), there are two regions where there is low, and possibly negative, time-averaged lift generation. The first region is defined by low plunge amplitudes (low $2h_a/c_m$), low AoA (high α_a), and advanced rotation (high φ). The second region is defined by low AoA (high α_a), and delayed rotation (low φ) and is where the 3D kinematics also generates low time-averaged lift values.

A variable's sensitivity is directly related to the gradients along the respective design variables, whereas a more quantitative measure is the global sensitivity analysis; these measures were examined in Trizila et al. [301]. The gradients along the α_a and φ axes are much more significant than that along the normalized stroke amplitude. (Note: Lua et al. [316] found that the effect of *Re* is noticeably smaller than that of α_a on the mean lift.) Furthermore, Figure 3.21 shows that advancing the phase lag is beneficial in 2D except when at high α_a; in 3D there is no such exception within the bounds studied. Comparisons with Sane and Dickinson's experiments [292] show qualitative agreement in the trends in time-averaged lift as a function of α_a and φ within the common ranges, with the noticeable difference in setups being one study [301] used pure translation to represent the plunge whereas the experimental study flapped about a pivot point.

The difference between 2D and 3D time-averaged lift may raise a question about areas of the design space for which simplified 2D aerodynamic analyses can sufficiently approximate their analogous 3D counterparts versus those areas where there are substantial 3D effects, which would preclude such a comparison. Figure 3.21c suggests that the difference in the time-averaged lift coefficient larger than 0.1 is due to three-dimensionality.

Following Shyy et al. [296] and Trizila et al. [301], we now highlight four cases corresponding to those presented in Figure 3.21: (i) synchronized hovering, high AoA; (ii) advanced rotation, low AoA; (iii) delayed rotation, high AoA, low plunging amplitude; and (iv) delayed rotation, low AoA, and low plunging amplitude. Each of these cases represent a region where the lift for the low *AR* wing is significantly different in 2D than in 3D, as indicated in Figure 3.21. In addition to these cases, we look at another region where 2D lift is similar to 3D lift and discuss the trade-off between lift and power.

3.4.1.1 Regions of Significant 3D Effects

SYNCHRONIZED HOVERING, HIGH AOA – PRESENCE OF A PERSISTENT JET. As shown in region 1 in Figure 3.21d, the kinematics in this region corresponds to nearly synchronized hovering (i.e., including cases with slightly delayed or advanced rotation), low angular amplitude (high AoA), and larger plunging amplitudes. Figure 3.22 shows the time history of lift and vertical velocity contour plots at three time instants: in the forward stroke from the 2D simulation, in the symmetry plane of the 3D computation, and near the wingtip. It can be seen that there are two local maxima per stroke in 2D. The first peak is associated with the wake capture at the beginning of the

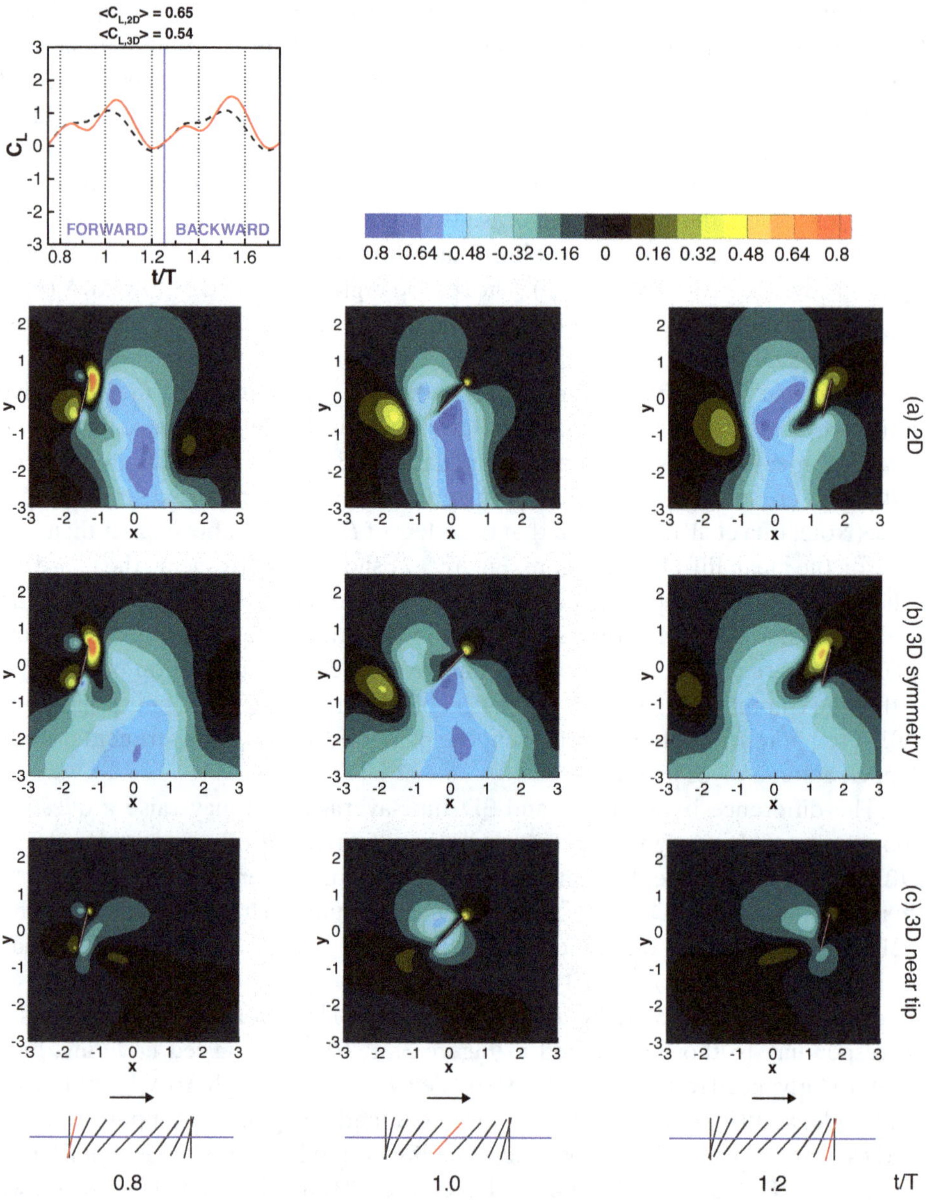

Figure 3.22. Instantaneous force history (2D: solid, red; 3D: broken, black) and vertical velocity contour plots at three time instants in the forward stroke for the synchronized rotation and high AoA ($2h_a = 3.0c_m$, $\alpha_a = 45°$, and $\varphi = 90°$): (a) 2D computation; (b) in the symmetry plane of 3D computation; (c) near the wing tip ($z/c = 1.8$).

stroke at $t/T = 0.8$. Between the two peaks, there is a local minimum referred to as a "wake" valley, which is caused by a combination of decreasing AoA and interaction with a pocket of downward momentum that takes the form of a persistent jet. As reported during the experimental studies of Freymuth [175], the jet develops as a result of a reverse Karman vortex street interacting with the downward momentum created by the wing as it translates. As the wing passes the jet, vortices are shed with an orientation that reinforces the downward momentum previously created by

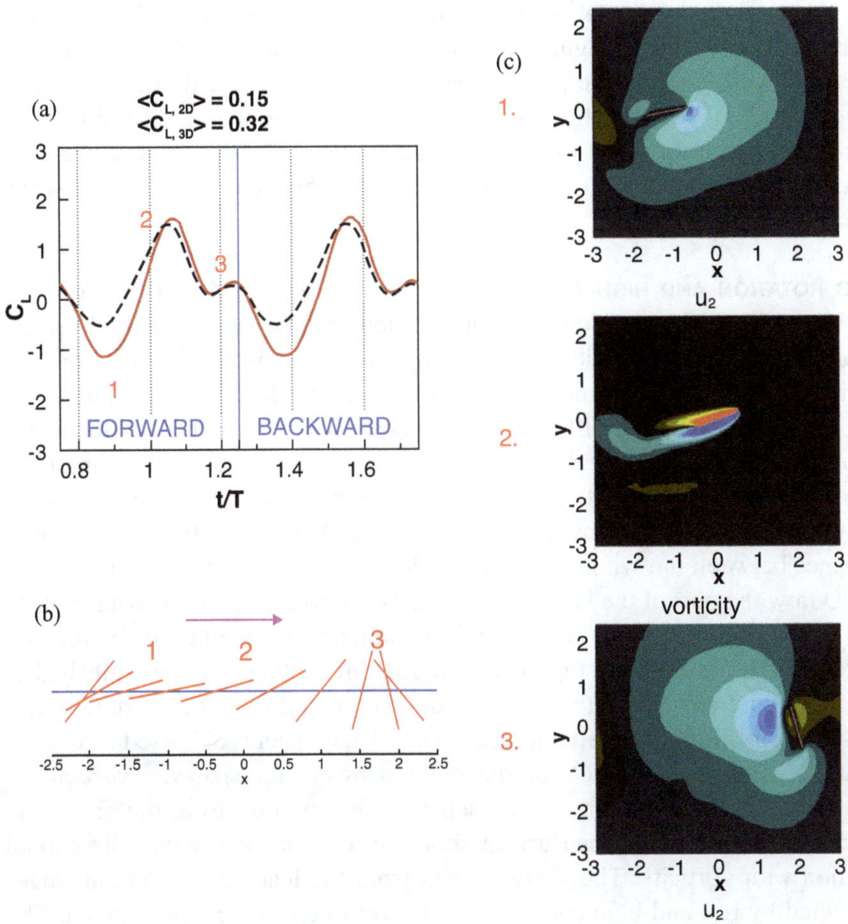

Figure 3.23. Time history of lift coefficients in a representative case in region 2; the advance rotation and low AoA ($2h_a = 4.0c_m$, $\alpha_a = 80°$, and $\varphi = 120°$), with the associated flow features. (a) Lift, (C_L), during a motion cycle. Red-solid, 2D computation; black-dashed, 3D computation. (b) Kinematic schema of the flat plate motion. (c) Representative flow features at (1) $t/T = 0.9$, vertical velocity (u_2) contours; (2) $t/T = 1.0$, vorticity contours; (3) $t/T = 1.2$, vertical velocity contour.

the wing. These vortices sustain the downward momentum, and they further entrain surrounding fluid to create a flow feature with which the wing then interacts during subsequent strokes. In 3D, the weaker downward momentum pocket does help generate lift compared with 2D, which can be confirmed in the instantaneous lift history at $t/T = 0.9$. In 2D, the persistent jet shows larger vertical velocity magnitude and is narrower. However, the weaker LEV has a negative effect on the lift. In the 2D case, the LEV is largely attached, and anchoring the LEV has no benefit. Overall, for this case, the 2D lift is better than the 3D counterpart. More generally, cases in region 1 as shown in Figure 3.21d have larger lift in 2D.

ADVANCED ROTATION AND LOW AOA – HIGHER 3D LIFT. In region 2 as shown in Figure 3.21d, the kinematics is characterized by advanced rotation and a high angular amplitude. Figure 3.23 shows the time histories of lift and the associated flow features for an advanced rotation case with a generally low AoA ($2h_a = 4.0c_m$, $\alpha_a = 80°$, and $\varphi = 120°$). Right after the stroke reversal, the flat plate moves into the wake generated in the previous stroke. Because of the downwash in this wake (see also Fig. 3.23) and the low and decreasing AoA (Fig. 3.23b), lift drops. Note that the pocket of downward momentum encountered for these kinematics is not a persistent jet: the 3D case does not suffer the same drop in lift as the 2D case.

DELAYED ROTATION AND HIGH AOA – TIP VORTEX CAN ENHANCE LIFT. Region 3 is defined by the kinematics with delayed rotation, low angular amplitude (high AoA), and shorter plunging amplitudes. This region shows a significant impact from the TiVs. Figure 3.24 illustrates a delayed rotation case with high AoA ($2h_a = 2.0c_m$, $\alpha_a = 45°$, and $\varphi = 60°$). The difference in the flow physics encountered due to 3D phenomena is noticeable. The main characteristics of the vortices, including their sizes, strengths, and movement, are distinctly different between 2D and 3D results. Not only is there a strong spanwise variation in the 3D flow but there is also little resemblance between the symmetry plane of the 3D and the 2D computations.

In 2D flow the pair of the large-scale vortices is noticeably closer to each other and the airfoil than in the 3D flow. The instantaneous lift coefficient for the two cases examined is illustrated in Figure 3.24; the 3D lift coefficient is generally higher than its 2D counterpart. With these kinematics patterns, the TiVs can interact with the LEV to form a lift-enhancement mechanism. This aspect is discussed next.

Figure 3.25 shows an iso-Q contour colored by ω_3, the spanwise vorticity in forward stroke. In this fashion we can separate the rotation from the shear, via Q, which can be used as a measure of rotation; we can then obtain directional information with vorticity. The vortices shed from the leading and trailing edges are identified by red and blue colors, respectively, whereas the TiV is green. The role of the TiV in the hovering cases studied is particularly interesting. For the case presented in Figure 3.25 (delayed rotation) – the Q iso-surface colored with ω_3-vorticity, along with the spanwise distribution of C_L due to pressure – the effects of the tip vortices become apparent. First, a low-pressure region at the wingtip favorably influences the lift. Second, the TiV anchors the large-scale vortex pair near the tip. At mid-span, however, the vortex pair has separated from the wing. This in turn drives the spanwise variation seen in the flow structures and force history.

Compared with an infinite wing, the TiVs cause additional mass flux across the span of a low-aspect-ratio wing, which helps push the shed vortex pair, from the leading edges and trailing edges, at mid-span away from one another. Furthermore, there is a spanwise variation in effective AoA induced by the downwash, which is stronger near the tip. Overall, the TiVs allow the vortex pair in their neighborhood of the tip to be anchored near the wing surface, which promotes a low-pressure region and enhances lift. The end result is an integrated lift value that departs considerably from the 2D value.

From the discussions in this section, it is clear that the kinematic motions have a significant impact on TiV formation and LEV/TEV dynamics. Interestingly, for many of the kinematic motions examined, the TiV force enhancement could be confined to lift benefits; that is, the resulting drag did not increase proportionally.

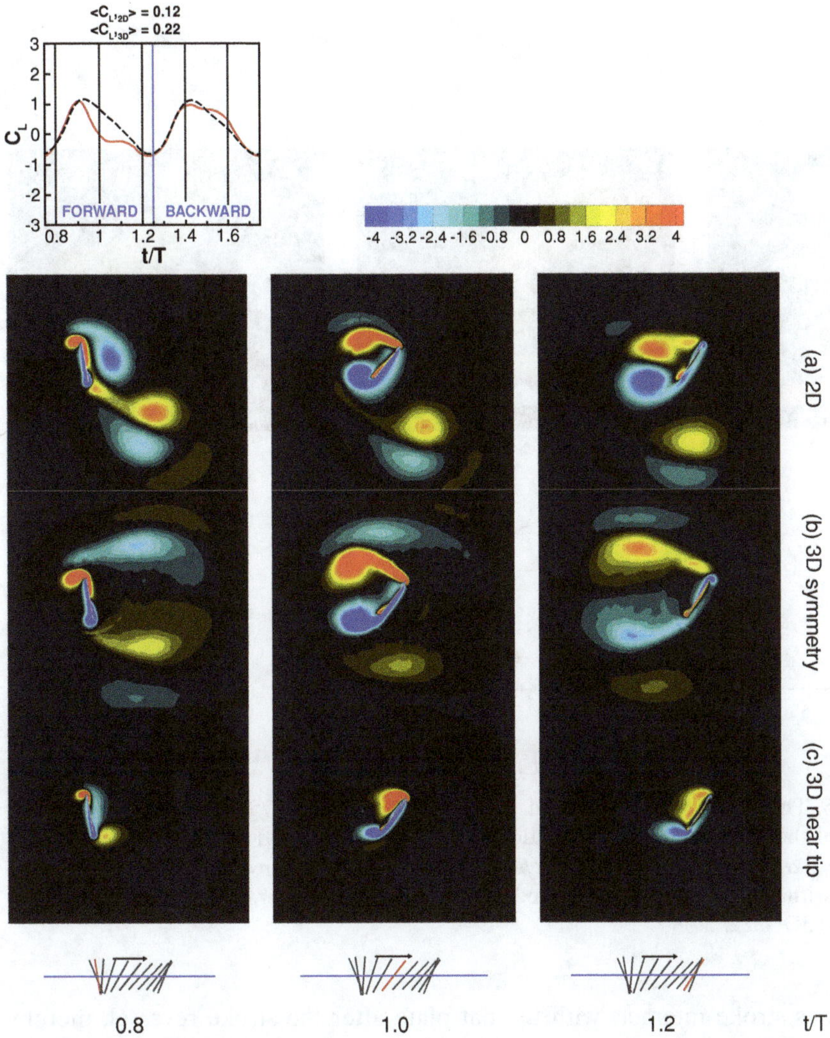

Figure 3.24. Force history (2D: solid red, 3D: dash black) for a flapping cycle and z-vorticity contour plots at three time instants in the forward stroke for the delayed rotation and high AoA ($2h_a = 2.0c_m$, $\alpha_a = 45°$, and $\varphi = 60°$): (a) 2D computation; (b) in the symmetry plane of 3D computation; (c) near the wingtip ($z/c = 1.8$).

DELAYED ROTATION, LOW AOA, LOW AMPLITUDE – ABSENCE OF WAKE CAPTURE IN 3D. The kinematics in region 4 is characterized by delayed rotation, large angular amplitude (or low AoA), and shorter plunging amplitude (see Fig. 3.21d). Note that the phase lag between the plunging and pitching motion is similar to that of previous sections, but the amplitude of both pitch and plunge differs. Figure 3.26 shows the time histories of lift from the 2D and the 3D computations, along with a schema for the kinematics as a representative case for region 4: $2h_a = 2.0c_m$, $\alpha_a = 80°$, and $\varphi = 60°$. The largest discrepancy between 2D and 3D is seen around $t/T = 0.9$. Because the rotation is delayed, after the stroke reversal at $t/T = 0.75$ the flat plate creates rotational starting vortices to increase the lift, with its first peaks around $t/T = 0.9$. However, as shown in Figure 3.26, in the 2D case, the TEV shed in

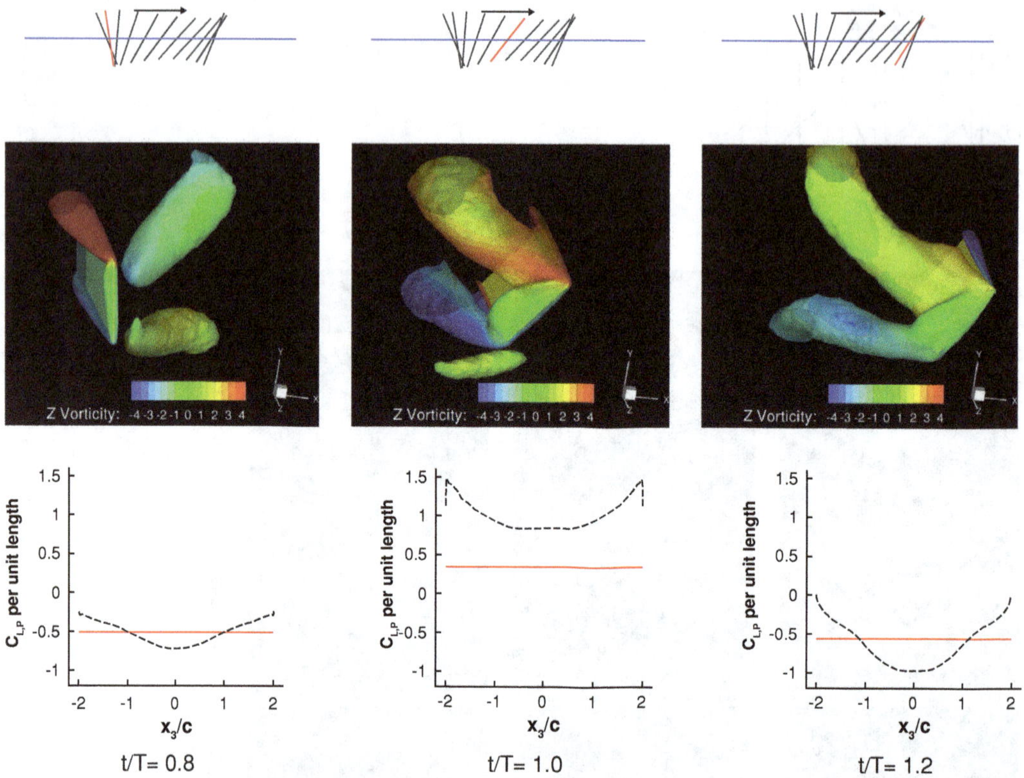

Figure 3.25. The lift per unit span and iso-Q surfaces ($Q = 0.75$) colored by z-vorticity over half of the wing using the kinematic parameters, the delayed rotation, and high AoA ($2h_a = 2.0c_m$, $\alpha_a = 45°$, and $\varphi = 60°$) at $t/T = 0.8$, 1.0, 1.2. The spanwise variation in forces is examined with the 2D equivalent marked for reference. Time-averaged lift coefficient for i) 2D: 0.13, ii) 3D: 0.22.

the previous stroke interacts with the flat plate after the stroke reversal, thereby enhancing the lift by the wake-capture mechanism ($t/T = 0.9$). In contrast, for the 3D case, the shed LEV and the TEV repel from each other and from the path of the flat plate, such that after the stroke reversal, the wake capturing is absent. The first lift peak in the time history in Figure 3.26 is then only due to the rotational effects. So the diverging behavior of the vortices, observed for all delayed rotation cases, and the interaction of the vortices from the leading and trailing edges with the TiVs, play a central role as important 3D effects as described in the previous section.

3.4.1.2 Region of Similarity: 2D versus 3D

There are interesting combinations of wing kinematics that result in similar lift in both 2D and 3D. For some cases, even the instantaneous forces agree well.

Figure 3.27 shows the flow fields of such a case corresponding to these kinematic parameters: $2h_a = 3.0c_m$, $\alpha_a = 80°$, and $\varphi = 90°$ at $t/T = 0.8$, 1.0, and 1.2. The variation along the spanwise direction is modest, making the 2D and 3D simulations substantially similar. The 2D flow field and the corresponding 3D flow on the symmetry plane are strikingly consistent. The high angular amplitudes lead to low AoAs

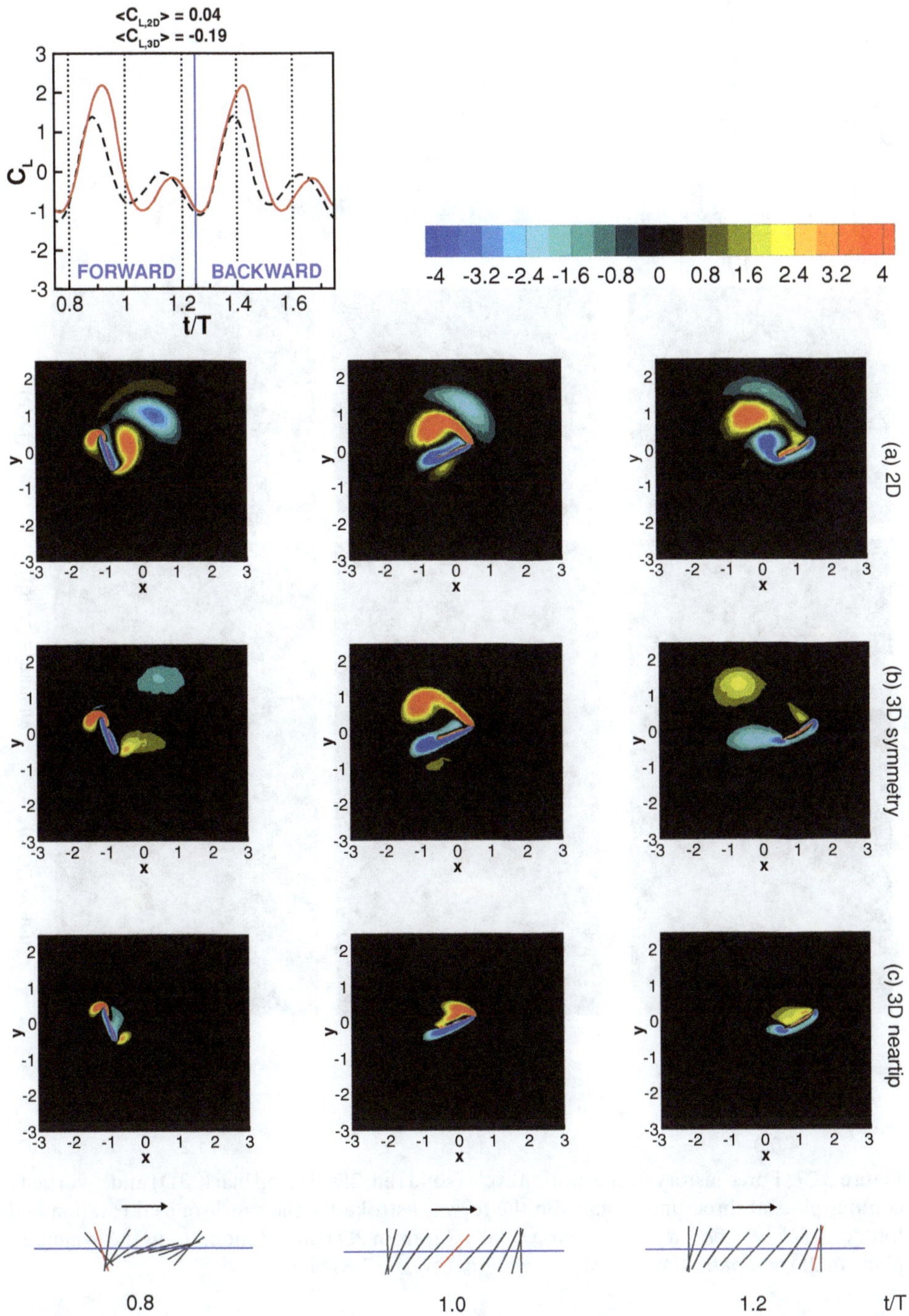

Figure 3.26. Force history (2D: red, 3D: black) for a motion cycle and z-vorticity contour plots at three time instants in the forward stroke for the delayed rotation and low AoA ($2h_a = 4.0c_m$, $\alpha_a = 80°$, and $\varphi = 60°$): (a) from 2D computation; (b) in the symmetry plane of 3D computation; (c) near the wingtip ($z/c = 1.8$) plane.

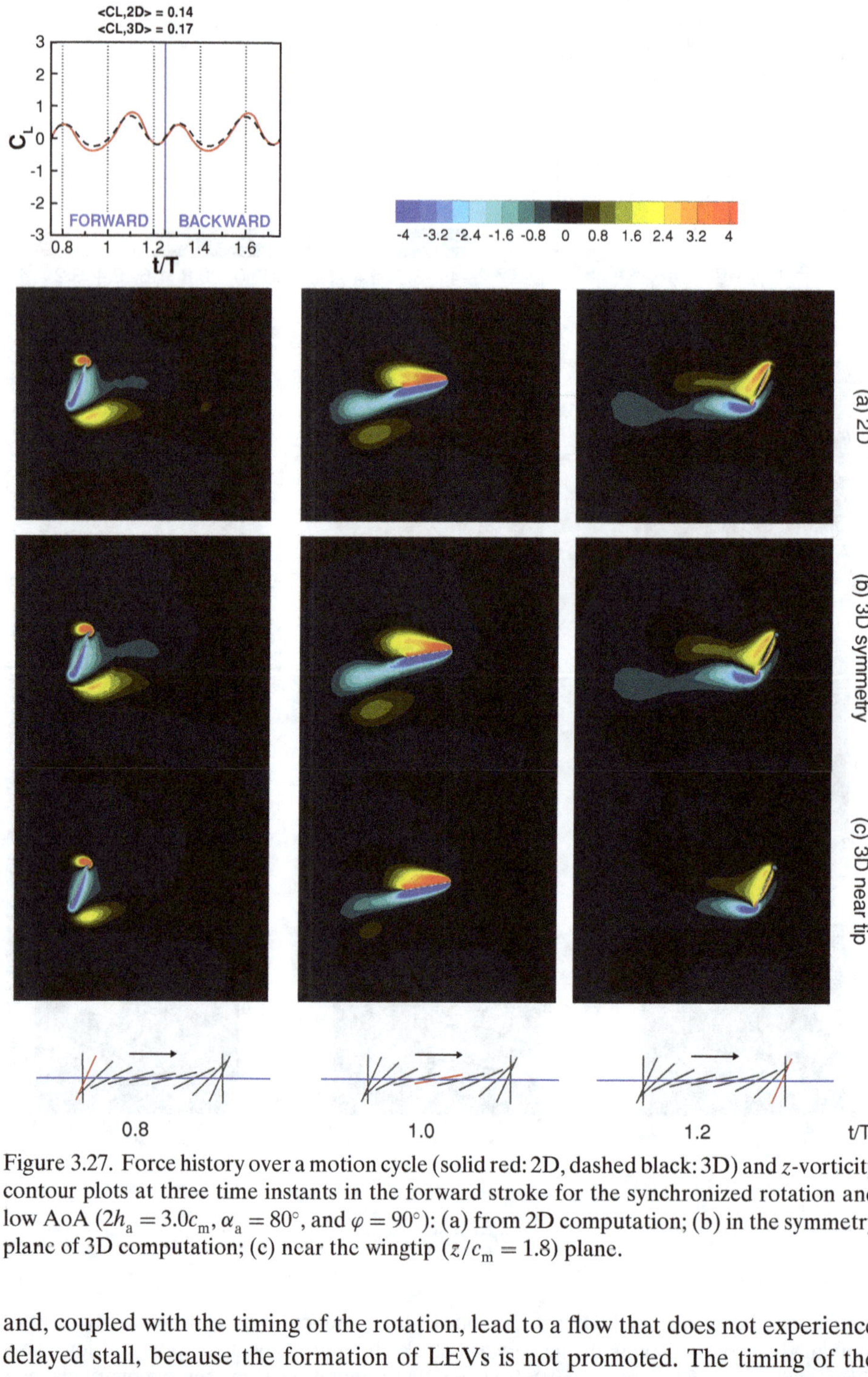

Figure 3.27. Force history over a motion cycle (solid red: 2D, dashed black: 3D) and z-vorticity contour plots at three time instants in the forward stroke for the synchronized rotation and low AoA ($2h_a = 3.0c_m$, $\alpha_a = 80°$, and $\varphi = 90°$): (a) from 2D computation; (b) in the symmetry plane of 3D computation; (c) near the wingtip ($z/c_m = 1.8$) plane.

and, coupled with the timing of the rotation, lead to a flow that does not experience delayed stall, because the formation of LEVs is not promoted. The timing of the rotation for this example puts the flat plate at its minimum AoA (10°) at the moment of maximum translational velocity, while the translational velocity is zero when the flat plate is vertical. It is seen from the flow field that both the TiV and the LE/TE vortex formation are largely suppressed. The net effect is a fairly uniform spanwise lift distribution, closely resembling the 2D case with the same kinematics.

Neither the 2D nor 3D results in this case promote downward induced jet formation. As summarized in Figure 3.27, the 2D and 3D lift coefficients are similar in the instantaneous and time-averaged senses. One implication of this similarity is the usefulness of the 2D simulation for generating quantitative data on a 3D counterpart. However, not all cases in this region display this similar instantaneous behavior, and sometimes the time-averaged lift similarity results from the partial canceling out of different features.

3.4.1.3 Trade-off between Lift and Power

In a multi-objective investigation, it is often the case that different goals are in competition when selecting design variables. One tool used to evaluate the trade-offs between objective functions is called the *Pareto front* [317], which consists of non-dominated points and can be thought of as the set of best possibilities. Non-dominated points are those points for which one could not improve all objective functions simultaneously.

Trizila et al. [301] presented the Pareto front for the competing objectives of maximizing time-averaged lift and minimizing the power requirement in the 2D and 3D hovering flat plate case highlighted in previous sections, as illustrated in Figure 3.28. Points on the Pareto front therefore involve those for which increases in lift are accompanied by increases in power, and vice versa. The resulting Pareto front itself is very comparable between 2D and 3D (see Fig. 3.28). The primary differences are that the peak lift values attained in 2D exceed their 3D counterparts and the density of the design variable combinations near the Pareto front is higher in the 3D case. The paths through the design space are plotted below their respective Pareto fronts in Figure 3.28. Note that the jaggedness of the path is due to the resolution of the tested points and the fine balance in objective functions for design variables in that region. The high-lift region follows the lower bound of the angular amplitude, suggesting that future iterations should decrease the lower bound for higher lift solutions. Overall, the design variable combinations on the optimal front are consistent qualitatively. The high time-averaged lift values are obtained by a combination of advanced rotation and low angular amplitude in the 2D and 3D cases. The general trends remain largely the same.

3.4.2 Effects of Wind Gust on Hovering Aerodynamics

One of the main difficulties in realizing a functional MAV is its inherent sensitivity to the operating environment because of its size and weight. Wind gusts create intrinsic unsteadiness in the flight environment [318] (see also discussions of the fixed wing in Section 2.2.4) As discussed in Chapter 1, the characteristic flapping time scale of insects and hummingbird – $O(10^1)$ to $O(10^2)$ of Hz – is much shorter than the time scale of a typical wind gust (around $O(1)$ Hz). Hence, from the flapping wing time scale, many wind gust effects can be treated in a quasi-steady manner. However, the vehicle control system (as in the case of a biological flyer) operates at lower frequencies, and their time scales are comparable to those of anticipated wind gusts. Therefore, there is a clear multi-scale problem between unsteady aerodynamics, wind gust, and vehicle control dynamics. Since a small bird or insect flaps much faster than commonly experienced wind, the aerodynamics

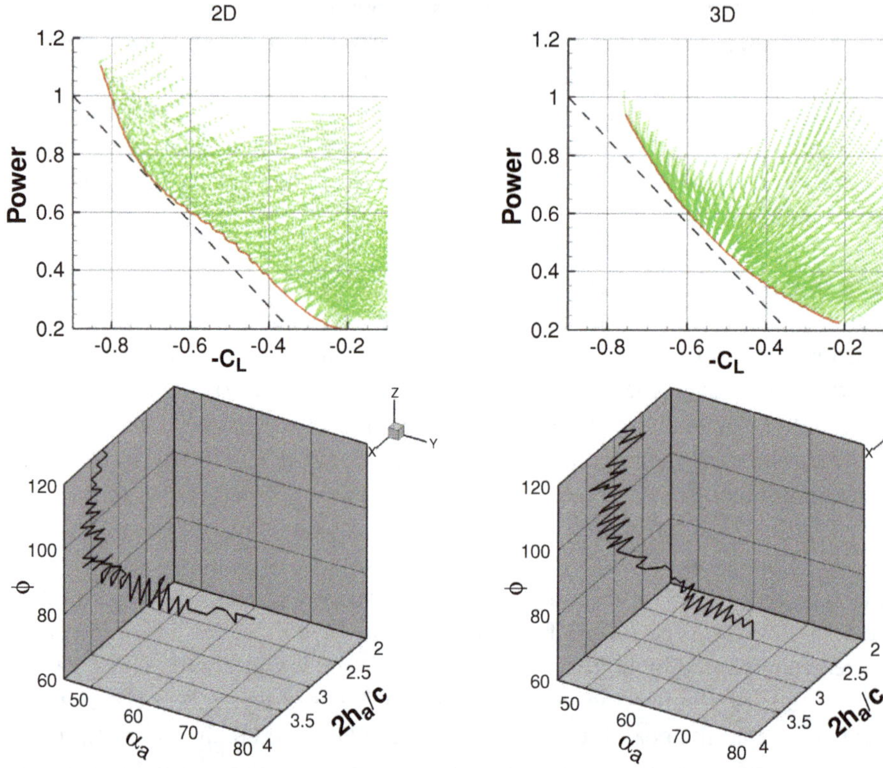

Figure 3.28. Pareto fronts illustrating the competing objectives of time-averaged lift and power requirements in 2D (left) and 3D (right) and the design variable combinations that provide those fronts. The dashed line is for reference.

associated with flapping wings can be pragmatically modeled by a constant free-stream.

Trizila et al. [301] investigated environmental sensitivity in regard to varying kinematic schemes and free-stream strengths and orientations; the same kinematic patterns highlighted in previous sections are adopted again in this discussion. The free-stream strength was fixed at 20 percent of the maximum translational velocity of the wing. Based on parameters associated with fruit flies [$Re = O(10^2)$ and wing speed is approximately 3.1 m/sec], the free-stream fluctuation would be approximately 0.6 m/s, a relatively light wind gust. The directions of the free-stream varied between heading down, right, or up.

The 2D cases were much more sensitive to the free-stream than their 3D counterparts. Instantaneous lift associated with all three kinematic patterns (synchronized rotation with either low or high AoA and delayed rotation with high AoA) was very sensitive to the horizontal free-stream and much less sensitive to the downward heading free-stream. The downward free-stream generally decreased lift by suppressing vortex generation, while making the forward strokes and backstrokes more symmetric as the vortical activity was washed away from the airfoil more quickly. Overall, the general nature of the force history was kept intact. In contrast, the upwind free-stream had the opposite effect. Namely, the vortex interactions were sustained for a longer period of time, because the free-stream held the wake closer to the airfoil and the increased AoA also served to accentuate the unsteady aerodynamics. This upward free-stream may or may not have had a significant impact on the force history, which was dependent on the kinematics. The horizontal free-stream had the largest impact over the range of kinematic motions studied, sometimes more than doubling the lift felt for a free-stream strength of 20 percent of the maximum translational velocity, a relatively tame environmental situation leading to a significant change in hovering performance.

Figure 3.29 shows the lift histories and vorticity contours, illustrating the vortex formation and interactions during the LEV-dominated portion of the stroke (Fig. 3.29a-c) and during the wake-capture-dominated portion of the stroke (Fig. 3.29d-f) at their respective maximal lifts for a 20 percent strength headwind hover scenario without free-stream. The 3D LEV-dominated portion of the stroke is highlighted with ω_3 contours at two spanwise locations with a 20 percent free-stream. Immediately apparent is the large impact on both the instantaneous and time-averaged lift. To clarify, the lift coefficients are still normalized by the velocity related to wing motion (e.g., the mean/maximum translational velocity); that is, the normalization is independent of the free-stream. Flow fields are plotted during the headwind portion of the stroke (backstroke) and show that the headwind case exhibits a more developed and stronger LEV, as well as stronger vorticity shedding from the TE. The increased vortical activity created by the headwind, which then interacts with the airfoil in a favorable manner, qualitatively explains the increase in performance during the backstroke.

However, the significant lift peak in the presence of tailwind is due to a different lift-generation mechanism. This peak occurs after stroke reversal as the airfoil interacts with the previously shed wake, or the wake-capture-dominated portion of the stroke cycle. The hover case temporarily drops off in lift (see Fig. 3.29d), whereas the 20 percent free-stream case, now a tailwind, continues to increase in lift. Vorticity contours at their respective local maximums in lift (see Figs. 3.29e and 3.29f) show a few striking differences, noticeably the strength and position of the previously shed vorticity. The presence of headwind during the backstroke creates stronger vortices. On the return stroke, the strengths of these vortices, in addition to their position relative to the airfoil, significantly help promote vortex growth (see Fig. 3.29f). This interaction, resulting in a temporary enhancement, eventually plays itself out, and a decline in lift ensues in what used to be dominated by the LEV but now amounts to slower relative translational velocity.

Looking at the 2D force histories (Fig. 3.30a) again, one sees that the response of a free-stream depends not only on the kinematics but also on its orientation. Figure 3.31 illustrates the lift response to a free-stream for 2D (Fig. 3.31a-c) and 3D (Fig. 3.31d-f) computations. The free-stream magnitude is 20 percent of the maximum plunging velocity and is headed in three different directions for the three hovering kinematics investigated in the previous two sections (3.4.1.1 and 3.4.1.2). The dotted lines in Figure 3.31a-f are the lift responses of the hovering cases (i.e., no free-stream). For some situations, the qualitative nature of the flow does not change much over the course of the entire cycle, nor are the forces too sensitive; see the vertical free-stream in Figure 3.31a or the downward free-stream in Figure 3.31c. However, the horizontal free-stream has an appreciable impact for all of the kinematic patterns considered in Trizila et al. [301]. Yet the upward and downward

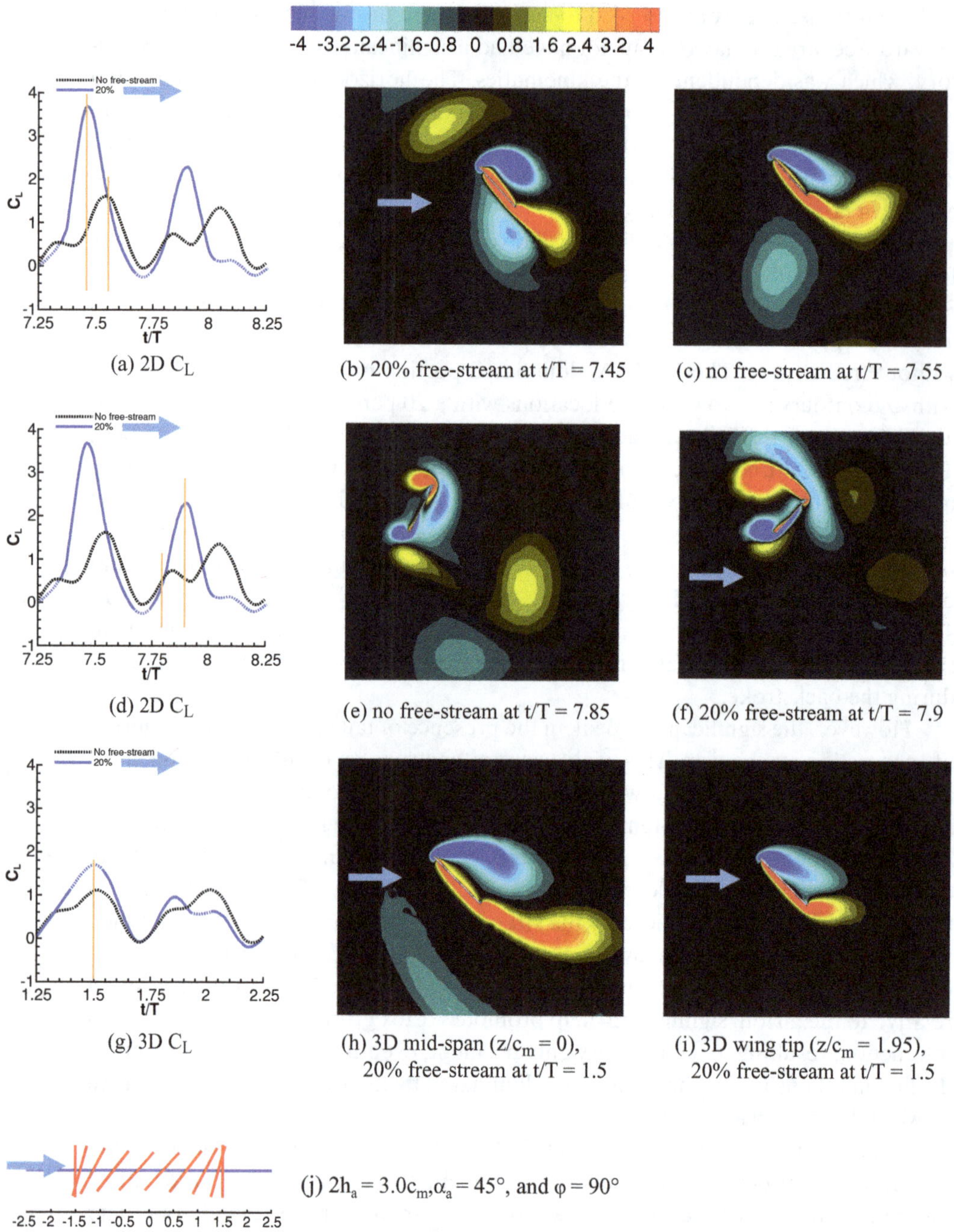

Figure 3.29. Force history and vorticity contours illustrating the vortex formation and interaction during the LEV dominated portion of the stroke (a–c) or wake capture dominated portion of the stroke (d–f) at their respective maximal lift for a 20% strength headwind free-stream and no free-steam. The 3D LEV-dominated portion of the stroke is highlighted with z-vorticity at two spanwise locations with a 20% free-stream in g, h, and i. Wing motion is shown in j.

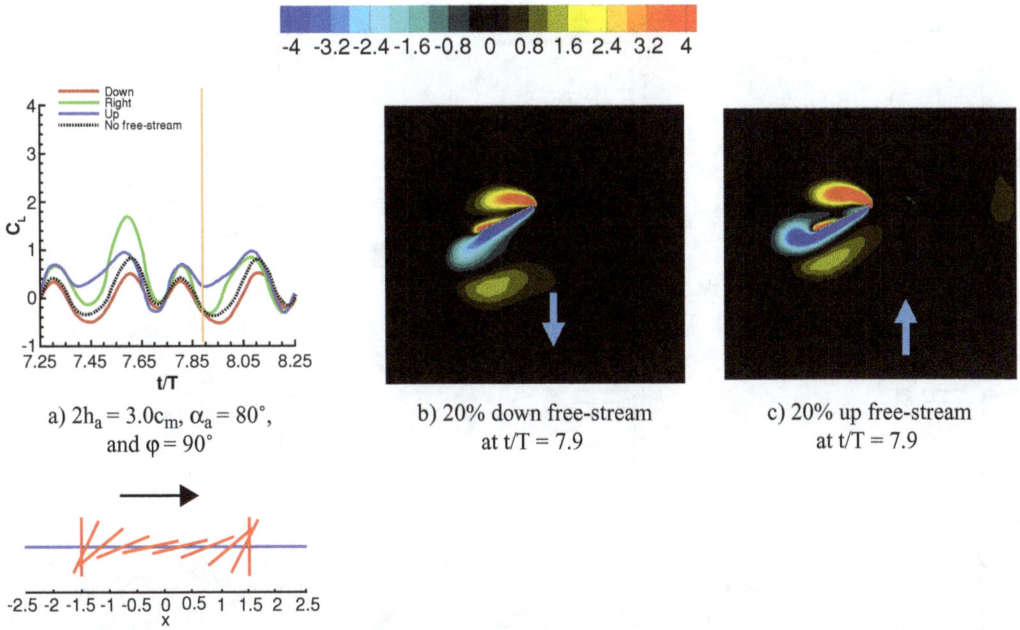

Figure 3.30. Force history and vorticity contours illustrating the vortex formation between stroke reversal and their respective maximums in lift for (b) 20% downward freestream and (c) 20% upward stream.

free-streams do not necessarily elicit similar responses in opposite directions, as highlighted in Figure 3.30. This brings into question the relevance of using effective AoA in these situations, because the force history may respond more noticeably to the upward free-stream than the downward free-stream; see Figure 3.31a-c, where a 20 percent free-stream imposed from different directions significantly changes the qualitative behavior of the resulting force history.

For all of the synchronized rotation cases (which have positive AoAs at all times), a 20 percent downward free-stream does indeed decrease the lift. The downward free-stream also suppresses the vortical flow strength as the effective AoA is lowered. However, some cases have a much more pronounced sensitivity to the upward free-stream. Figure 3.30a illustrates again the force histories for a 20 percent free-stream at various orientations relative to the hover case, as well as the vorticity flow field (Figs. 3.31a and 3.31c) for the 20 percent upward and downward free-stream cases at a time where the difference in force history between the two is pronounced. What is seen in Figure 3.30b and Figure 3.30c is the increase in LEV and TEV formation, as well as a more pronounced interaction with the wake, as the upward free-stream promotes the growth of the vortex structures and holds the wake in the vicinity for a longer period of time. The non-linear response in lift as the free-stream lowers or raises the effective AoA is a product of these factors.

The 3D cases, in contrast, are much less sensitive to the free-stream (see Fig. 3.29 g-i and Fig. 3.31 d-f. Note, however, that the scale for the force histories was chosen so that they could be directly compared with the 2D case and the free-stream could be quite influential. The impact is non-negligible for a 20 percent strength free-stream, but overall, the nature of the flow is very similar for most cases.

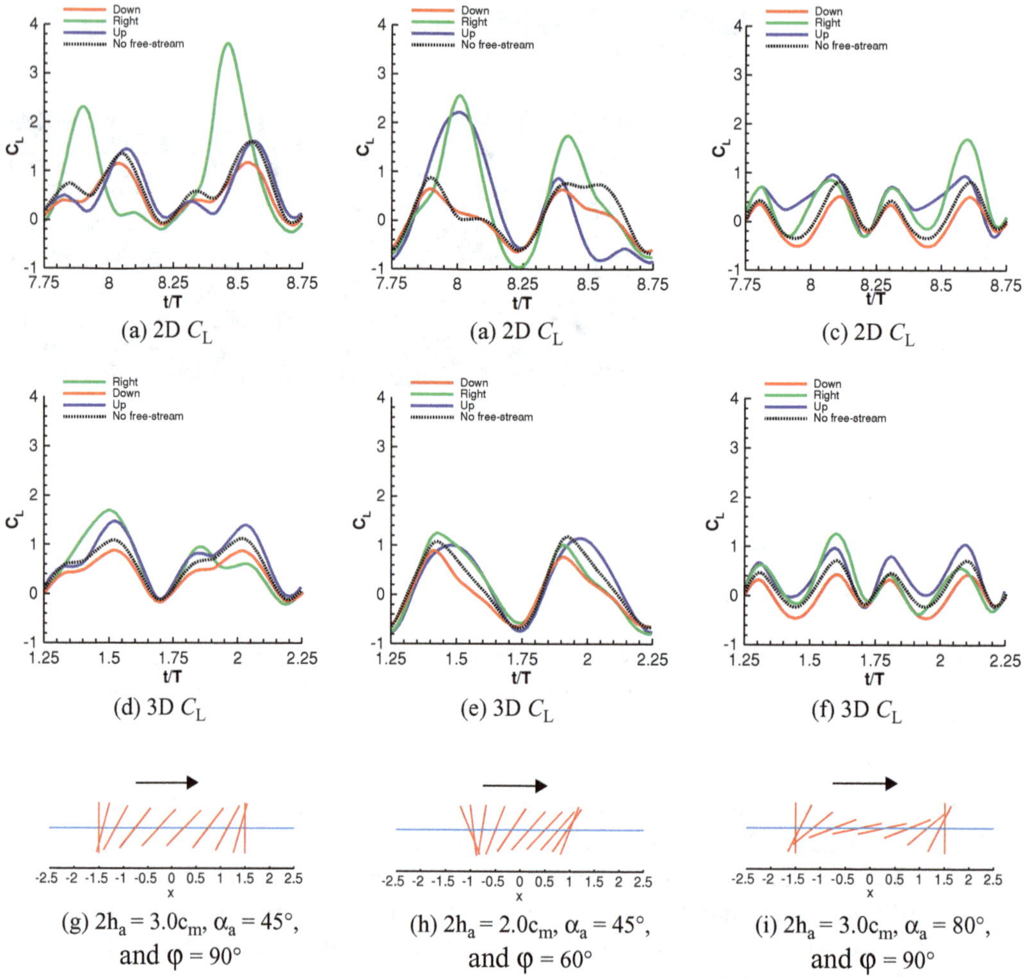

Figure 3.31. 2D (a–c) and 3D (d–f) C_L in response to a free-stream with a magnitude of 20% of maximum plunging velocity heading in three distinct directions (down: red, right: green, and up: blue) for three hovering kinematics (g, i). The black dotted line is the reference hovering case.

The downward free-stream once again degrades lift, and the upwind free-stream enhances it, albeit to a lesser degree than in the 2D cases.

Figure 3.32 highlights the vortex interactions at the beginning, middle, and ending of the strokes for a 20 percent free-stream tailwind. The blue arrow in vorticity contours for 2D and at mid-span and the wingtips for 3D denote free-stream directions. The 3D wing is unable to generate vortices of the same magnitude as the analogous 2D counterpart. This inability directly affects the wing's benefit from LEV interactions, as well as subsequent interactions with the previously shed wake. The spanwise variation of vorticity exhibited shows a decrease in LEV generation from mid-span to tip, and although the tip vortices are prominent, they do not make up for the weakened LEV formation and wake interaction, as experienced in 2D. Figure 3.29 illustrates the vorticity flow fields during maximal lift during the headwind, resulting in a 2D lift (Fig. 3.29a) that is almost twice as large as its 3D counterpart (Fig. 3.29g).

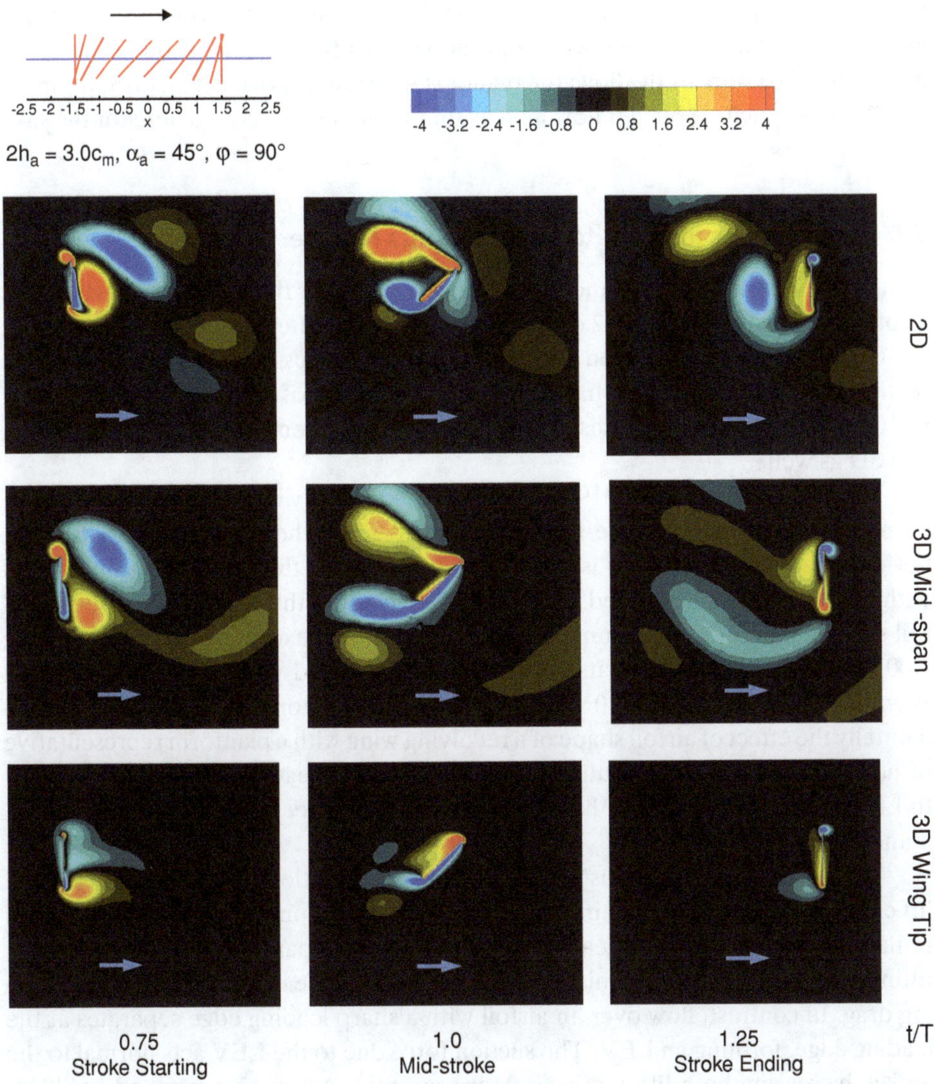

Figure 3.32. Vorticity contours during beginning, mid, and end-stroke for a 2D flat plate and at mid-span ($z/c = 0$) and wingtip ($z/c = 1.95$) for a 3D flat plate with $AR = 4$ with a 20% free-stream tailwind.

This discrepancy in sensitivity to free-stream between 2D and 3D shows up across the range of kinematic motions. A limited subset of kinematic motions show very similar force histories in the time-averaged sense, as well as instantaneously (see Section 3.4.1.2) when not under the influence of an external free-stream. The kinematics in this region of the design space shares low AoAs across much of the flapping cycle and synchronized rotation, limiting the high angular velocities. A lower angular velocity, in turn, tends to limit vortex size, strength, formation, and influence. As the free-stream is introduced (see Figs. 3.31c and 3.31f for 2D and 3D force histories, respectively, in the presence of a 20 percent free-stream), the response is not uniform across the span of the finite wing. The downward free-stream (20 percent strength) tends to suppress the vortex dynamics, and as such, the 2D and 3D force histories remain quite similar. In contrast, the horizontal free-stream, most notably during the headwind and the upward free-stream, shows differences due to the 3D nature of the flow; the reader is referred to Trizila et al. [315] for more details, which include further flow field examinations not described here for the sake of conciseness.

3.5 Fluid Physics in $O(10^4$ to $10^5)$ Reynolds Number Regime

Natural flyers such as insects and birds have wings thinner than those of conventional airplanes, such as NACA 0012 or SD7003 airfoils. The effects of kinematics at Re of $O(10^2)$ in terms of St and k on aerodynamics were discussed in Section 3.4. In this section the effects of airfoil shape on the performance of flapping airfoils at Re of $O(10^4)$ play a central role in discussing the effects of kinematics, Reynolds number, and 3D as well.

Lentink and Gerritsma [319] investigated numerically the effects of airfoil shape on aerodynamic performance in forward flight. Even though the Reynolds number considered in their study is $O(10^2)$, they concluded that a plunging thin airfoil with aft camber outperformed other airfoils, including the more conventional airfoil shapes with thick and rounded leading edges. One exception was the plunging N0010 airfoil, which due to its largest frontal area had good performance. At a Reynolds number of 5.0×10^3, Usherwood and Ellington [320] examined experimentally the effect of airfoil shape of a revolving wing with a planform representative of hawkmoth wings. The results showed that detailed leading-edge shape and airfoil twist and camber do not have a substantial influence on aerodynamic performance.

Sane [312] used Polhamus's [321] leading-edge suction analogy to explain the lift characteristics of thinner airfoils. The flow around rounded leading-edge airfoils remains attached, creating a leading-edge suction force parallel to the airfoil chord, tilting the resulting aerodynamic force forward toward the incoming flow, and reducing drag. In contrast, flow over an airfoil with a sharp leading edge separates at the leading edge, forming an LEV. The suction force due to the LEV acts normal to the airfoil, increasing both lift and drag. More recently, Ashraf, Young, and Lai [322] numerically investigated the effects of airfoil thickness on combined pitching and plunging airfoils at $k = 1$ and Reynolds numbers varying from 2.0×10^2 to 2×10^6. They found that at lower Reynolds numbers thin airfoils outperformed thick airfoil, whereas at higher Reynolds numbers the thick airfoil show better performance.

Initiated by the Research Technology Organization (RTO) of NATO, multiple institutions from several countries have investigated the unsteady flow around a pitching and plunging airfoil at $Re = O(10^4)$ using different methodologies. Ol et al. [323] numerically and experimentally considered the interplay between a SD7003 airfoil undergoing either combined pitching and plunging or purely plunging motions and the resulting aerodynamics at Reynolds numbers 1×10^4, 3×10^4, and 6×10^4 at a fixed Strouhal number and reduced frequency: $St = 0.08$ and $k = 0.25$. The two different kinematics produced a shallow stall (under combined pitching-plunging) and a deep stall (under purely plunging) of the instantaneous flow about the airfoil. Kang et al. [324] further reported results on the flow physics of a 2.3 percent thickness flat plate in 2D and 3D configurations and an SD7003 airfoil. In particular, they

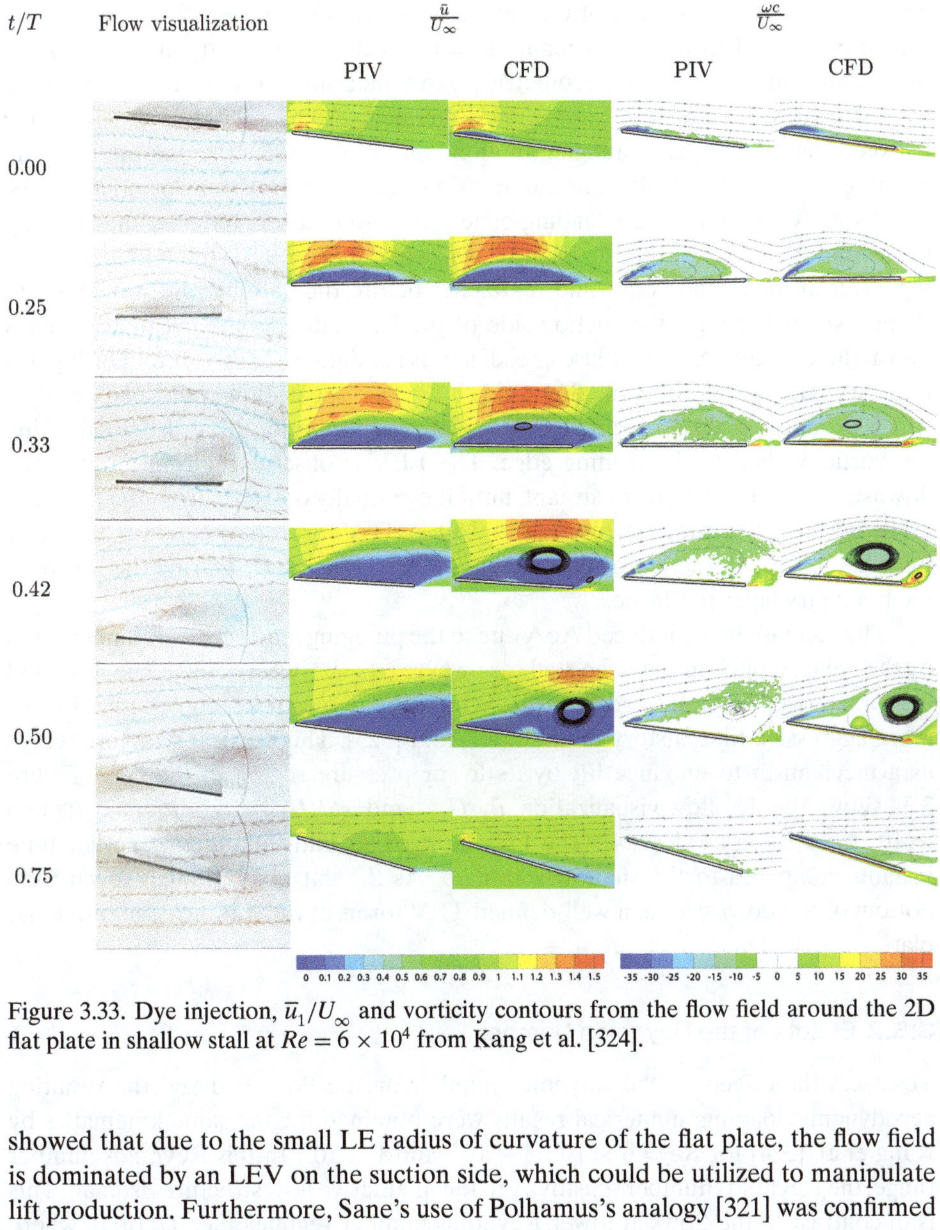

Figure 3.33. Dye injection, $\overline{u}_1/U_\infty$ and vorticity contours from the flow field around the 2D flat plate in shallow stall at $Re = 6 \times 10^4$ from Kang et al. [324].

showed that due to the small LE radius of curvature of the flat plate, the flow field is dominated by an LEV on the suction side, which could be utilized to manipulate lift production. Furthermore, Sane's use of Polhamus's analogy [321] was confirmed for the shallow stall motion.

3.5.1 Flow around a Flat Plate in Shallow and Deep Stall at $Re = 6 \times 10^4$

Figure 3.33 shows dye flow visualization, $\overline{u}_1/U_\infty$, and $\omega_3/(U_\infty/c)$ contours from Kang et al. [324] for the flow about a flat plate in shallow stall at $Re = 6 \times 10^4$. They gave the shallow stall kinematics as

$$h(t/T) = h_0 c_m \cos(2\pi t/T), \tag{3–19}$$

$$\alpha(t/T) = \alpha_0 + \alpha_a \cos(2\pi t/T + \varphi), \tag{3–20}$$

where $h(t/T)$ is the location of the center of rotation ($x_p/c_m = 0.25$) of the airfoil measured normal to the free-stream, $h_0 = 0.5$ is the normalized amplitude of the plunge motion, $\alpha(t/T)$ is the geometrical AoA measured relative to the incoming free- stream with velocity U_∞, $\alpha_0 = 8°$ is the mean AoA, $\varphi = 90°$ is the phase lag between the pitching and plunging motion, and $\alpha_a = 8.43°$ is the amplitude of the pitching motion. The resulting non-dimensional numbers are $k = 0.25$ and $St = 0.08$.

As shown in Figure 3.33, leading-edge separation is observed over the majority of the motion cycle. At the top of the downstroke ($t/T = 0.00$) the boundary layer separates at the leading edge and reattaches before the half-chord, forming a thin shear layer that covers the suction side of the flat plate. As the flat plate plunges down, the effective AoA increases, reaching its maximum at $t/T = 0.25$. During this portion of the downstroke, the flow field shows flow separation at the leading edge, which produces a closed recirculation region and formation of an LEV capturing the vorticity shed at the leading edge. The LEV is observed during most of the downstroke, convecting downstream, until it eventually detaches from the flat plate at the bottom of the downstroke, $t/T = 0.50$. A TEV forms at $t/T = 0.33$, 0.42, and 0.50 due to the interaction of the LEV and the trailing edge. During the upstroke the boundary layer reattaches.

The increase in the induced AoA due to the plunging motion is not compensated by the delayed pitch in the deep stall case where $\alpha_a = 0°$ in Eq. (3.20). The flow field is characterized by separated flow throughout the downstroke, consistent with the more aggressive time history of effective AoA [323]. This vortical structure serves as a mechanism to enhance lift by its lower pressure region in the core. Figure 3.34 shows the dye flow visualization, $\bar{u}_1/U_\infty$, and $\omega_3/(U_\infty/c)$ contours at $Re = 6 \times 10^4$ from Kang et al. [324]. The LEV is stronger and separates at earlier time instants compared to the shallow stall case. As the flat plate plunges toward the bottom of the downstroke, a well-defined TEV forms at the trailing edge of the flat plate.

3.5.2 Effects of the Reynolds Number

To assess the effects of the Reynolds number on the flow field and the resulting aerodynamic loading, numerical results were obtained for the same kinematics by Kang et al. [324] for $Re = 6 \times 10^4$, 3×10^4, and 1×10^4. In this Reynolds number range, the Reynolds number sensitivity to the qualitative flow structures is small. This is in contrast to the flows at lower Reynolds number regimes (i.e., $O(10^2)$), where the viscosity plays a more important role than at the current Reynolds number, or for the more conventional SD7003 airfoil with a larger radius of curvature at the leading edge [323].

For the deep stall case, Figure 3.35 shows the time history of force coefficients at $Re = 1 \times 10^4$, 3×10^4, and 6×10^4. Both lift and drag coefficients are negligibly affected by the Reynolds number variation. For all Reynolds number cases, the lift coefficient reaches its maximum near $t/T = 0.25$, decreases, and starts to increase from $t/T = 0.75$.

At these moderately high Reynolds numbers, forces due to pressure that arise from large-scale vortical effects, such as LEVs, dominate over viscous forces. The dependence of the small Reynolds number on the $\overline{u}_1/U_\infty$ profiles shown in

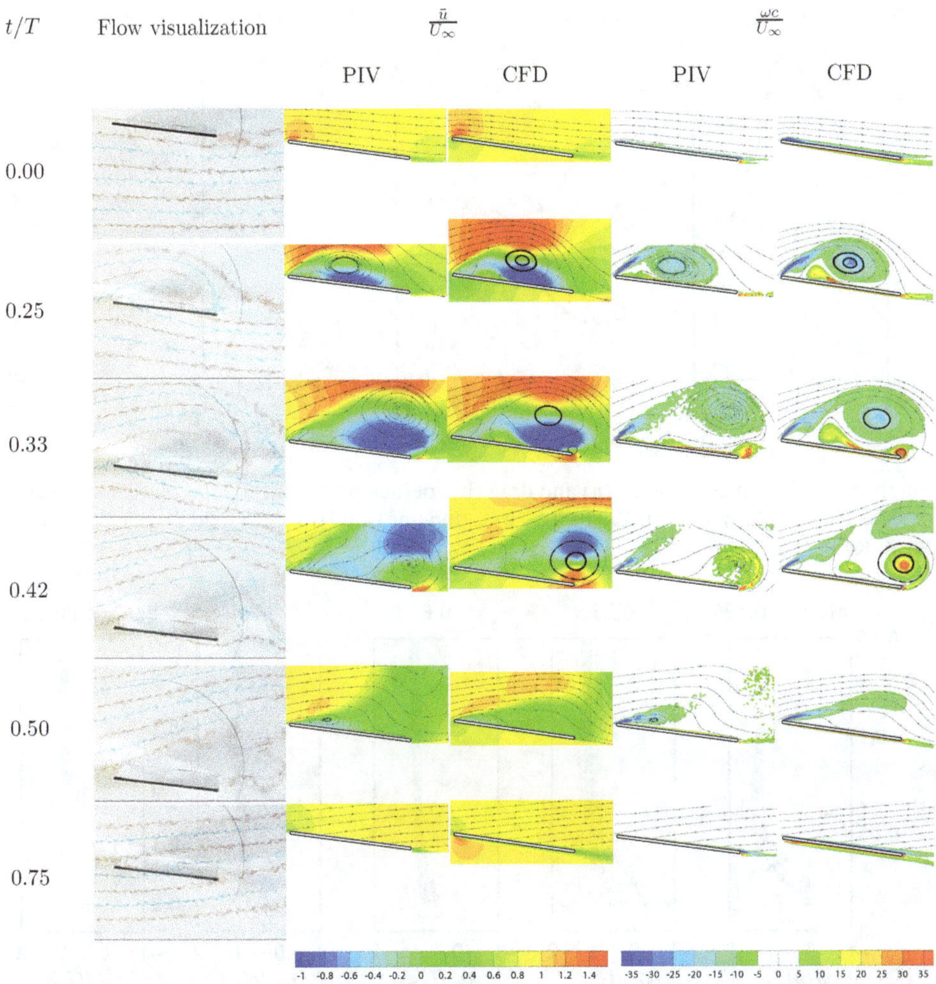

Figure 3.34. Dye injection, $\overline{u}_1/U_\infty$ and vorticity contours from the flow field around the 2D flat plate in deep stall at $Re = 6 \times 10^4$ from Kang et al. [324].

Figure 3.36 and the qualitative similarity in the large-scale flow features ($t/T = 0.50$ in Fig. 3.36) suggest that the resulting aerodynamic forces are insensitive to the change in the Reynolds numbers. Figure 3.37 shows the time histories of the force coefficient from the numerical computations. At $Re = 1 \times 10^4$ the maximum lift coefficient is around $t/T = 0.25$, which is slightly lower ($C_{L,\,\text{max, 10K}} = 1.86$) than for $Re = 6 \times 10^4$ ($C_{L,\,\text{max, 60K}} = 1.92$) and 3×10^4 ($C_{L,\,\text{max, 30K}} = 1.90$). Similarly the time histories of the drag coefficient coincide during the downstroke, whereas for $Re = 1 \times 10^4$ the drag is slightly greater between $t/T = 0.50$ to 1.00 than at $Re = 3 \times 10^4$ and 6×10^4.

3.5.3 Airfoil Shape Effects: Sane's Use of Polhamus's Analogy

Figure 3.38 shows the normalized vorticity field around a SD7003 in deep stall, following the same kinematics as introduced in Section 3.5.1 at $Re = 6.0 \times 10^4$. The phase-averaged field obtained using a 3D implicit filtered large-eddy simulation

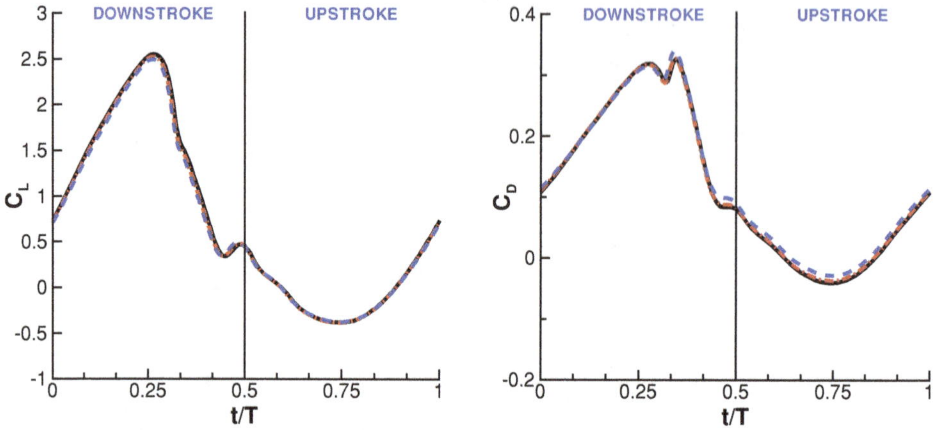

(a) lift coefficient (b) drag coefficient

Figure 3.35. Time history of lift (a) and drag (b) coefficients for the 2D flat plate in deep stall at $Re = 1 \times 10^4$ (dashed line), 3×10^4 (dash-dotted line), and 6×10^4 (solid line), respectively. From Kang et al. [324].

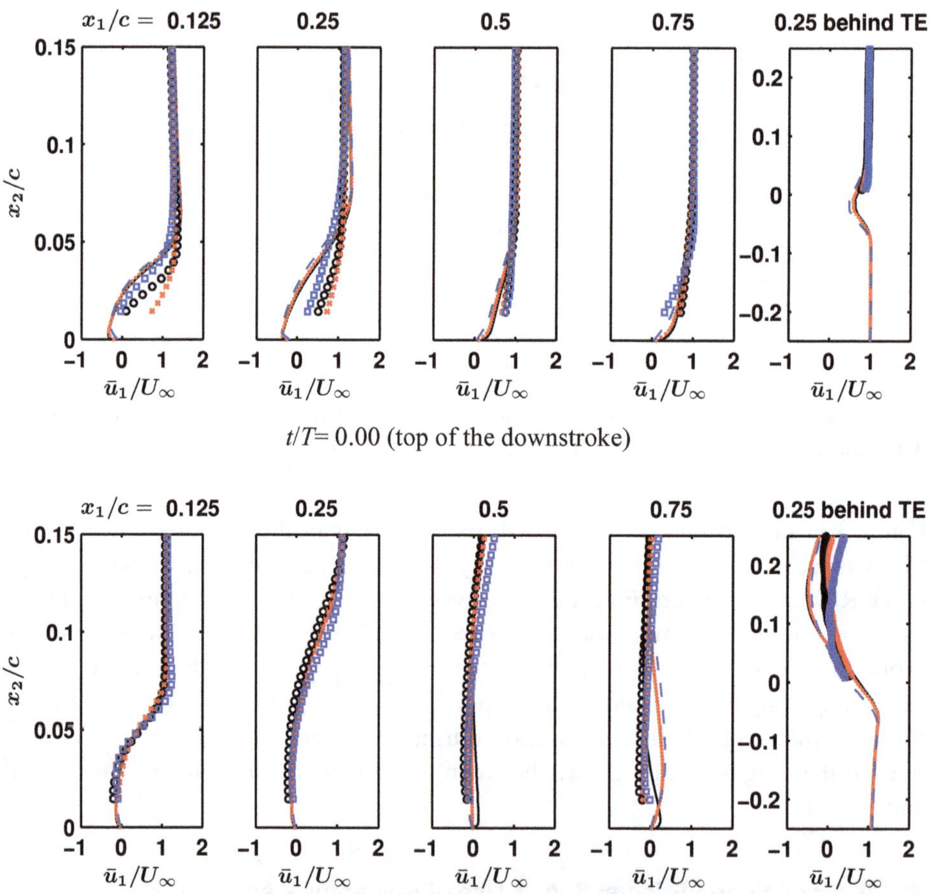

t/T= 0.50 (bottom of the downstroke)

Figure 3.36. $\bar{u}_1/U_\infty$ profiles from the numerical and experimental results at constant x_1/c_m = 0.125, 0.25, 0.50, 0.75, and 0.25 behind the trailing edge at t/T = 0.00 and 0.50 at $Re = 1 \times 10^4$ (numerical: dashed line; experimental: square), 3×10^4 (numerical: dash-dotted line; experimental: cross), and 6×10^4 (numerical: solid line; experimental: circle) for the 2D flat plate in shallow stall. From Kang et al. [324].

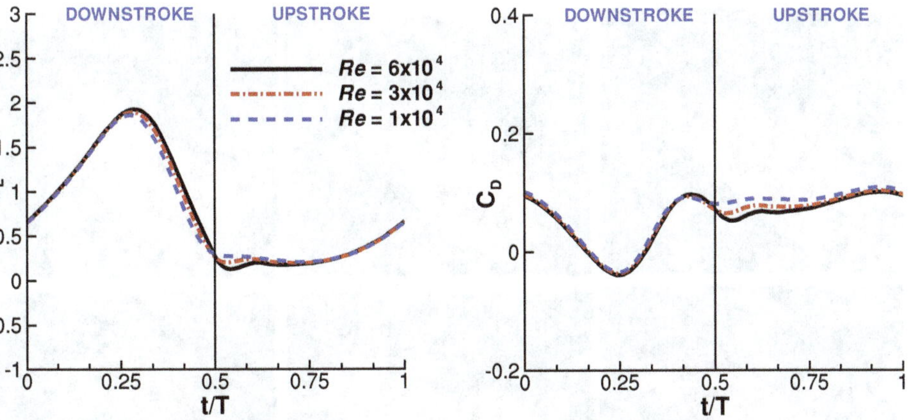

(a) lift coefficient (b) drag coefficient

Figure 3.37. Time history of (a) lift and (b) drag coefficients for the 2D flat plate in shallow stall for three different Reynolds numbers. From Kang et al. [324].

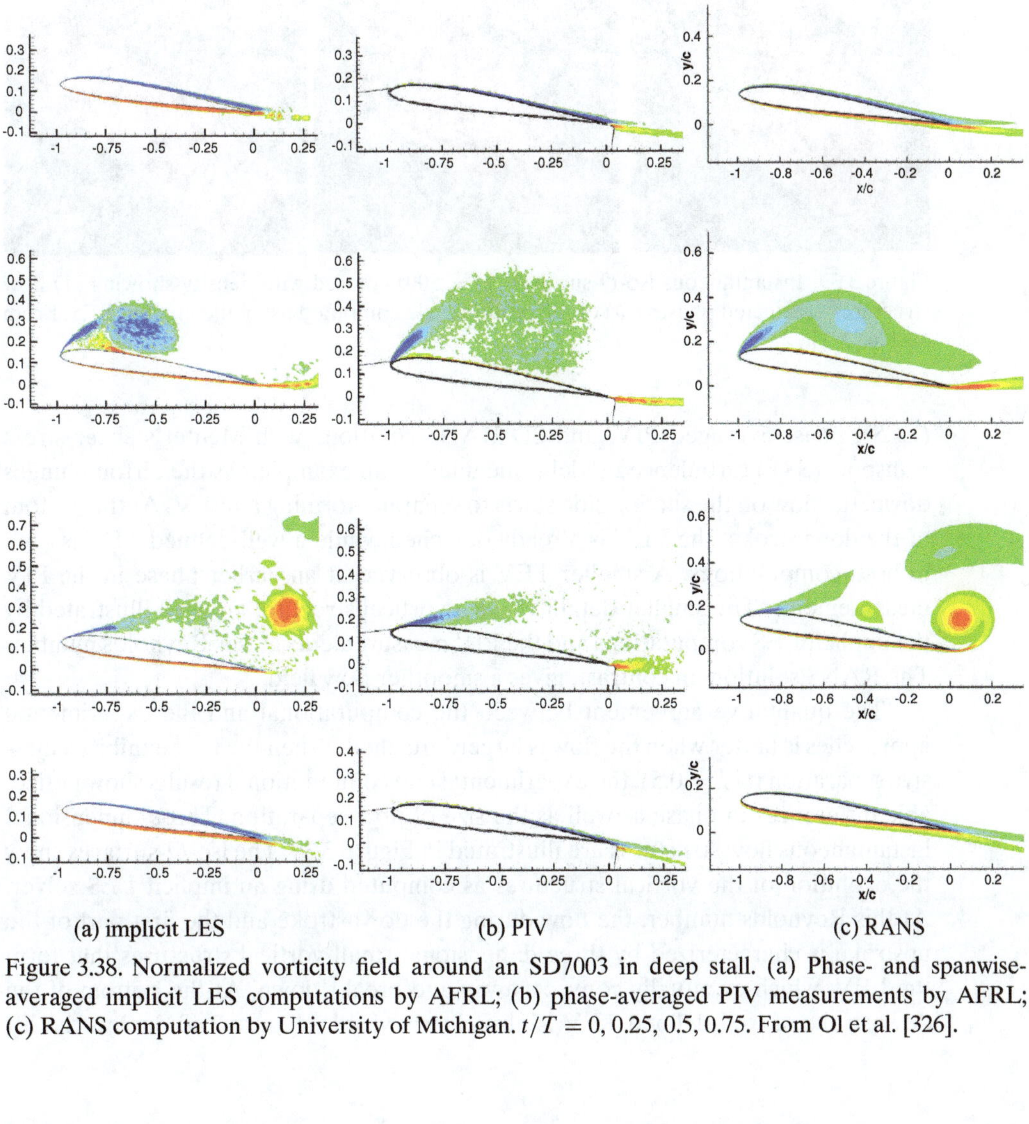

(a) implicit LES (b) PIV (c) RANS

Figure 3.38. Normalized vorticity field around an SD7003 in deep stall. (a) Phase- and spanwise-averaged implicit LES computations by AFRL; (b) phase-averaged PIV measurements by AFRL; (c) RANS computation by University of Michigan. $t/T = 0, 0.25, 0.5, 0.75$. From Ol et al. [326].

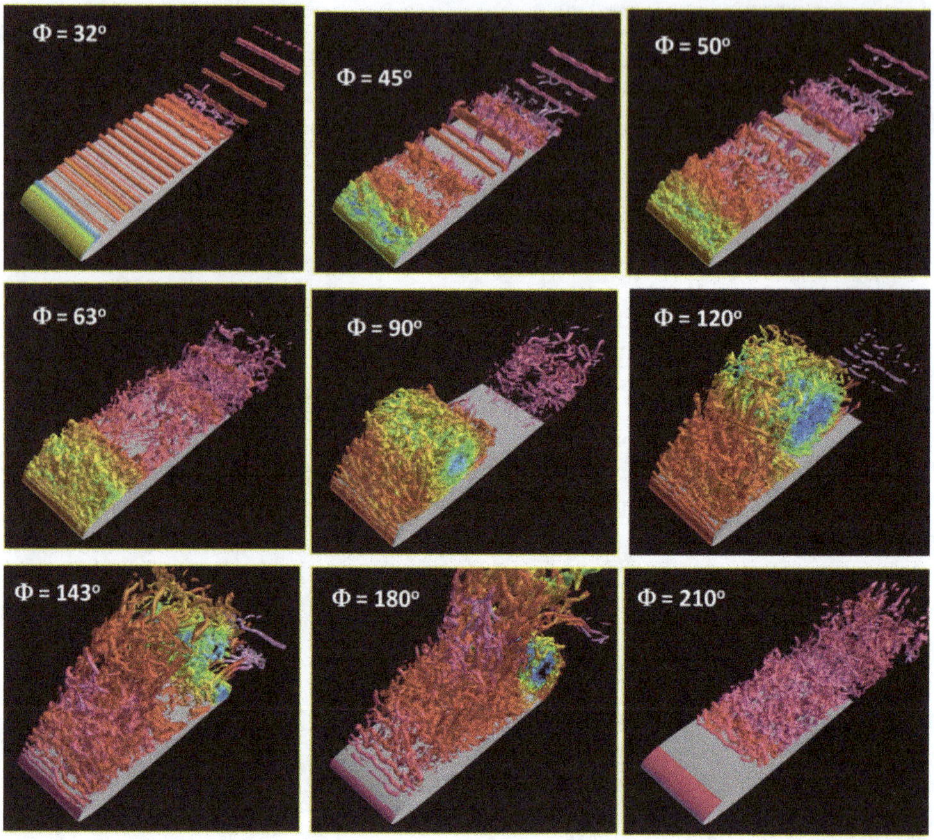

Figure 3.39. Instantaneous iso-Q-surfaces ($Q = 500$) colored with density showing 3D flow structures at selected phases (Φ) of the motion cycle computed using the implicit LES. From Visbal [327].

(LES), phase-averaged PIV, and 2D RANS solutions with Menter's shear stress transport (SST) turbulence model is included as an example. As the airfoil plunges down, the flow on the suction side starts to separate, forming an LEV. At the bottom of the downstroke, the LEV is already detached, while a well-defined TEV is seen in both computations. A smaller TEV is observed at an earlier phase in the PIV measurements. The smaller details of each vortical structure are well illustrated by the implicit LES computations and the PIV measurements at this Reynolds number. The RANS solution, in contrast, gives a smoother flow field.

The qualitative agreement between the computational and the experimental approaches is better when the flow is largely attached. When the flow exhibits a massive separation ($t/T = 0.5$), the experimental and computational results show noticeable differences in phase as well as the size of flow separation. Three-dimensional instantaneous flow structures are illustrated in Figure 3.39. The iso-Q-surfaces show the evolution of the vortical structures as computed using an implicit LES solver. At this Reynolds number, the flow during the downstroke and the first part of the upstroke is characterized by three-dimensional small vortical structures that form the LEV, which eventually convects away and breaks down. At the bottom of the downstroke, a well-defined TEV is also clearly visible. Figure 3.40 depicts the lift

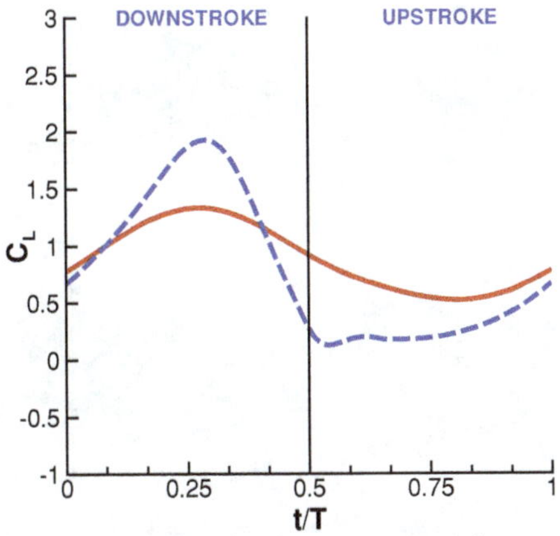

(a) lift, shallow stall, $Re= 6\times10^4$

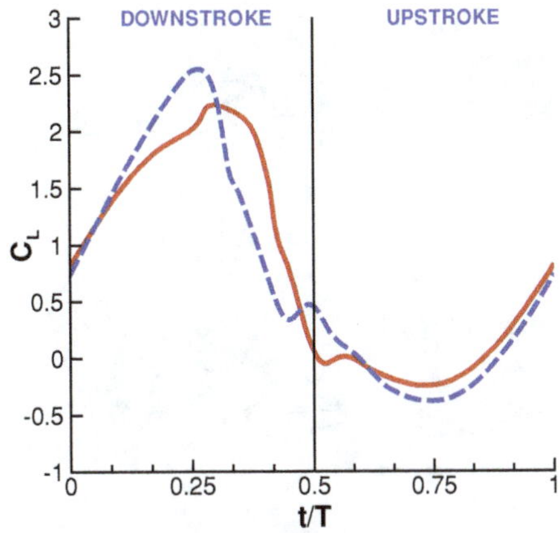

(b) lift, deep stall, $Re= 6\times10^4$

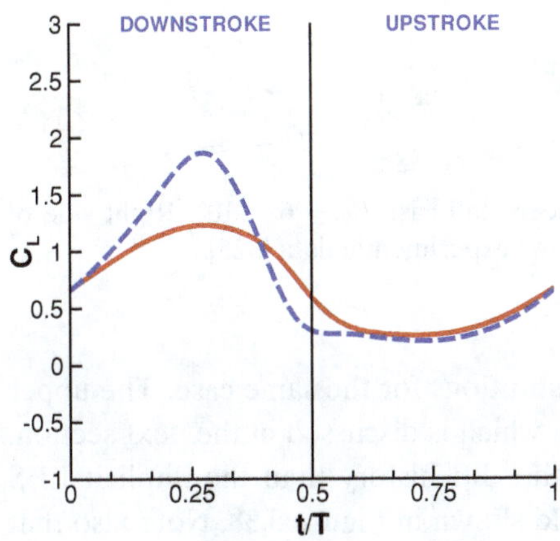

(c) lift, shallow stall, $Re= 1\times10^4$

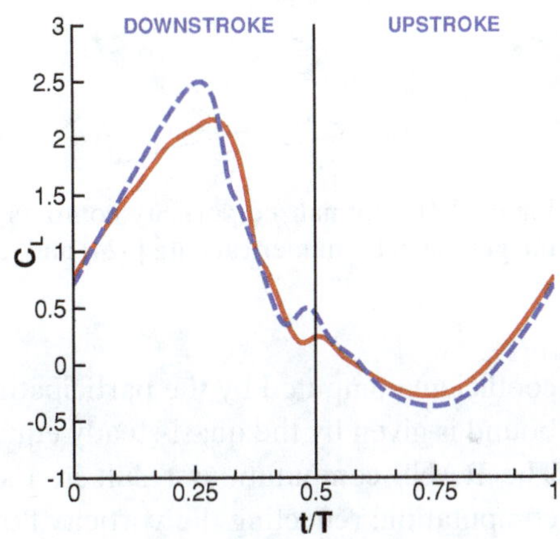

(d) lift, deep stall, $Re= 1\times10^4$

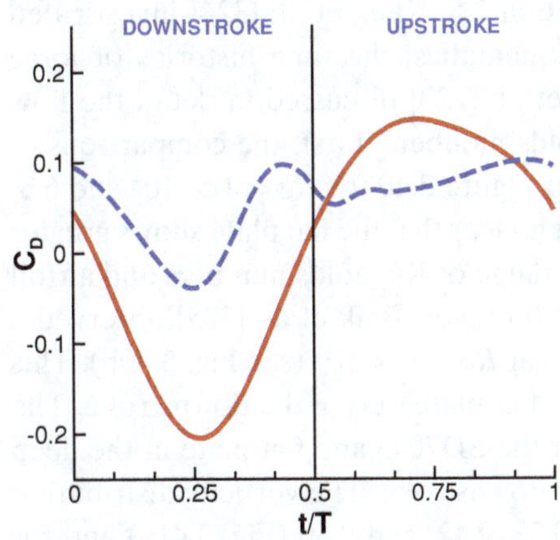

(e) drag, shallow stall, $Re= 6\times10^4$

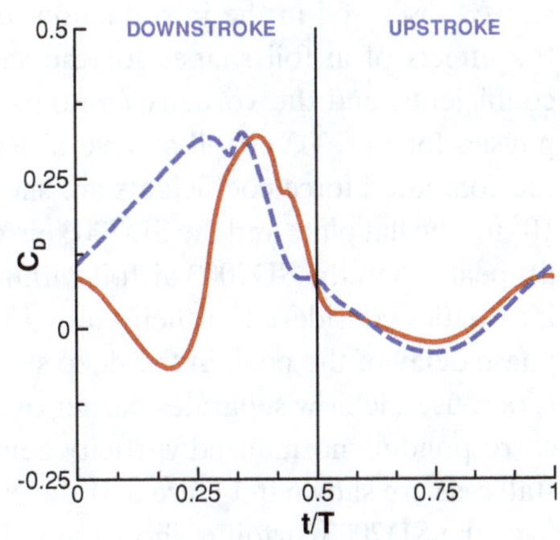

(f) drag, deep stall, $Re= 6\times10^4$

Figure 3.40. Time histories of forces on the 2D flat plate (dashed line) and SD7003 airfoil (solid line). From Kang et al. [324].

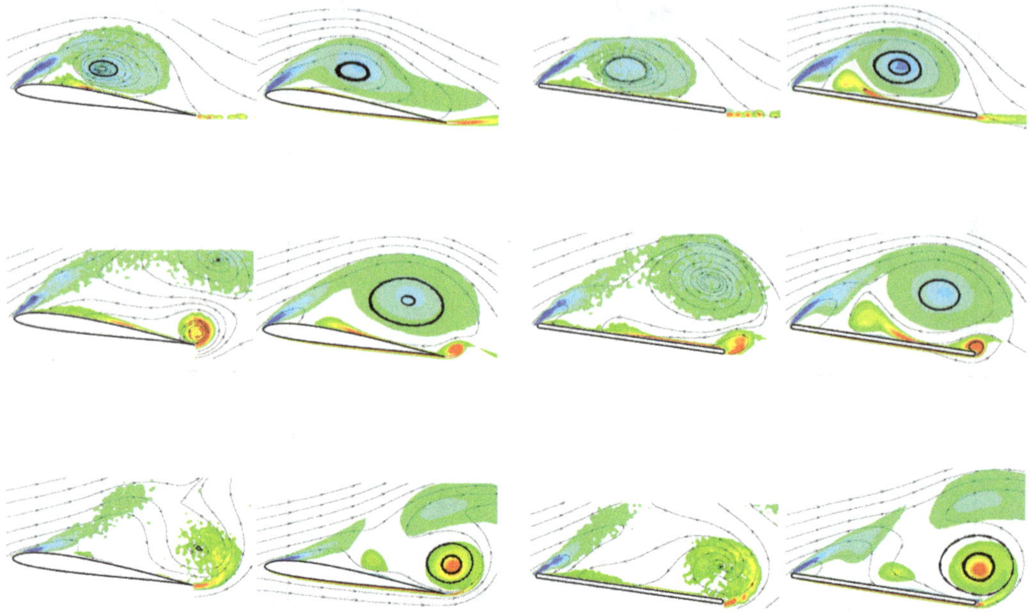

Figure 3.41. Normalized vorticity contours for deep stall case $Re = 6 \times 10^4$. Right side of images indicates numerical data [324] and left shows experimental data [325].

coefficient computed by the participating institutions for the same case. The upper bound is given by the quasi-steady equation, which is discussed in the next section. The RANS computations result in a smoother lift history than the implicit LES computation, reflecting the vorticity flow field shown in Figure 3.38. Note also that Theodorsen's lift formula approximates the unsteady lift well (see Section 3.6).

As discussed in the introduction to Section 3.5, Kang et al. [324] investigated the effects of airfoil shapes for the same kinematics, the time histories of force coefficients, and the vorticity contours. Ol et al. [323] discussed in detail the flow physics for SD7003 airfoil at several Reynolds numbers. First, the comparisons of the computed force coefficients are shown in Figure 3.40 for $Re = 1 \times 10^4$ and 6×10^4 for the flat plate and the SD7003 airfoil. It is clear that the flat plate shows greater lift peaks than the SD7003 airfoil within the range of Reynolds numbers and airfoil kinematics considered in their study [324]. Moreover, Baik et al. [325] observed a phase delay of the peak in the deep stall case at $Re = 6 \times 10^4$ (see Fig. 3.40b). This is because the flow separates earlier over the flat plate during the downstroke. The corresponding normalized vorticity fields for the SD7003 and flat plate in the deep stall case are shown in Figure 3.41 during the downstroke. The vorticity distribution over the SD7003 airfoil is shown at $t/T = 0.33$, 0.42, and 0.50 (Fig. 3.41a) and for the flat plate at earlier time instants of $t/T = 0.25$, 0.33, and 0.42 (Fig. 3.41b). Due to the smaller radius of curvature at the leading edge of the flat plate, the flow around the leading edge experiences a stronger adverse pressure gradient, separating at a lower effective AoA. The leading-edge shear layer and the LEV on the suction side of the flat plate have similar topology to that of the flow field of SD7003 with a phase lag of approximately 0.083. As the LEV and the TEV detach from the airfoils, the difference in lift between these two airfoils diminish.

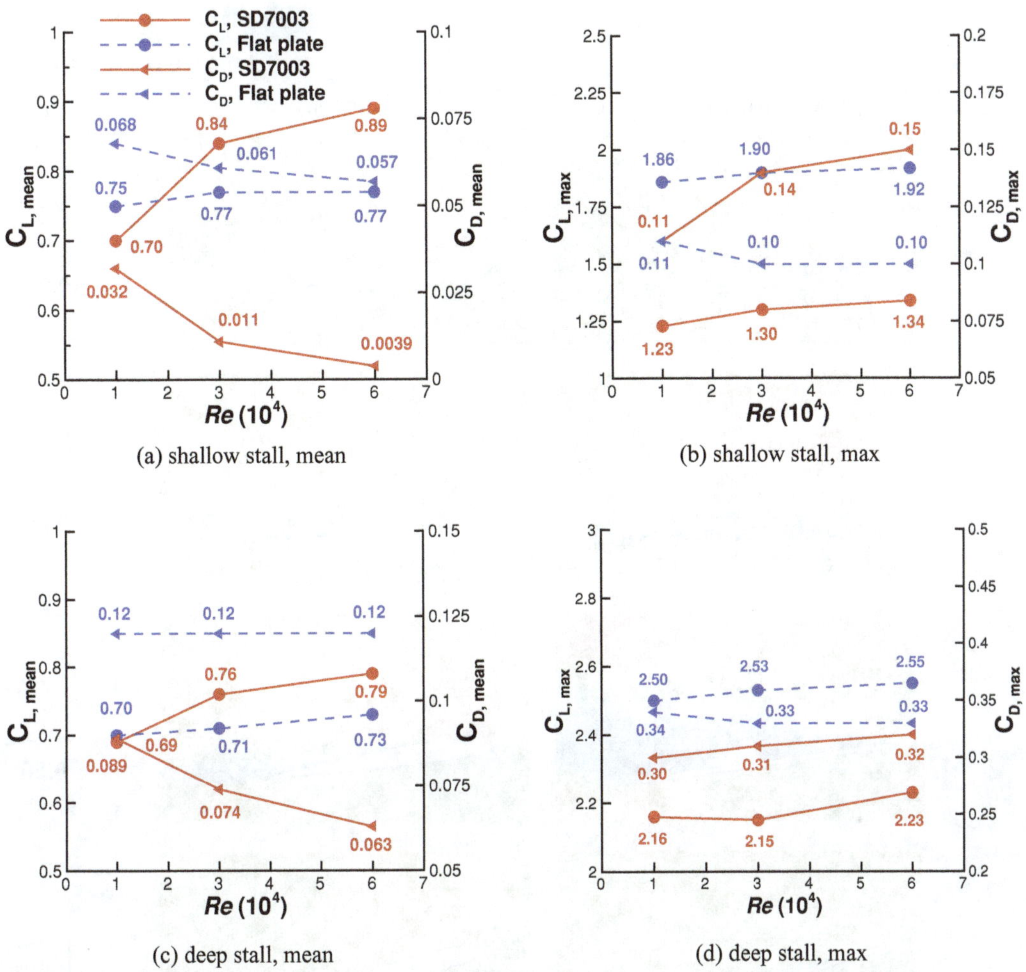

Figure 3.42. Mean and maximum lift and drag coefficients as a function of Reynolds number. From Kang et al. [324].

The mean and maximum force coefficients for both airfoil shapes for several Reynolds numbers are summarized in Figure 3.42. The geometric effect due to the sharper leading edge of the flat plate overwhelms the variations in motion kinematics and Reynolds number: the maximum lift is obtained by the flat plate for both kinematics. Furthermore, the force coefficients of the flat plate are insensitive to the Reynolds number. It is also interesting to note that the mean drag coefficient is lower for the SD7003 airfoil because of its streamlined body and that the mean lift coefficient is higher for the SD7003 airfoil for $Re = 3 \times 10^4$ and 6×10^4, when the flow is attached over most of the motion period.

Sane [312] discussed the effects of delayed stall and LEV on force generation using Polhamus's analogy [321]. For airfoils without leading-edge separation, the flow over the leading edge accelerates and creates a suction force that is parallel to the chord. This suction force tilts the resulting force forward in the direction of thrust at low AoAs. For a flat plate the flow separates at the leading edge, and an LEV forms on the suction side. The LEV increases the force normal to the flat plate,

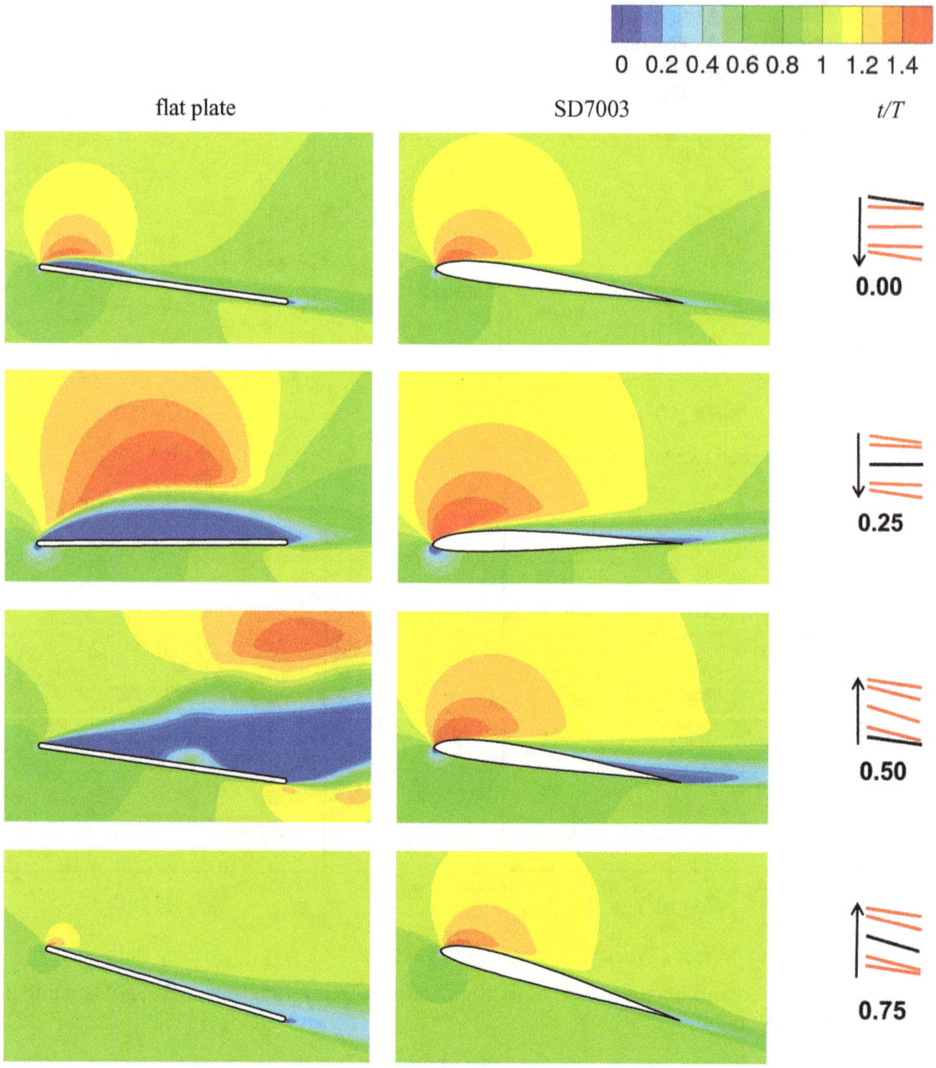

Figure 3.43. $\bar{u}_1/U_\infty$ contours for the 2D flat plate and airfoil in shallow stall at $Re = 6 \times 10^4$. From Kang et al. [324].

resulting in increased lift and drag for positive AoA. In this effort, for the shallow stall case, the SD7003 airfoil generates lower lift peak (Fig. 3.40a) but greater thrust compared to the flat plate due to attached flow and leading-edge suction (Fig. 3.40e). Figure 3.43 shows the normalized streamwise velocity contours around the 2D flat plate and the SD7003 airfoil at $Re = 6 \times 10^4$ for the shallow stall case. During the downstroke the flow separates at the leading edge, forming an LEV that covers the whole suction surface of the flat plate, consistent with the higher lift peak and greater drag during the downstroke shown in Figure 3.40a and Figure 3.40e, while the flow over the SD7003 remains mostly attached. When the prescribed motion is more aggressive for the deep stall case, the flow over the SD7003 airfoil separates near the leading edge during the downstroke (Fig. 3.41), and the resulting lift and drag time histories are similar to those of the flat plate (Fig. 3.40b, f) for the period in which the flow is separated. There is still a significant difference in drag (Fig. 3.40f) during the first portion of the downstroke, because the flow over the SD7003 airfoil separates later than over the flat plate. Hence, Sane's application of Polhamus' analogy is valid for the shallow stall motion and to a lesser extent for the deep stall motion. For motions with greater effective AoA, the flow over the blunt airfoil may separate, and then the resulting force on the blunt airfoil is qualitatively similar to that of a flat plate.

3.5.4 2D versus 3D Flat Plate in Shallow Stall

As discussed in Sections 2.3 and 3.4, the difference between 2D and 3D flow structures in terms of force generation has important implications for the applicability of 2D computations to approximate a 3D flow field. For example, at Re of 10^2 for a delayed rotation kinematics, the TiV anchored the vortex shed from the leading edge, thereby increasing the lift compared to a 2D computation under the same kinematics (see Section 3.4.1.1). In contrast, under different kinematics with a smaller AoA and synchronized rotation, the generation of TiVs was small, and the aerodynamic loading was captured by the analogous 2D computation (see Section 3.4.1.2). At $Re = O(10^3\text{–}10^4)$ Visbal [328] studied the unsteady separation process in a 3D flow field around a purely plunging flat plate with $AR = 2$, $k = 1.0$, and $St = 0.16$. The implicit LES computations employed by Visbal [328] show the evolution of both the TiV and the LEV. The TiV forms as the flat plate plunges downward, breaks down, and collapses during the first part of the upstroke. Meanwhile the LEV, which is fairly uniform at the early stage of downstroke, experiences intense axial flow toward the center of the wing, eventually evolves into an arch-type vortex, and sheds as a ring vortex, suggesting the 3D nature of the flow at higher reduced frequency and Strouhal number.

Kang et al. [324] investigated the 3D effects for the flat plate with AR of 2 in shallow stall at $Re = 4 \times 10^4$. Figure 3.44 illustrates the $\bar{u}_1/U_\infty$ contours from the computations at $Re = 4 \times 10^4$ and the experimental results obtained at $Re = 3 \times 10^4$. The flow field is dominated by large leading-edge separation due to the geometric effects in the downstroke as discussed in the previous section. In 3D, the separation is mitigated and the flow reattaches before mid-chord. The velocity field at the center of the downstroke, $t/T = 0.25$, is further depicted in Figure 3.45 from the 3D computation and experimental measurements at 75 percent span location. The velocity profiles also show the boundary-layer separation at the leading edge and reattachment around $x_1/c_m = 0.25$. It should be noted that the experimental measurements were obtained at $Re = 3 \times 10^4$. The reasonable correlation shown in Figure 3.44 and Figure 3.45 may indicate that the Reynolds number effect is limited, similar to the observation made in Section 3.4.1.2 for the 2D flow around the flat plate.

The absence of the strong leading-edge separation for the flat plate with $AR = 2$ manifests itself in the aerodynamic force felt on the wing. The time histories of the lift coefficient for both 2D and 3D flat plates in shallow stall are plotted in Figure 3.46. In the upstroke where both flat plates evince attached flow, the lift coefficient shows comparable magnitudes. However, during the downstroke the peak of the lift generated on the 2D flat plate is 2.3 times greater than on its 3D counterpart.

At the center of the downstroke, $t/T = 0.25$, the effective AoA is at maximum, and the development of the TiV is also significant as illustrated by the iso-surfaces

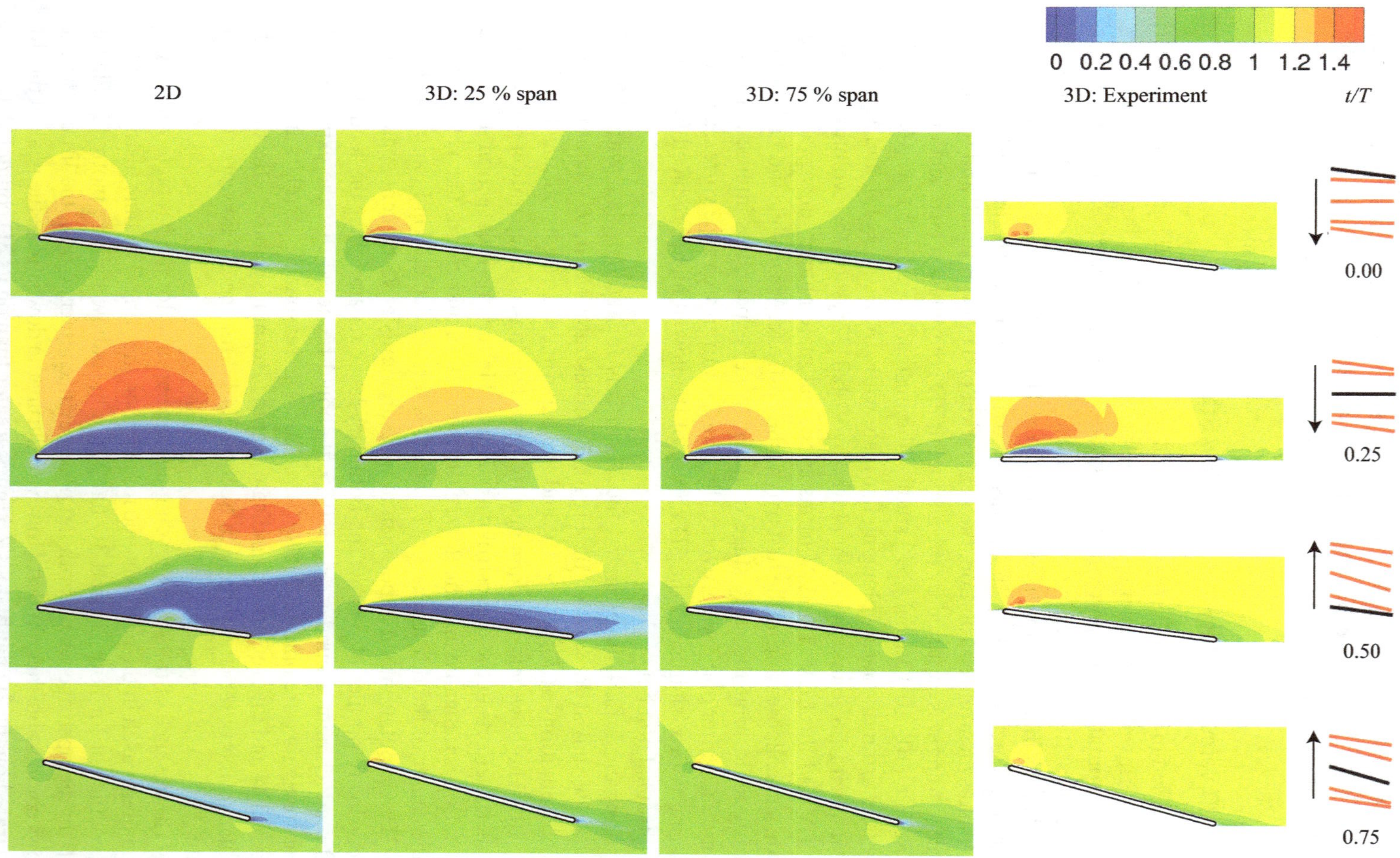

Figure 3.44. $\overline{u}_1/U_\infty$ contours for the 2D and 3D ($AR = 2$, 25% and 75% span) flat plates in shallow stall at $Re = 4 \times 10^4$. Experimental data are obtained at $Re = 3 \times 10^4$ and at 75% span. From Kang et al. [324].

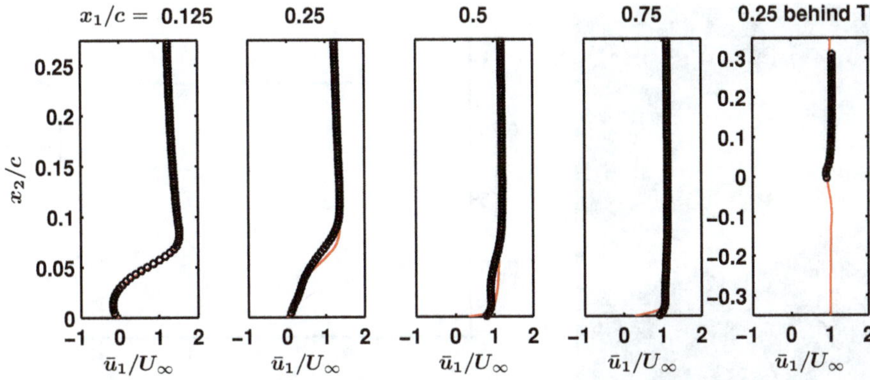

Figure 3.45. $\overline{u}_1/U_\infty$ profiles at $t/T = 0.25$ at $Re = 4 \times 10^4$for the 3D flat plate with $AR = 2$ in shallow stall at 75% span. Experimental data were performed at $Re = 3 \times 10^4$. From Kang et al. [324].

of Q in Figure 3.47. This TiV interacts with the LEV developed at the leading edge of the flat plate. Because of the downwash induced from the presence of the TiV, the effective AoA at the leading edge near the tip is smaller than in 2D, which results in a spanwise non-uniform LEV formation that is stronger away from the tip. Consequently, the spanwise lift distribution in 3D is smaller than in 2D as shown in Figure 3.47. Close to the tip of the flat plate a small peak in spanwise lift is shown. A reason for this local maximum is the lower pressure region associated with the presence of the TiV [297]. In contrast, at the center of the upstroke the effective AoA is at minimum, and the flow field has negligible 3D effects such as TiV generation and interactions between TiV and LEV, so that the spanwise lift distribution of 3D is similar to that of 2D.

3.6 Approximate Analysis for Non-Stationary Airfoil

3.6.1 Force Prediction for a Pitching and Plunging Airfoil in Forward Flight

It was in the early 20th century that the aerodynamic force generation on a moving airfoil was treated using the linearized aerodynamic theories. Wagner [329]

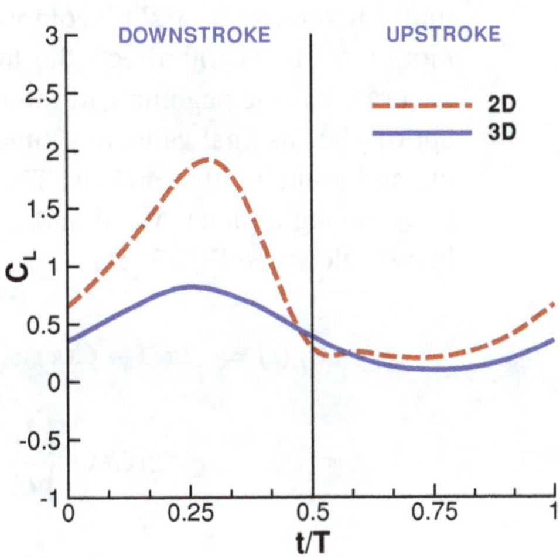

Figure 3.46. Lift coefficient for the 2D and 3D ($AR = 2$) flat plates in shallow stall at $Re = 4 \times 10^4$. From Kang et al. [324].

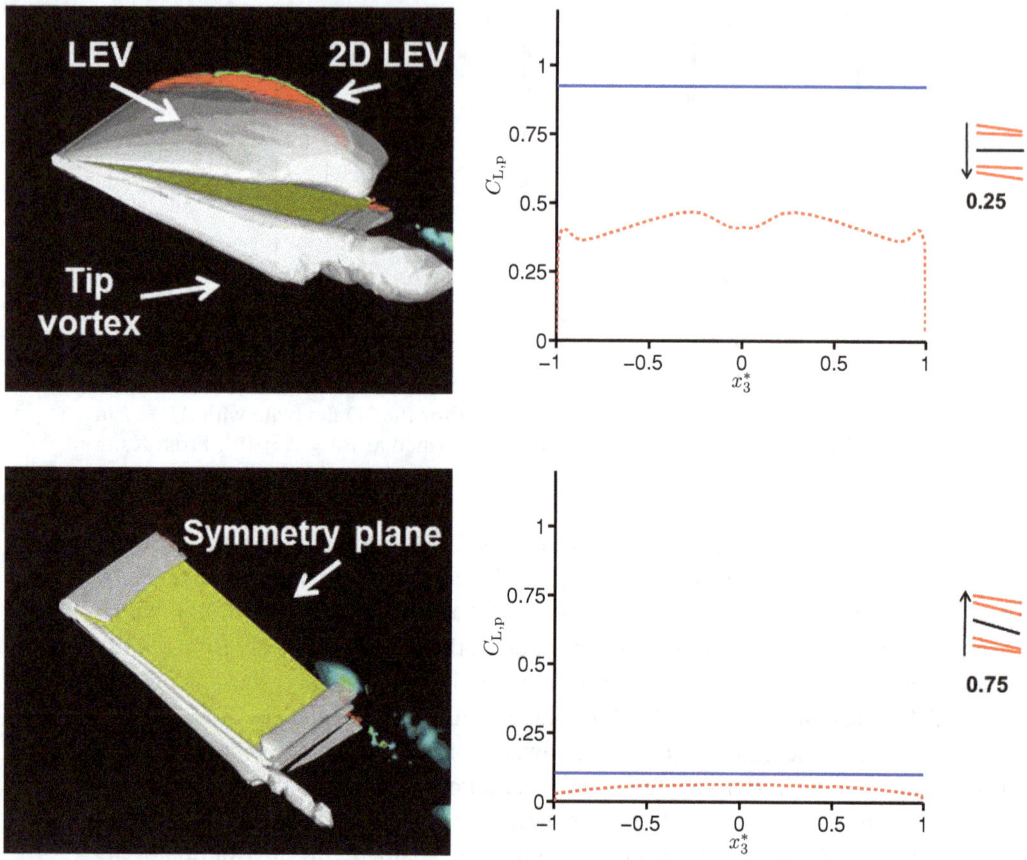

(a) iso-Q-surface at $Q = 4$ (white) and Q contours from the 2D computations on the symmetry plane.

(b) spanwise lift distribution due to pressure. Red dot: 2D; Blue solid: 3D.

t/T

Figure 3.47. Iso-Q surfaces (a) and spanwise lift distribution (b) to illustrate the difference in the flow structures at the center of the downstroke ($t/T = 0.25$) and the center of upstroke ($t/T = 0.75$). From Kang et al. [324].

calculated the lift generation on a thin airfoil moving impulsively from rest to a uniform velocity in a 2D incompressible fluid. Vortex wake generated due to the motion of the airfoil affects the force acting on the airfoil. Half of the final lift is generated at the beginning of the motion, and the instantaneous lift asymptotically approaches its final value as a function of increasing time. For harmonically pitching and plunging thin airfoils, Theodorsen [330] derived an expression for the lift by assuming a planar wake and a trailing-edge Kutta condition in incompressible inviscid flow; see Eq. (3–21):

$$\begin{aligned} C_L(t) &= 2\pi(1 - C(k))\alpha_0 + \frac{\pi c}{2}\left\{\frac{\dot{\alpha}}{U_\infty} + \frac{\ddot{h}}{U_\infty^2} - \frac{c(2x_p - 1)\ddot{\alpha}}{2U_\infty^2}\right\} \\ &\quad + 2\pi C(k)\left\{\frac{\dot{h}}{U_\infty} + \alpha + c(1.5 - 2x_p)\frac{\dot{\alpha}}{2U_\infty}\right\}, \end{aligned} \tag{3–21}$$

The pitch and plunge motions are described by the complex exponentials, $\alpha(t) = \alpha_0 + \alpha_a e^{(2\pi ft+\Psi)i}$ and $h(t) = h_a e^{2\pi fti}$. The phase lead of pitch compared to plunge is denoted by ψ. $C(k)$ is the complex-valued *Theodorsen function* with magnitude ≤ 1; it accounts for the attenuation of lift amplitude and the time lag in lift response from its real and imaginary parts, respectively. The first term is the steady-state lift, and the second term is the non-circulatory lift due to acceleration effects. The third term models circulatory effects. The same expressions were obtained by Küssner [331] and von Karman and Sears [332].

For a harmonically plunging thin rigid flat plate in a free-stream, the lift coefficient can be derived assuming inviscid incompressible flow (see Eq. (5–347) as in Bisplinghoff, Ashley, and Halfman [333]):

$$C_L = 2\pi^2 St\, k\, \cos(2\pi ft) + 4\pi^2\, St\, \sin(2\pi ft), \tag{3–22}$$

assuming quasi steady-state flow where the influence of the wake vorticities is neglected. The first term in Eq. (3–22) is the non-circulatory term that is consistent with the added mass force derived in Section 3.6.4. The second term in Eq. (3–22) is the circulatory term, which can be expressed in a more familiar form, $2\pi\alpha_e$, by recognizing that $2\pi St\, \sin(2\pi ft) \approx \alpha_e$ where α_e is the effective AoA for purely plunging motions. Note also that Eq. (3–22) is a simplification of Eq. (3–21) for purely plunging motion with $C(k) = 1$ equivalent to $k \to 0$.

Based on the formulas derived by Theodorsen for the lift and the work by Karman and Burgers [334], Garrick [335] investigated the thrust generation by a harmonically pitching and plunging thin airfoil in uniform free-stream. For a special case of a pure plunge motion the thrust generation C_T was

$$C_T = \pi^3 St^2 (F^2 + G^2), \tag{3–23}$$

where $F(k)$ and $G(k)$ are the real and the imaginary parts, respectively, of the Theodorsen function $C(k)$ – see Eq. (3–21) – predicting greater thrust for higher Strouhal numbers. For a NACA 0012 airfoil Heathcote and Gursul [336] compared the experimentally measured thrust generation with Garrick's formula [335] and used numerical computation to solve the Navier-Stokes equations [337]. It was shown that Garrick's formula overestimates the experimentally measured thrust, which is in good agreement with the Navier-Stokes computations.

3.6.2 Simplified Aerodynamics Models

Following the helicopter theory, a simplified analysis for flapping flight can be established based on the actuator disk model. An actuator is an idealized surface that continuously pushes air and imparts momentum to downstream by maintaining a pressure difference across itself (i.e., the lift is equal to the change in fluid momentum). Assuming that insect wings beat at high enough frequencies so that their stroke planes approximate an actuator disk, the wake downstream of a flapping wing can be modeled as a jet with a uniform velocity distribution [65] [338]. Although the momentum theory accounts for both axial and rotational changes in the velocities at the disk, it neglects the time dependency in wing shape, kinematics, and associated unsteady lift-producing mechanisms.

By using the Bernoulli equation for steady flow to calculate induced velocity at the actuator disk and the jet velocity in the far wake downstream (i.e., the downwash), Weis-Fogh [339] derived the induced downwash velocity w_i for a hovering insect at the stroke plane as

$$W_i = \sqrt{\frac{W}{2\pi\rho R^2}}, \tag{3–24}$$

where W is the insect weight, ρ the air density, and R the wing length. From the experimental measurements of the beetle *Melolontha vulgaris* Weis-Fogh [339] assumed that the downwash velocity in the far wake is twice that at the disk (i.e., $w = 2w_i$), even though he pointed out that w_i varies through a half-stroke and that stroke plane amplitude Φ is rarely 180°.

Instead of using a circular disk, Ellington [340] proposed a *partial* actuator disk of the area $A = \Phi R^2 \cos(\beta)$ that flapping wings cover on the stroke plane, as depicted in Figure 3.48, and modified the expression for the induced power P_{ind}, such that

$$P_{\text{ind}} = \sqrt{\frac{W}{2\rho\Phi R^2\cos(\beta)}} = W\sqrt{\frac{1}{2\rho}\left(\frac{W}{A}\right)}, \tag{3–25}$$

where β is the stroke plane angle and W/A is the disk loading that controls the minimum power requirement. He also noted that, because of the time-varying nature of flapping, a *pulsed* actuator disk seems more representative of hovering flight. He showed that the circulation of the vortex rings in the far wake downstream is related to the jet velocity:

$$\Gamma = \frac{w^2}{2f_s}, \tag{3–26}$$

where f_s is the shedding frequency.

Rayner [79] [341] proposed a method representing the wake of a hovering insect by a chain of small-cored coaxial vortex rings (one produced for each half-stroke). Although the approach could determine the lift and drag coefficients, it did not account for the effects of stroke amplitude and stroke plane angle. Sunada and Ellington [293] developed a method that models the shed vortex sheets in the wake as a grid of small vortex rings, with the shape of the grid modeled by wing kinematics so that all forward speeds can be handled.

Overall, the relatively simple approaches presented in this section are of limited usefulness because they only include stroke-plane angle and disk loading. The models do not allow, for example, estimation of lift forces for a given wing kinematics or wing geometry.

One way to estimate the force generation is to measure the vorticity ω in the flow field around the flapping wing. Then, for steady inviscid irrotational (potential) flow, the lift per unit span on such a wing can be approximated as

$$L = \rho U_\infty \Gamma, \tag{3–27}$$

where U_∞ is the free-stream velocity and Γ is the circulation over the flow field S, defined as

$$\Gamma = \int \boldsymbol{\omega} \cdot d\boldsymbol{S}. \tag{3–28}$$

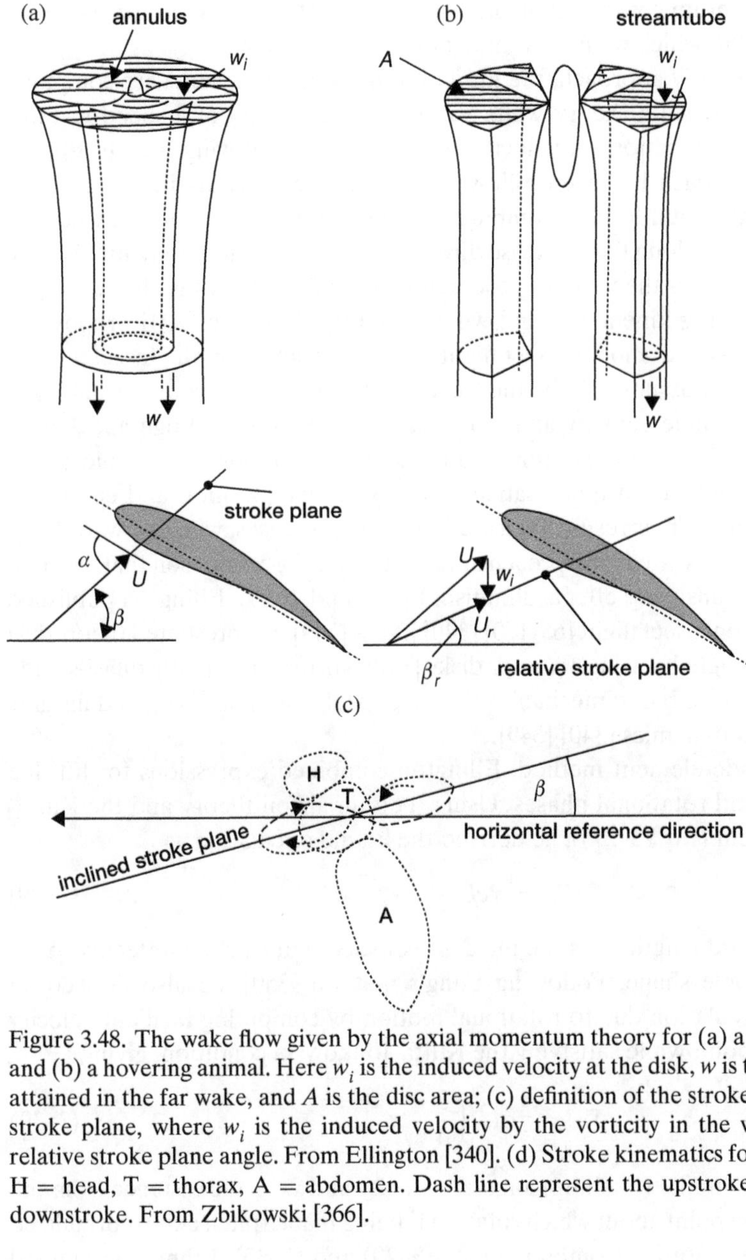

Figure 3.48. The wake flow given by the axial momentum theory for (a) a hovering propeller and (b) a hovering animal. Here w_i is the induced velocity at the disk, w is the vertical velocity attained in the far wake, and A is the disc area; (c) definition of the stroke plane and relative stroke plane, where w_i is the induced velocity by the vorticity in the wake and β_r is the relative stroke plane angle. From Ellington [340]. (d) Stroke kinematics for a hovering insect; H = head, T = thorax, A = abdomen. Dash line represent the upstroke, dotted line is the downstroke. From Zbikowski [366].

Because obtaining a direct pressure measurement over a moving wing is difficult, applying Eq. (3–27) has been a popular way to estimate the lift generation. However, the so-called wake momentum paradox arose when the lift generation of slow-flying pigeons [342] and jackaws [343] appeared to be only 50 percent of the force that is required to sustain their weight. Later using high-resolution wake vorticity measurements and by accounting for all vortical structures Spedding et al. [344] showed that the resulting wake structures provide sufficient momentum for weight support. Another caveat is that Eq. (3–27) is only valid for steady-state flow field around a stationary wing. When the wing and its wake change in time, the unsteady term in the momentum conservation needs to be correctly accounted for, as shown by Noca et al. [345], which we discuss in detail in Section 3.6.4.

In the quasi-steady approach, the lift and drag force coefficients are computed based on the steady-state theory while varying the geometry and speed in time. To account for the variations in velocity and geometry from wing base to tip, the blade-element approach has been followed to discretize the wing into chordwise, thin wing strips; the total force is computed by summation of the forces associated with individual strips along the spanwise direction [65] [294]. Integrating lift over the entire stroke cycle gives the total lift production of the flapping wings. For example, considering such wing kinematics and wing geometry, Osborne [294] proposed a quasi-steady approach to model insect flight: the forces acting on the insect wing at any point in time are assumed to be the steady-state values that would be achieved by the wing at the same velocity and AoA. Later, in 1956, Weis-Fogh and Jensen [346] laid out the basis of momentum and blade-element theories as applicable to insect flight and carried out quantitative analyses on wing motion and energetics available at the time. Their results indicated that, in most cases, when forward flight is considered, the quasi-steady approach appears to hold for the reason that, as flight velocity increases, unsteady effects diminish. In the mid-1980s, Ellington published a series of papers on insect flight [65] [70] [340] [347]–[349]. He presented theoretical models for insect flight by using actuator disks [340]; vortex wake [340]; quasi-steady methods [65]; rotation-based mechanisms of clap, peel, and fling [340]; and insights into unsteady aerodynamics [340] [349].

From the blade-element method, Ellington combined expressions for lift due to translational and rotational phases. Using the thin airfoil theory and the Kutta-Joukowski theorem (Eq. (3–27)), he derived the bound circulation as

$$\Gamma_t = \pi c U \sin \alpha_e, \tag{3–29}$$

where c is the chord length, U is the incident velocity, and α_e is the effective AoA corrected for profile shape. Following Fung's method [350], he also derived an expression for circulation due to rotational motion by computing incident velocity at the 3/4 chord point while satisfying the Kutta-Joukowski condition, giving

$$\Gamma_r = \pi \dot{\alpha} c^2 \left(\frac{3}{4} - \hat{x}_0 \right), \tag{3–30}$$

where $\dot{\alpha}$ is the rotational (pitching) angular velocity and $\hat{x}_0$ is the distance from the leading edge to the point about which rotation is being made (pitch axis), normalized with respect to the chord c. Combining Eqs. (3–29) and (3–30), Ellington obtained the quasi-steady lift coefficient:

$$C_L = 2\pi \left[\sin \alpha_e + \frac{\dot{\alpha} c}{U} \left(\frac{3}{4} - \hat{x}_0 \right) \right]. \tag{3–31}$$

Equation (3–31) is equivalent to Theodorsen's formula for lift: Eq. (3–21) in the quasi-steady limit (i.e., $C(k) = 1$) without the added mass components. Although Osborne [294] suggested that the added mass may play an important role in flapping wing flight, Ellington [348] argued that the additional time-averaged lift due to the added mass (virtual mass) vanishes for periodic motions. The additional drag due to added mass is also zero because there is no net wing acceleration parallel to the wing stroke during a motion period. Recently, Kang et al. [351] normalized the integral form of the Navier-Stokes equation and proposed that the added mass effects are important for the high reduced frequency and high Reynolds number flows. In particular, the added mass force should not be neglected when investigating the performance of a flexible flapping wing in a high-density medium, such as in water. Furthermore, to determine lift and power requirements for hovering flight, Ellington [349] sought estimates for the mean lift coefficient through the flapping cycle and derived a non-dimensional parameter-based expression:

$$\langle C_L \rangle = \frac{8\,\langle L \rangle \cos^2(\beta_r)}{\rho f^2 \Phi^2 R^2 \hat{r}_2^2(S) \langle (d\hat{\phi}/dt)^2 \rangle S \cos^2(\beta)}, \tag{3–32}$$

where $\langle L \rangle$ is the mean lift through a half-stroke, ρ is the air density, f is the wing-beat frequency, Φ is the stroke angle, $\langle (d\hat{\phi}/dt)^2 \rangle$ is the mean-squared flapping angular velocity, S is the wing area, β is the stroke plane angle, β_r is the relative stroke plane angle (see Fig. 3.48c), and r_2 is the second moment of the wing area.

Numerous versions of the quasi-steady approach can be found in the literature; in general, the model predictions are not consistent with the physical measurements, especially when the hovering flight of insects is considered. For example, lift coefficients obtained under those conditions yield (i) 0.93–1.15 for dragonfly *Aeschna juncea* [229] [352], (ii) 0.7–0.78 for fruit fly *Drosophila* [353] [354], and (iii) 0.69 for bumblebee *Bombus terrestris* [212]. However, lift coefficients estimated by direct force measurements in flying insects are significantly larger than those predicted by the quasi-steady methods, ranging from 1.2 to 4 for various insects including the hawkmoth *Manduca sexta*, bumblebee *Bombus terrestris*, parasitic wasp *Encarsia formosa*, dragonfly *Aeschna juncea*, and fruit fly *Drosophila melanogaster* [339] [349] [355] [356].

Because quasi-steady methods are unable to predict flapping wing aerodynamics accurately, empirical corrections have been introduced. Walker and Westneat [357] presented a semi-empirical model for insect-like flapping flight, which includes, for example, Wagner's function [350], which is devised to account for the lift enhancement caused by an impulsively starting airfoil. They used a blade-element method to discretize the flapping wing and then computed forces on the wing elements, in which the forces comprise a circulation-based component and a non-circulatory apparent mass contribution. Sane and Dickinson [216] refined a quasi-steady model to describe the forces measured in their earlier experiments on the *Robofly*, a mechanical, scaled-up model of the fruit fly *Drosophila melanogaster* [228]. They decomposed the total force F into four components, namely,

$$F = F_t + F_r + F_a + F_w, \tag{3–33}$$

where the subscripts t and r are for translational and rotational quasi-steady components, respectively; a is for added mass; and w is for wake capture. In the blade-element approach, a *Robofly* wing is divided along the spanwise direction into chordwise strips, and the forces on each strip are computed individually and integrated along the span. The translational quasi-steady force F_t is computed from empirically fitted equations from a 180° sweep with fixed AoAs as

$$C_L = 0.225 + 1.58\ \sin(2.13\alpha - 7.2),$$

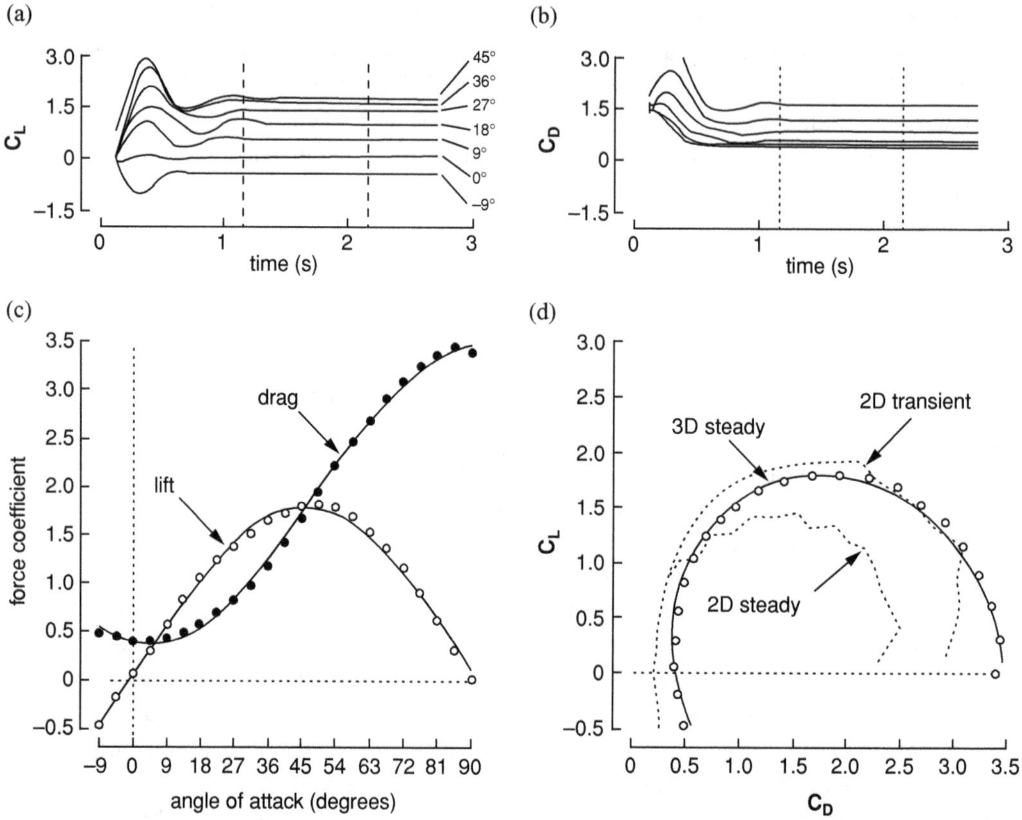

Figure 3.49. Lift coefficient measurements of a translating *Robofly* wing as a function of angle of attack. Dickinson et al. [201] used these measurements to propose a quasi-steady model for the translating component. $Re = 100$–200.

$$C_D = 1.92 - 1.55\ \cos(2.04\alpha - 9.82), \tag{3–34}$$

see also Figure 3.49. To determine the rotational quasi-steady force F_r, Sane and Dickinson [216] set the forces due to added mass and wake capture to zero (i.e., $F_a = F_w = 0$) by removing any accelerations and avoiding wake reentry, respectively. They measured the force F acting on a wing undergoing a constant translation and rotation for one forward stroke only and obtained the rotational force F_r by subtracting the empirically predicted translational force F_t from the measured total force F. For the quasi-steady treatment of the rotational force they used Eq. (3–34) and replaced the free-stream velocity with the instantaneous translation velocity, U_t, as $F_r = C_{\text{rot}}\rho U_t \Gamma_r$, with C_{rot} being a coefficient that depends on the angular velocity. Furthermore, they computed the added mass F_a by applying the added mass term from the linearized aerodynamic theories (e.g., Eq. (3–21)), for each blade element and integrating along the span. Knowing these three components, they evaluated the wake-capture force F_w by subtracting the components F_t, F_r, and F_a from the total measured force F. The wake-capture force arises from the interaction with the wakes and vortices shed in the previous strokes. At the beginning of the stroke, the enhanced momentum increases lift as described in Section 3.3.1. In contrast, during the mid-stroke the effects of the wing-wake interaction are negative due to the downwash, which resides upstream of the wing generated by the wakes shed in the previous strokes [358] [359]. As the wing starts to interact with this downward wake, the effective angle of attack reduces.

Kang and Shyy [359] combined the geometric angle of attack and this downwash to calculate an effective angle of attack, which is fed in the lift estimation of the translational component, Eq. (3–34). The translational lift corrected for the downwash indeed predicted a loss of lift during this part of the stroke, thereby giving a better approximation of the lift calculated by solving the Navier-Stokes equations. Further discussion can be found in Section 4.5.1. Wing-wake interaction depends on the history and non-linear aerodynamic effects, and an a priori estimation of this force due to wing-wake interaction has still not been found.

A model for unsteady lift generation for insect-like flapping wings was proposed by Pendersen and Zbikowski [360]. The model is modular, giving a better insight into various effects on aerodynamic force generation, and it includes added mass effects, the quasi-steady assumption, an LEV effect, and the wake effect. The model's predicted lift and drag forces were compared with the measurements of Dickinson et al. [201], and despite its simplifications, the model captures reasonably well the lift evolution, but over-predicts the force peak values.

Although such semi-empirical methods can be tuned to provide good agreement with experimental measurements, their predictable capabilities are questionable because they cannot adequately account for the relevant unsteady, vortical fluid physics. In particular, the unsteady effects are important not only during the translational phases of the stroke (upstroke and downstroke) but also during the rotational phase near the end of each stroke when the wings are quickly rotated around their spanwise axes [247] [275] [229] [361]. Nevertheless, the quasi-steady model does provide some insight into flapping flight in insects and birds and offers quick estimates of unsteady aerodynamic coefficients.

Recently, Ansari et al. [362] [363] modified and extended the circulation approach for modeling insect-like flapping wings in the hover. It is based on the original approach of von Karman and Sears [332] together with the non-linear extensions proposed by McCune et al. [364] and Tavares and McCune [365], but with further significant extensions as well. Ansari et al. [362] [363] showed not only the validity of their unsteady aerodynamic model by comparing it with the experiments of Dickinson et al. [201] but also the limitation of the model (i.e., an inviscid, potential, and essentially 2D model).

3.6.3 Some Remarks on Simplified Models

The simplified models – such as Theodorsen's formula for lift, Eq. (3–21) for forward flight, the revised quasi-steady model proposed by Sane and Dickinson [292], or Eq. (3–33) for hover – are very powerful for obtaining a quick estimate of lift generation or for design and optimization purposes, but they do not guarantee the accuracy of the solution. A solution from a Navier-Stokes equation solver would be more accurate, but it would take hours or even days to obtain one single 3D simulation. It is hence of great importance to assess when or why the simplified models give reasonably accurate results.

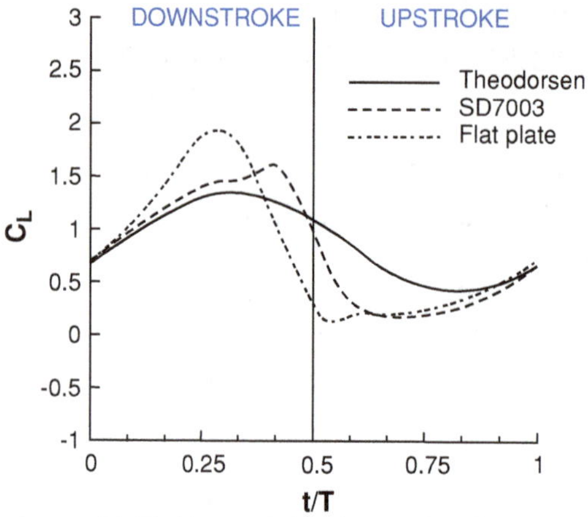

(a) Pitching and plunging (0.68, 0.89, 0.77)

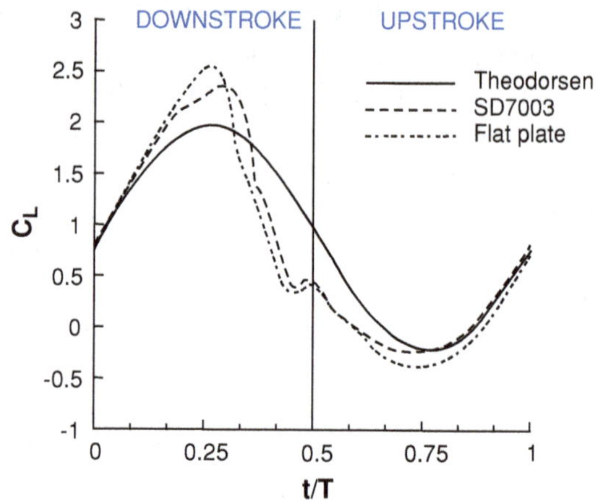

(b) Purely plunging (0.78, 0.79, 0.73)

Figure 3.50. Lift coefficient time histories from Navier-Stokes computations and Theodorsen, Eq. (3–21). The time-averaged lift values are given in the parentheses: Eq. (3–21), SD7003, and flat plate. $Re = 6 \times 10^4$.

For the shallow stall and deep stall kinematics discussed in Section 3.5, the lift prediction by the Theodorsen formula Eq. (3–21) is plotted against the lift obtained by numerical computations for a 2D SD7003 airfoil and flat plate in Figure 3.50. For the shallow stall case, Theodorsen's result matches the computations reasonably during the downstroke. The agreement improves in the second part of the upstroke. Although Eq. (3–21) is derived for a thin flat plate, the lift prediction is closer to that of SD7003 airfoil. This difference stems from the formation of LEV for the case of a flat plate as shown in Figure 3.33 during the downstroke. As the LEV convects downstream, the lower pressure region in the vortex core enhances the lift. Subsequently the LEV detaches and the flat plate loses the leading-edge suction. At $t/T = 0.5$ the discrepancy is the largest where the flow over the SD7003 experiences an open separation [323]. Since Theodorsen's solution assumes a planar wake and Kutta condition at the trailing edge, the wake structure at $t/T = 0.5$ violates this condition, causing the discrepancy in the lift coefficient. Overall, Theodorsen's solution approximates the lift coefficient from the numerical computation better when the wake is planar. In the deep stall, as discussed in Section 3.5 flow over both the SD7003 and flat plate separates early in the downstroke, leading to LEV formation. The resulting flow structures and the time history of lift are similar. Again, Theodorsen's prediction gives a reasonable estimation, but is less accurate when an LEV is formed.

For the hovering flat plates, the quasi-steady lift predicted by Eq. (3–34) is compared to the numerically computed lift of hovering flat plates at $Re = 100$ for the cases highlighted in Section 3.4. Although the case setup is different (e.g., Dickinson et al. [201] used a revolving 3D wing, whereas Trizila et al. [301] used plunging flat plates), such a comparison highlights the usefulness and the limitations of applying quasi-steady models for design or control purposes. For these cases Eq. (3–34) captures the general trends well, but over-predicts the lift peak during the second part of the stroke, which is reflected in the time-averaged lift coefficients: the quasi-steady values are greater than the computed values. Most notable differences are found during the first part of the stroke where the returning flat plate interacts with the wakes shed in the previous stroke, as explained in Section 3.6.2. For the delayed rotation and both the synchronized rotation cases (Fig. 3.51a, c, d), the first lift peak that is due to wake capture is not captured at all by Eq. (3–34). In fact because the airfoil pitches down while accelerating forward, the rotational component, F_r, dominates and yields negative lift. For the advanced rotation case shown in Figure 3.51b, because the interaction with the downward wake [359] is not included in the quasi-steady model Eq. (3–34), the negative portion of the lift between $t/T =$ 0.75 and 1.00 is not captured. The biggest difference in the predicted and computed time-averaged lift coefficient is found for the cases with these kinematics. The global trend is illustrated in Figure 3.52 where the time-averaged lift coefficients for all cases considered by Trizila et al. [301] are plotted with the x-coordinate being the quasi-steady model prediction given in Eq. (3–34) and the y-coordinate the Navier-Stokes computation from Trizila et al. [301]. It is clear that Eq. (3–34) over-predicts the mean lift, and for some cases the error can be huge (e.g., the case shown in Fig. 3.51b).

Gogulapati and Friedmann [367] extended Ansari's unsteady aerodynamic model to forward flight and flexible wings and also incorporated a vorticity decay model to include the effects of viscosity. They compared their results to the numerical results shown in Section 3.4.1 and showed that the correlation indeed improves when the viscosity effects are included. They also investigated the role of LEV shedding. For the case where the 3D flow field is similar to the 2D flow field (Section 3.4), the vortical activity from the leading edge is small, as shown in Figure 3.27. For this case, Gogulapati and Friedmann's model showed that the best correlation to the Navier-Stokes computations is obtained when an attached flow is assumed: the aerodynamic model is adjusted not to shed any vorticity from the leading edge. In contrast, for the delayed rotation with high AoA and low plunging amplitude case (Section 3.4.1.1), results with LEV shedding agree the best with the Navier-Stokes computations (see Fig. 3.53). For combined pitch and plunging motions in hover it was shown that the forces predicted by the approximate model match Navier-Stokes solutions reasonably well for a Zimmerman wing with the flapping amplitudes of 10° to 15° and the pitching amplitude of 5° and 10°. The motion frequency was fixed at 10 Hz, resulting in the reduced frequency of 1 to 1.5 based on the mean wingtip speed and Re of $O(10^3)$. The differences between this approximate model and the Navier-Stokes model were attributed to 3D effects at play, such as TiV generation, LEV-TiV interaction [296], or spanwise flow. Also for flapping Zimmerman wings in forward flight, the flapping amplitude was kept at 35° and the frequency at 10 Hz. By varying the advance ratio and hence the incoming free-stream velocity, they correlated the agreement between their result and the Navier-Stokes solutions to the advance ratio, which is inversely proportional to the reduced frequency. The agreement was more favorable for the higher reduced frequencies (see Fig. 3.54). A plausible reason for this effect is that, as shown in Section 3.6.4 for high reduced frequency and high Reynolds number flows, the added mass force dominates, which is related to the acceleration of the wing. As the reduced frequency decreases, the influence of the vortices in the flow field increases, which is harder to capture by potential theory-based approximate models.

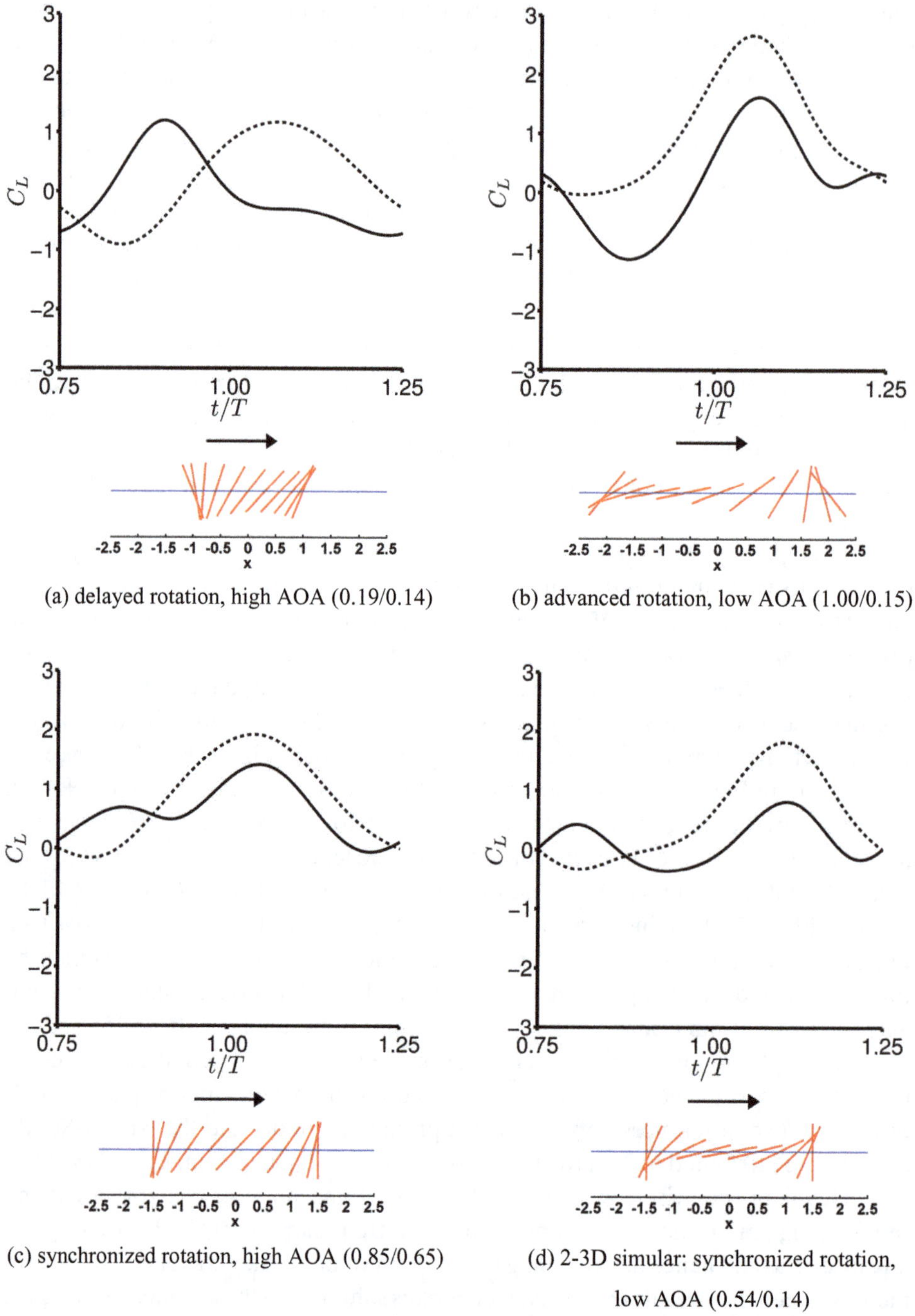

(a) delayed rotation, high AOA (0.19/0.14)

(b) advanced rotation, low AOA (1.00/0.15)

(c) synchronized rotation, high AOA (0.85/0.65)

(d) 2-3D simular: synchronized rotation, low AOA (0.54/0.14)

Figure 3.51. Lift coefficient during the forward stroke. —, Navier-Stokes equations (Re = 100); ---, quasi-steady model as shown in Eq. (3–34). The time-averaged lift coefficients from the quasi-steady model and the computation are indicated in the parentheses.

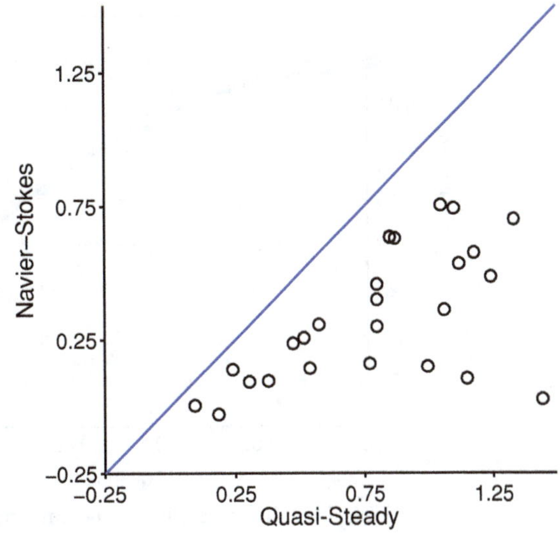

Figure 3.52. Time-averaged lift coefficients predicted by the quasi-steady model given in Eq. (3–34) on the x-axis and the Navier-Stokes computations on the y-axis for hovering flat plates at $Re = 100$ [301].

Recently, Ol and Granlund [368] considered several flapping rigid-wing experiments in water: (i) a flat plate free to pivot around its leading edge under a periodic translational motion, (ii) linear pitch-ramp motion with varying pivot points, and (iii) a combined pitch-plunge. They conjectured that the aerodynamic force responses are not quasi-steady; that is, the lift is proportional to an effective angle of attack with the proportionality constant being some coefficient (e.g., 2π), but it can be modeled by including the second and the first derivatives of the effective angle of attack. Despite the inherent limitations of linearized aerodynamic models as discussed in this section, research in this area has been popular recently because of these models' applicability to control applications or design optimizations. As is shown in Section 3.6.2, the quasi-steady model can be significantly improved if one can estimate the

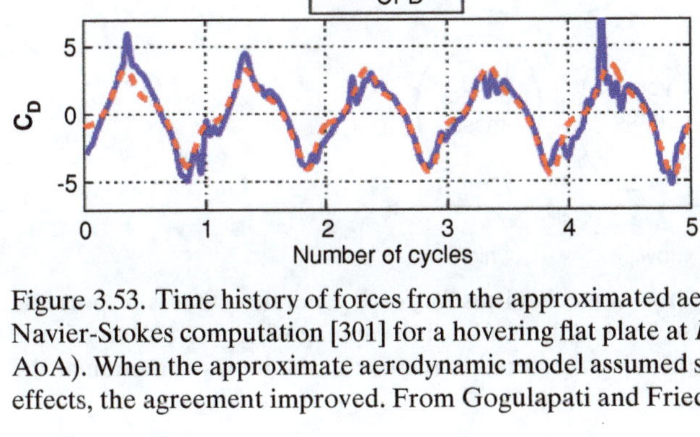

Figure 3.53. Time history of forces from the approximated aerodynamics model [367] and the Navier-Stokes computation [301] for a hovering flat plate at $Re = 100$ (delayed rotation, high AoA). When the approximate aerodynamic model assumed separated flow, including viscous effects, the agreement improved. From Gogulapati and Friedmann [367].

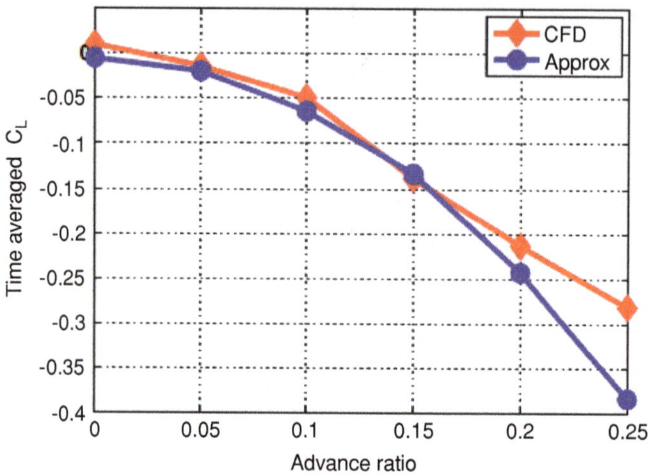

Figure 3.54. Time-averaged lift coefficients from the approximated aerodynamics model (blue circle) and the Navier-Stokes computations (red diamond) for a rigid Zimmerman wing in forward flight at $Re = 4.6 \times 10^3$, flap amplitude of 35°, and frequency of 10 Hz. The advance ratio is inversely proportional to the Strouhal number and also to the reduced frequency for fixed flap amplitude. From Gogulapati and Friedmann [367].

effective angle of attack, instead of the nominal value. For low Reynolds number, flapping wing aerodynamics, the viscous effect is significant, and the interaction between the wing and the surrounding large-scale vertical flows created by the wing motion in the previous and present strokes can noticeably affect the instantaneous actual angle of attack. Thus, in addition to the historical effects of the fluid physics, careful consideration of the quasi-steady framework is required before adopting it as a predictive tool.

3.6.4 Scaling of the Forces Acting on a Moving Body Immersed in Fluid

The aerodynamic theories described in the previous sections dealt with the linearized aerodynamic theories by assuming thin flat plates or by taking the quasi-steady

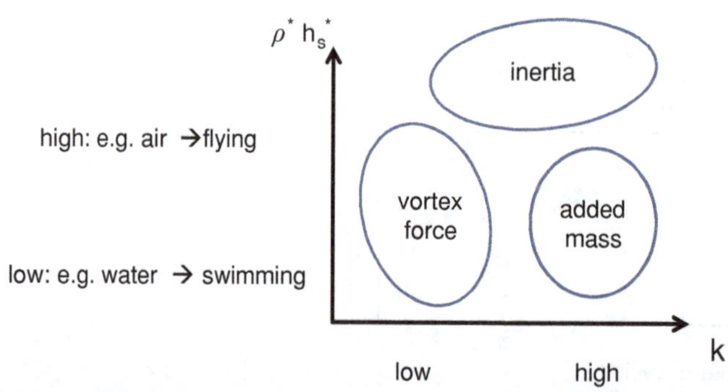

Figure 3.55. Schematic of the dominant mechanisms for force generation that is responsible for the wing deformation (see also Table 3.2) of a flexible flapping wing. Inertia force is the force acting on the wing due to the wing acceleration relative to the imposed motion at the wing root.

assumption in inviscid flows. Exact solutions for the unsteady viscous flows described by the Navier-Stokes (see Eq. (3–18)), are unknown due to their non-linearity. Recently, numerical computations (e.g., [200] [296] [369]) and detailed experimental measurements (e.g., [199] [201] [323]) have revealed intriguing unsteady flow physics related to the flapping wing aerodynamics (see, e.g., Section 3). However, some key questions, such as the relation between the aerodynamic performance of flapping wings and the non-dimensional parameters introduced in Section 3.2, are still challenging.

Based on a control volume analysis of incompressible viscous fluid around a moving body [370] [371], Kang et al. [351] normalized the resulting hydrodynamic impulse term, which relates the vortices in the flow field to the force acting on the moving body, and the acceleration-reaction term, as shown in Eq. (3–35),

$$C_F = C_{F,\text{impulse}} + C_{F,a} \sim St\left\{\frac{1}{Re\,k}O(1) + \frac{St}{k}O(1)\right\} + St\,k\,O(1) \qquad (3\text{–}35)$$

as a first-order approximation. Note that the first term is independent of the motion frequency and the second term is proportional to $2\pi f$. The force due to the hydrodynamic impulse scales with St. However, if the viscous time scale, $c_m^2\rho/\mu$, is much greater than the motion time scale, $1/(2\pi f)$, such that $Re\,k \gg 1$, then the first term in Eq. (3–35) becomes negligible. Moreover, when the plunge amplitude $h_a/c_m \sim St/k$ is small, the second term in Eq. (3–35) will make only a small contribution to the total force felt on the wing. In general, however, complex fluid dynamic mechanisms, such as the wing-wake interaction, or the wake-wake interactions would additionally affect the vorticity distribution in the flow field.

The last term of Eq. (3–35) is the acceleration-reaction term indicating the force due to the acceleration of the wing. When a body accelerates in a fluid, the fluid kinetic energy changes. The rate of work done by pressure moving the body yields the acceleration-reaction force. In an inviscid fluid this force is proportional to the acceleration (see, e.g., [372]). The constant of proportionality has the dimension of mass; hence, the name "added mass." The added mass term is usually some fraction of the fluid mass displaced by the body. Determination of the added mass, which is a tensor because it relates the acceleration vector to the force vector, is not easy in general because the local acceleration of the fluid is not necessarily the same as the acceleration of the body [45]. However, for an accelerating thin flat plate with a chord length c_m normal to itself, the force acting normal to the flat plate can be obtained as

$$\rho_f F_a = \frac{d}{dt}\left(\rho_f \frac{\pi}{4} v_i c_\text{m}^2\right), \qquad (3\text{–}36)$$

where v_i is the vertical velocity component: the added mass of a vertically accelerating thin flat plate is equal to the displaced fluid cylinder with radius $c_\text{m}/2$. The added mass due to angular rotations can be obtained similarly, which results in the non-circulatory term as in Theodorsen, Eq. (3–21), or its quasi-steady approximation for the pure plunge motion, Eq. (3–22). Note also that the combination of non-dimensional numbers appearing in the previous formulas and in Eq. (3–35) is consistent.

This non-dimensionalization process reveals that the fluid dynamic force is proportional to St; hence, with increasing St, the force acting on the wing is expected to increase. Furthermore, if the motion is highly unsteady (i.e., k is high), the force due to the motion of the body, appearing as the acceleration-reaction component, dominates over the forces due to vorticity in the flow field.

A parametrization of special interest for the flapping wing community is the dependence of the force on the flapping motion frequency, ω. The current scaling shows that for forward flight with $U_{\text{ref}} = U_\infty$ the acceleration-reaction force has the highest order of frequency as $\sim (2\pi f)^2$. The resulting dimensional force is then proportional to the square of the motion frequency. Similarly, for hovering motions the current scaling shows that the non-dimensional force is independent of the motion frequency since the Strouhal number is a constant and the reduced frequency is only a function of flapping (plunging) amplitude. However, since $U_{\text{ref}} \sim (2\pi f)^2$ the resulting dimensional force is also proportional to the square of the motion frequency. Similar observations were reported by Gogulapati and Friedmann [367].

At a high Reynolds number, high reduced frequency regime, Visbal, Gordnier, and Galbraith [373] considered a high-frequency small amplitude plunging motion at $Re = O(10^4)$ over a 3D SD7003 wing ($\alpha_0 = 4$ deg, $k = 3.93$, $St = 0.06$, $Re = 1 \times 10^4$ and 4×10^4). They used the implicit LES approach to solve for the flow structures, including the laminar-to-turbulence transition and the forces on the wing. The flow field exhibits formation of dynamic stall features such as LEVs, breakdown due to spanwise instabilities, and transitional features; however the forces on the wing could still be well predicted by the Theodorsen [374] formula for lift. The time history of lift was "independent of Reynolds number and of the 3D transitional aspects of the flow field" [373]. Visbal and co-workers explained that the lift is dominated by the acceleration of the airfoil, which is proportional to the square of the motion frequency. This observation is also consistent with the fact that the scaling of the hydrodynamic impulse term is small compared to the acceleration-reaction term in Eq. (3–35) for the given non-dimensional parameters.

In contrast, at lower Reynolds numbers, Trizila et al. [301] showed at $Re = 100$ and k in the range of 0.25–0.5 that the formation and interaction of leading-edge and trailing-edge vortices with the airfoil and previous shed wake substantially affect the lift and power generation for hover and forward flight. Furthermore, three-dimensionality effects play a significant role. For instance, for a delayed rotation kinematics ($k = 0.5$, low angle of attack, $Re = 100$), the tip vortex generated at the tip of the $AR = 4$ flat plate would interact with the LEV, thereby enhancing lift compared to its two-dimensional counterpart; this is in contrast to the classical steady-state thin wing theory [301] [296] that predicts the formation of TiVs as a lift-reducing flow feature. This complex interplay between the kinematics, the wing-wake and wake-wake interactions, and the fluid dynamic forces on the wing at the given range of non-dimensional parameters is also consistent with the scaling analysis described in this section as summarized in Table 3.2. The scaling shows that, for low reduced frequency motions or low Reynolds number flows, the hydrodynamic impulse term, which indicates the interaction between the vortices and the wing becomes important. In contrast, when the reduced frequency increases, the acceleration-reaction term dominates over the hydrodynamic impulse term. Both components are proportional to the Strouhal number. An interesting consequence that needs to be investigated

Table 3.2. *Summary of the force scaling*

Force	Scaling	Note
Hydrodynamic	$St/(Re\,k) \sim (2\pi f)^0$	Viscous term
impulse	$St^2/k \sim (2\pi f)^1$	Vortex force due to motion of moving body
Added mass	$St\,k \sim (2\pi f)^2$	

more is that in the hovering flight condition where both the Strouhal number and reduced frequencies are independent of motion frequency, the normalized force will be independent of frequency for high Reynolds number flows.

3.6.5 Flapping Wing Model versus Rotating Wing Model

As discussed in the previous section, the helicopter blade model has been used to help explain the flapping wing aerodynamics; however, spanwise axial flows are generally considered to have a minor influence on the helicopter aerodynamics [289] [375]. In particular, helicopter blades operate at a substantially higher Reynolds number and lower AoA than flapping wings. The much larger AR of a blade also makes the LEV harder to anchor. These are key differences between helicopter blades and typical biological wings.

Usherwood and Ellington [320] investigated the aerodynamic performance associated with "propeller-like" rotation of the hawkmoth wing model ($Re = O(10^3)$). They reported that revolving models produced high lift and drag forces because of the presence of the LEV. Knowles et al. [376] performed computational studies involving with translating 2D and rotating 3D flat plate models at $Re = 5.0 \times 10^2$. Their results showed, that for 2D flows, the LEV is unstable, but the rotating 3D flat plate model at high AoA produces a conical LEV, as observed in the results presented earlier and by others [320] [377]. Moreover, they noted that if the Reynolds number increases above a critical value, a Kelvin–Helmholtz instability [378] occurs in the LEV sheet, resulting in the sheet breaking down on outboard sections of the wing.

Lentink and Dickinson [379] revisited existing hypotheses regarding stabilizing the LEV. Using a fruit fly wing model, they systematically investigated the effects of propeller-like motion on aerodynamic performance and LEV stability at $Re = 1.1 \times 10^2$–1.4×10^4 based on theoretical and experimental approaches. They stated that the LEV is stabilized by the "quasi-steady" centripetal and Coriolis accelerations that are present at low Rossby numbers (i.e., a half of the AR) and result from the propeller-like sweep of the wing. In addition, they suggested that the force augmentation through a stably attached LEV could represent a convergent solution for the generation of high fluid forces over a range of Reynolds numbers.

Jones and Babinsky [380] experimentally examined unsteady lift generation on rotating wings with AoA of 5° and 15° at a Reynolds number of 6.0×10^4. A transient high list peak, approximately 1.5 times the quasi-steady value, occurred in the first chord length of travel, and it was caused by the formation of a strong LEV. They showed that wing kinematics has only a small effect on the aerodynamic force produced by the waving wing. In the early stages of the wing stroke, the velocity

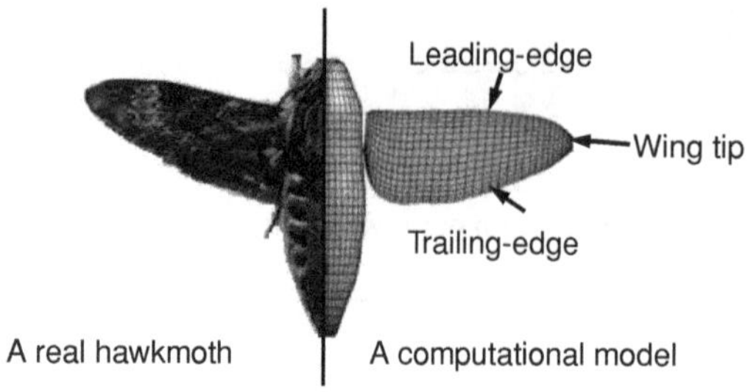

Figure 3.56. A morphological model of a hawkmoth, *Agrius convolvuli*, with a computational model superimposed on the right half. The hawkmoth has a body length of 5.0 cm, a wing length of 5.05 cm (mean wing chord length $c_m = 1.83$ cm), and an aspect ratio of 5.52. The image is from Aono, Shyy, and Liu [384].

profiles with low accelerations affect the timing and the magnitude of the lift peak, but at higher accelerations, the velocity profile is insignificant.

From the viewpoint of unsteady aerodynamics, the LEV as a lift-enhancement mechanism at a higher *Re* range ($O(10^5$–$10^6)$) may be questionable because a dynamic stall vortex on an oscillating airfoil is often found to break away and to convect elsewhere as soon as the airfoil translates [277].

3.7 Modeling of Biological Flyers in a Rigid-Wing Framework

There are a number of computational studies based on realistic wing configurations of natural flyers, such as a hornet [381], bumblebee [382] [383], hawkmoth [200] [247] [384], honeybee [225], drone fly [385], hover fly [386], fruit fly [387], and thrips [225]. In following, we present selected examples to highlight some of the features discussed in previous sections, as well as to gain insight into how flow structures are associated with the specific characteristics of wing motion.

3.7.1 Hovering Hawkmoth

Figures 3.56 and 3.57 show the morphological and wing kinematics models of a realistic hawkmoth model. Computations were performed by Liu and co-workers using "a biology-inspired dynamic flight simulator" [387] [388].

3.7.1.1 Vortical Structures and Lift Generation

Iso-vorticity-magnitude surfaces around a hovering hawkmoth at selected instants during a flapping cycle are shown in Figure 3.58. The contour color on the iso-vorticity-magnitude surfaces in Figure 3.58 indicates the magnitude of the normalized helicity density. The corresponding time histories of the vertical force on the wings and body are plotted in Figure 3.60.

FLOW STRUCTURES DURING DOWNSTROKE. In the first half of the downstroke as indicated at point (a) in Figure 3.57, a horseshoe-shaped vortex is generated by the initial wing motion of the downstroke (see Fig. 3.58A-1. Poelma et al. [270] showed a similar 3D flow structure around an impulsively started dynamically scaled flapping wing using PIV (see Fig. 3.59). The horseshoe-shaped vortex is composed of three vortices – an LEV, a TEV, and a TiV – and it grows in size as the translational and angular velocity of the wing increases. These vortices produce a low-pressure region in their core and on the upper surfaces of the wing (Figure 3.61, section 1) when they are attached. Lift forces show a peak at the corresponding time instant (point (b)

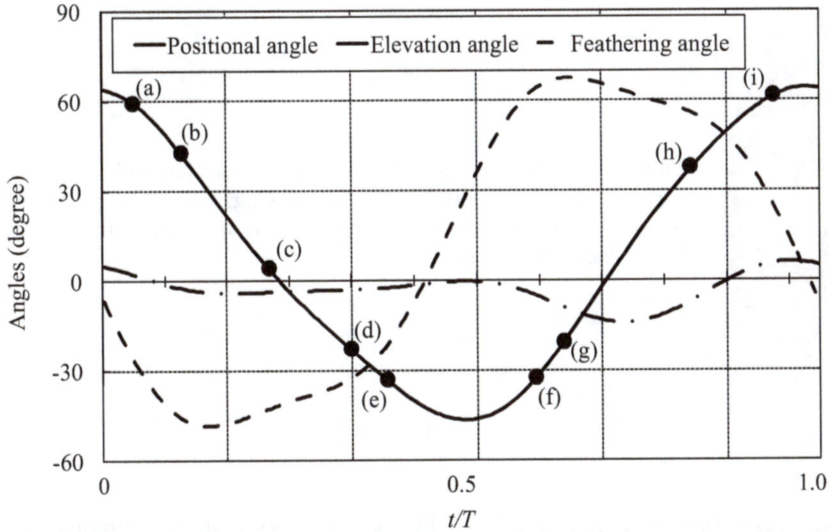

Figure 3.57. Time histories of positional angle ϕ, feathering angle α, and elevation angle θ of the hawkmoth wing over one complete flapping cycle. Points marked: (a) late pronation, (b) early downstroke, (c) middle downstroke, (d) late downstroke, (e) early supination, (f) late supination, (g) early upstroke, (h) late upstroke, and (i) early pronation. T denotes dimensionless period of one flapping cycle. From Aono, Shyy, and Liu [384].

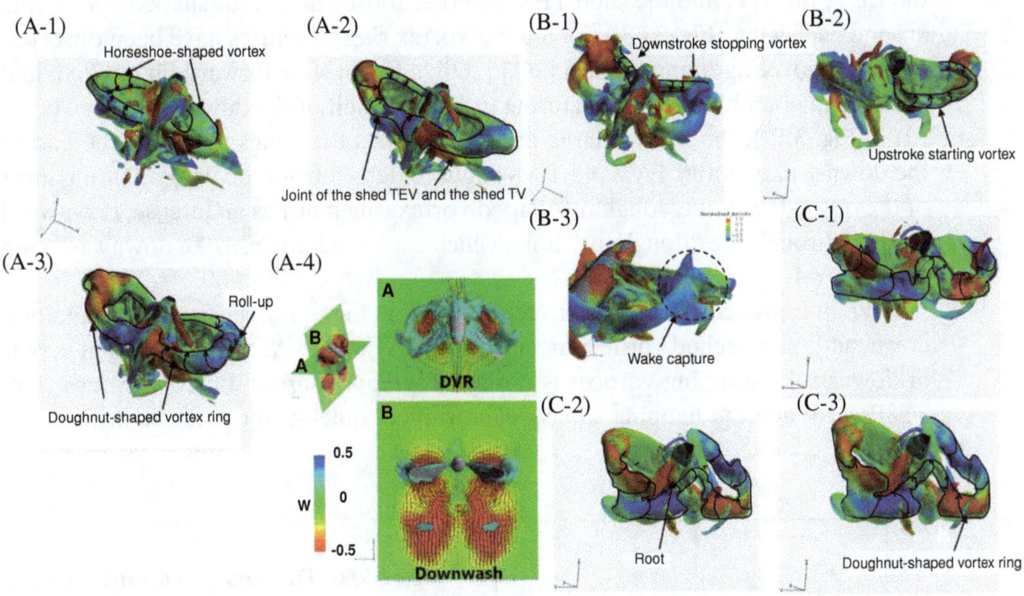

Figure 3.58. Visualization of flow fields around a hovering hawkmoth. Iso-vorticity-magnitude surfaces around a hovering hawkmoth during (a) the downstroke, (b) the supination, and (c) the upstroke, respectively. The color of iso-vorticity-magnitude surfaces indicates the normalized helicity density, which is defined as the projection of a fluid's spin vector in the direction of its momentum vector, being positive (red) if these two vectors point in the same direction and negative (blue) if they point in the opposite direction.

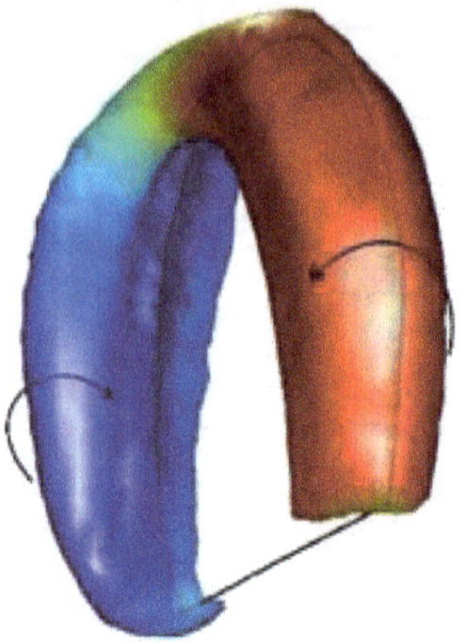

Figure 3.59. 3D horseshoe-shaped vortex indicated with an iso-Q-surface around an impulsively started wing. The contours on the wing represent the vorticity with its direction indicated by the arrows. From Poelma et al. [270].

in Fig. 3.57). As expected, coherent LEVs, TEVs, and TiVs together enhance the lift generation in hovering flight of the hawkmoth. During the mid-downstroke (see point (c) in Fig. 3.57), the TEVs shed mostly from the wings while on the body they stay attached. Moreover, the shed TEVs stay connected to the TiVs (Fig. 3.58A-2). Overall, the LEVs produce the largest and strongest area of low pressure on the wing surface (Fig. 3.61, section 2). Shortly afterward, the LEVs begin to break down at a location approximately 70–80 percent the span of the wing length. At the same time, the LEV, the TiV, and the shed TEV together form a doughnut-shaped vortex ring around each wing (Fig. 3.58A-3). Similar vortex ring structures have been observed around a hovering hummingbird [261] [266], a bat in slow forward flight [268], and a free-flying bumble bee [259]. During the second half of the downstroke (see point (d) in Fig. 3.57), the TiVs enlarge and weaken. As the wings approach the end of the downstroke, both LEVs and TiVs begin to detach from the wings. During most of the downstroke, the doughnut-shaped vortex ring pair has an intense, downward jet flow through the "doughnut" hole, which forms the downstroke downwash (see Fig. 3.58A-4).

While approaching supination (see point (e) in Fig. 3.57), the flapping wing slows down, and the attached vortices are shed from the wings. At this time instant, a pair of downstroke stopping vortices is observed wrapping around the two wings (Fig. 3.58B-1). When the flapping wings begin to pitch quickly about the spanwise axis,

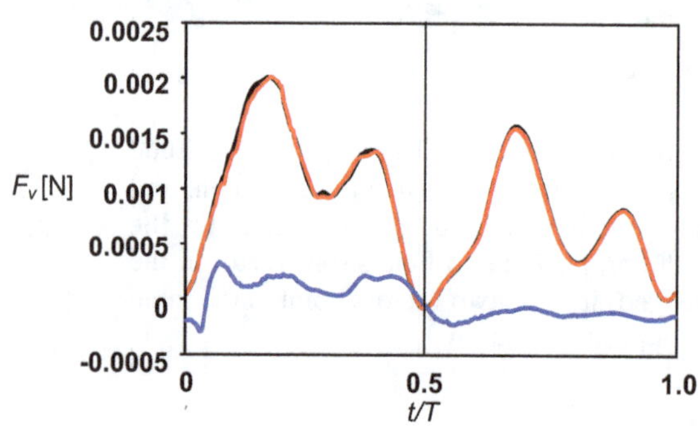

Figure 3.60. Time courses of vertical force of a hovering hawkmoth model over a flapping cycle. $t/T = 0$–0.5 corresponds to the downstroke. Black, red, and blue lines represent the aerodynamic forces acting on the right wing, left wing, and body, respectively. The weight of a hawkmoth is around 15.7×10^3 [N] [384].

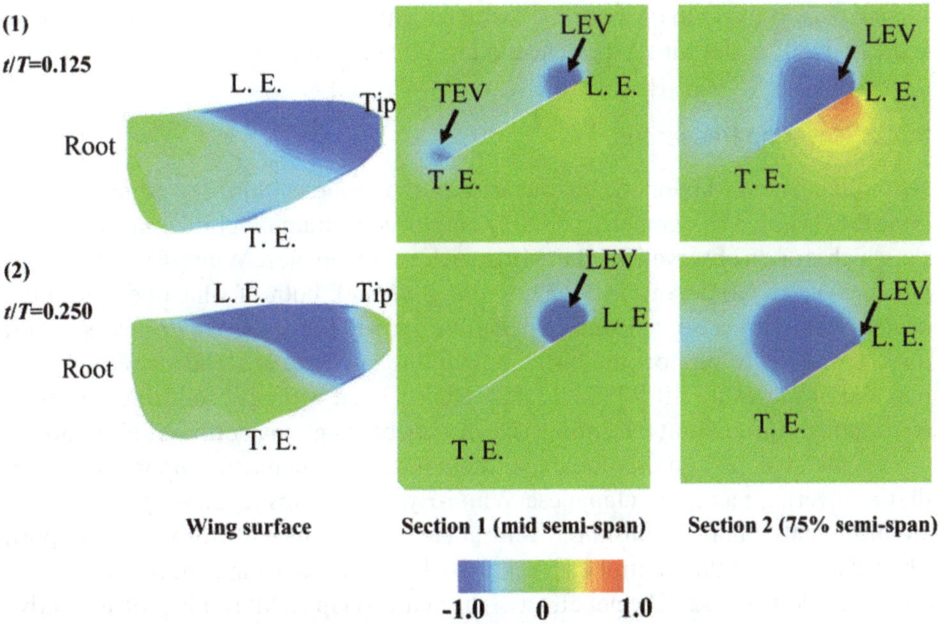

Figure 3.61. Pressure distributions at selected time instants corresponding to Figure 3.57 (b) and (c) during the downstroke. Left: Upper wing surface. Middle: Mid semi-span cross-section. Right: 75% semi-span cross-section from the wing root. Note that L.E. and T.E. indicate leading edge and trailing edge, respectively. Note that the pressure coefficients in the LEV are negative in all sub-figures.

a pair of upstroke starting vortices is detected around the wingtip and the trailing edge (Fig. 3.58B-2).

FLOW STRUCTURES DURING UPSTROKE. After supination (point (f) in Fig. 3.57), TEVs and TiVs associated with the beginning of the upstroke are generated when the flapping wings rapidly accelerate. The downstroke wakes of the circular vortex rings are subsequently captured (Fig. 3.58B-3), but only a minor impact on lift is seen (Fig. 3.60). As discussed in Section 3.3, the role of the unsteady flow structures on lift enhancement needs to be examined in specific context. As the upstroke starts, (point (g) in Fig. 3.57), the TEVs and TiVs are shed from the two wings (Fig. 3.58C-1). Together with the TEVs, the LEVs and TiVs form a horseshoe-shaped vortex pair wrapping each wing (Fig. 3.58C-2). Similar to what is seen during the downstroke, the horseshoe-shaped vortex grows and evolves into a doughnut-shaped vortex ring for each wing. It should be pointed out that, due to asymmetric variation of the AoA between the upstroke and downstroke, the LEVs generated in the upstrokes are smaller than those in the downstroke. Late in the upstroke (point (h) in Fig. 3.57), the doughnut-shaped vortex rings elongate and deform while maintaining their ring-like shape (Fig. 3.58C-3).

As for the aerodynamic force generation, two peaks in the lift force are predicted during each stroke for a hovering hawkmoth (see Fig. 3.60). Considering the correlation between the aerodynamic forces and the key unsteady flow features associated with flapping wings discussed in Section 3.3, the delayed stall of the LEV and contributions from the TEV and TiV are responsible for the first lift peak. The second peak is likely to be associated with a contribution from the rapid increase in vorticity [289] as the wing experiences a fast pitching motion.

3.7.2 Hovering Passerine

According to their kinematic characteristics, the modes of bird-hovering flight are typically classified as symmetric, notably employed by hummingbirds, and asymmetric, which is often observed in other birds, such as the Japanese White-Eye (*Zosterops japonicus*) and Gouldian Finch (*Erythrura gouldiae*), both of which belong to the order *Passeriforme*. For a hovering passerine, only the downstroke produces the lift force necessary to support the bird's weight; the upstroke is aerodynamically inactive and produces no lift [4] [389] [390]. Hence, even without producing lift forces continuously throughout the entire cycle of wing beating, a passerine is able to hover.

Chang et al. [391] presented experimental measurements to support the notion that a hovering passerine (Japanese White-Eye, *Zosterops japonicus*) can employ an unconventional mechanism of "ventral clap" to produce lift for weight support. They claimed that this ventral clap can first abate and then augment lift production during the downstroke. The net effect of the ventral clap on lift production is positive because the extent of lift augmentation is greater than the extent of lift abatement, as shown in Figure 3.62, where the PIV data illustrate the flow structures associated with the various phase of the flapping motion. As discussed in this section and by Trizila et al. [301], LEV and TEV flows as well as the downward jet flow are observable, contributing to the complicated balancing and net generation of aerodynamic forces.

In Chang et al.'s study [391], two methods based on wake topology were applied to evaluate the locomotive forces. The first is primarily associated with the Kutta–Joukowski lift theorem [392]–[394]. For a bird executing quasi-steady level flight, the total lift acting on its wings must be equal to its weight. The second method to evaluate the locomotive force is based on a vortex-ring model that obtains a time-averaged force from a calculation of the momentum change (i.e. impulse), divided by the generation period of a vortex ring shed by a flapping or beating appendage of a locomoting animal [342] [395] [396]. Although these methods are simple and convenient, as discussed in Section 3.6, such approaches encounter difficulties in offering an accurate account and they are not based on first principles.

3.7.3 Reynolds Number Effects on the LEV and Spanwise Flow: Hawkmoth, Honeybee, Fruit Fly, and Thrips in Hovering Flight

As discussed previously, the lift enhancement due to the delayed stall of the LEV is important in flapping wing flight [199] [245]. The formation of the LEV depends on the wing kinematics, the details of wing geometry, and the Reynolds number [397]. To examine the *Re* effect on LEV structures and spanwise flow for insect-like wing body configurations with appropriate kinematic motions, Shyy and Liu [397] and Liu and Aono [225] investigated the flapping wing fluid physics associated with the hawkmoth ($Re = 6.3 \times 10^3$, $k = 0.30$), honeybee ($Re = 1.1 \times 10^3$, $k = 0.24$), fruit fly ($Re = 1.3 \times 10^2$, $k = 0.21$), and thrips ($Re = 1.2 \times 10^1$, $k = 0.25$) in hover. Different representative kinematic parameters (flapping amplitude, flapping frequency, and type of prescribed actuation) and

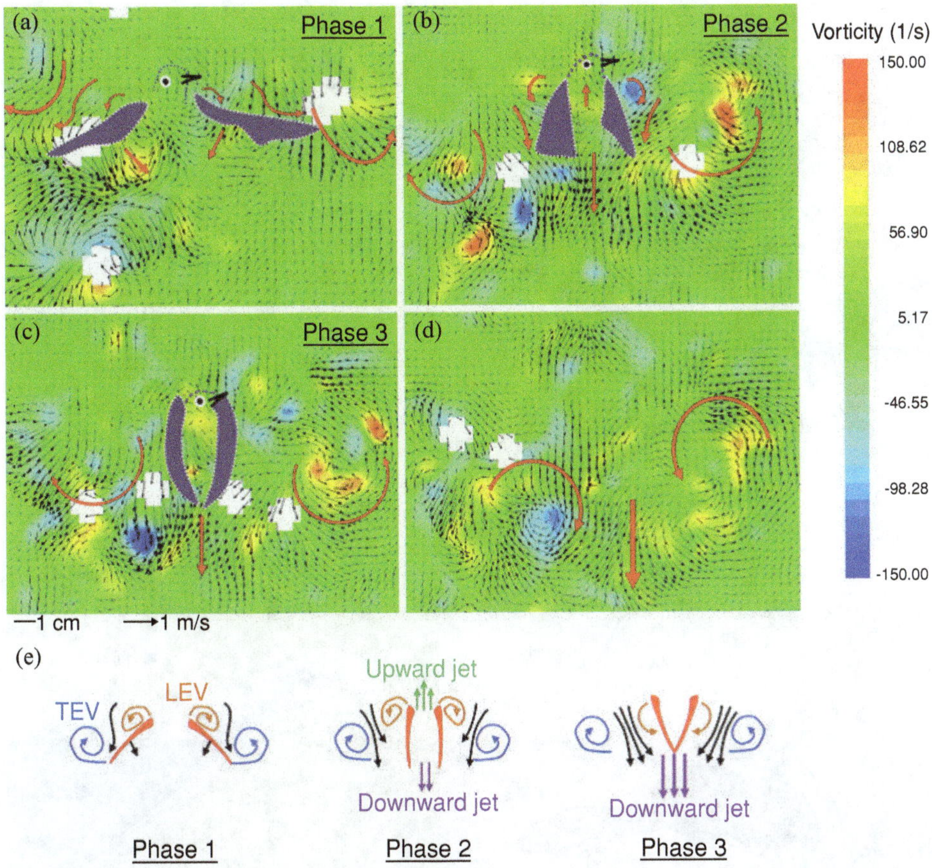

Figure 3.62. Hovering Japanese White-eye wake flow fields on a frontal plane. Color contours represent the vorticity distribution of the flow fields. (a–c) Near-wake flow fields pertaining to kinematic phases of the ventral clap; purple patches mark the position of the two wings, and purple dotted lines indicate an outline of the bird. (d) The far-wake flow field beneath a hovering Japanese White-eye after phase 3; thick red arrows depict the trends of the flow motion. (a–d) Thick red arrows show the trends of flow motions. (e) Schematic sketches summarizing the wake flow structures for the three kinematic phases. Black arrows indicate fluid jets; orange and blue spiral arrows represent, respectively, the LEV and TEV. Purple arrows signify the downward jet generated by the downstroking wings executing the ventral clap. From Chang et al. [391].

dimensionless numbers (Reynolds number, reduced frequency) are considered for each insect model [225]. The Reynolds number and reduced frequency are calculated based on mean chord and tip velocity. The morphological and kinematical model of a hawkmoth was already shown in Figure 3.56 and Figure 3.57. For other insect models, similar information and computational models can be obtained in Liu and Aono [225].

Figure 3.63 shows the velocity vector field on an end-view plane at 60 percent semi-span for these four insects. The LEV structure and the spanwise flow in the hawkmoth and the fruit fly cases (see Fig. 3.63A and C) are in good qualitative agreement with the corresponding experimental results reported in Birch et al. [283]. For the thrips ($Re = 1.2 \times 10^1$, $k = 0.25$), the LEV forms upstream of the leading edge, and spanwise flow is weakest among all cases. For the fruit fly ($Re = 1.3 \times 10^2$, $k = 0.21$), the LEV structure is smaller than that of the hawkmoth

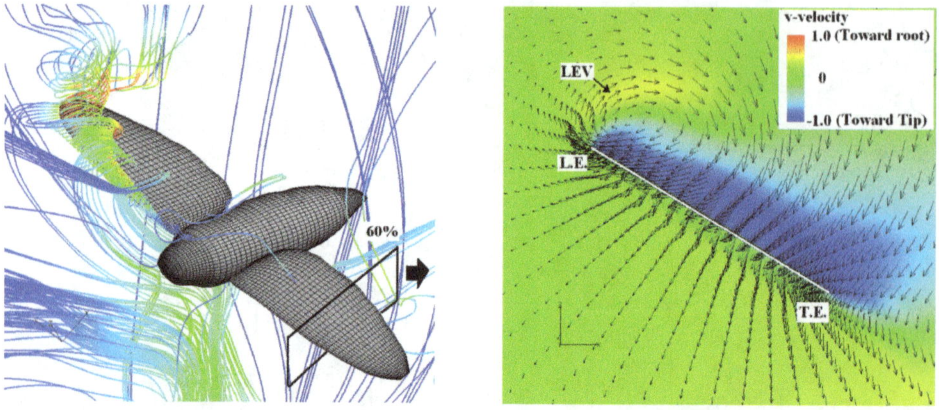

(A-1) Streamlines (A-2) Spanwise flow

(A) Hawkmoth model (*Re*= 6.3x10³, *k*=0.30)

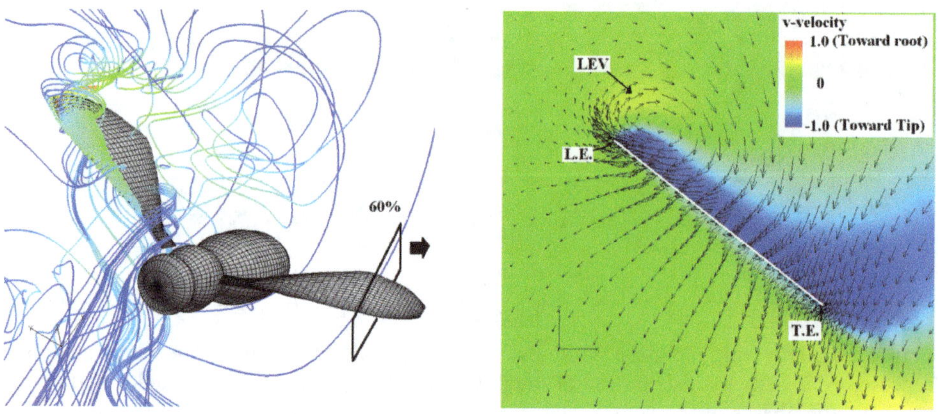

(B-1) Streamlines (B-1) Spanwise flow

(B) Honeybee model (*Re*= 1.1x10³, *k*=0.24)

Figure 3.63. Comparison of near-field flow fields between a hawkmoth, a honeybee, a fruit fly, and a thrips around the middle of the downstroke. Wing-body models of (a) a hawkmoth model, (b) a honeybee model, (c) a fruit fly model, and (d) a thrips model, respectively.

and the honeybee. The fruit fly LEV is tube-like and ordered, and spanwise flow is observed around the upper region of the trailing edge. The hawkmoth ($Re = 6.3 \times 10^3$, $k = 0.30$) and honeybee ($Re = 1.1 \times 10^3$, $k = 0.24$) cases yield much more pronounced spanwise flow inside the LEV and upper surface of the wing, which together with the LEV forms a helical flow structure near the leading edge (see Fig. 3.63A-1 and B-1). Figure 3.64 shows the spanwise pressure gradient contours on the wing of four typical insects at the middle of the downstroke. Compared to the hawkmoth and the honeybee, even though the wing kinematics and the wing-body geometries are different, fruit flies, at a *Re* of $1.0 \times 10^2 \sim 2.5 \times 10^2$, cannot create as steep pressure gradients at the vortex core; nevertheless, they seem to be able to maintain a stable LEV during most of the down- and upstroke. Although the LEVs on both the wings of the hawkmoth and honeybee experience a breakdown near the middle of the downstroke, the LEV on the fruit fly's wing remains attached during

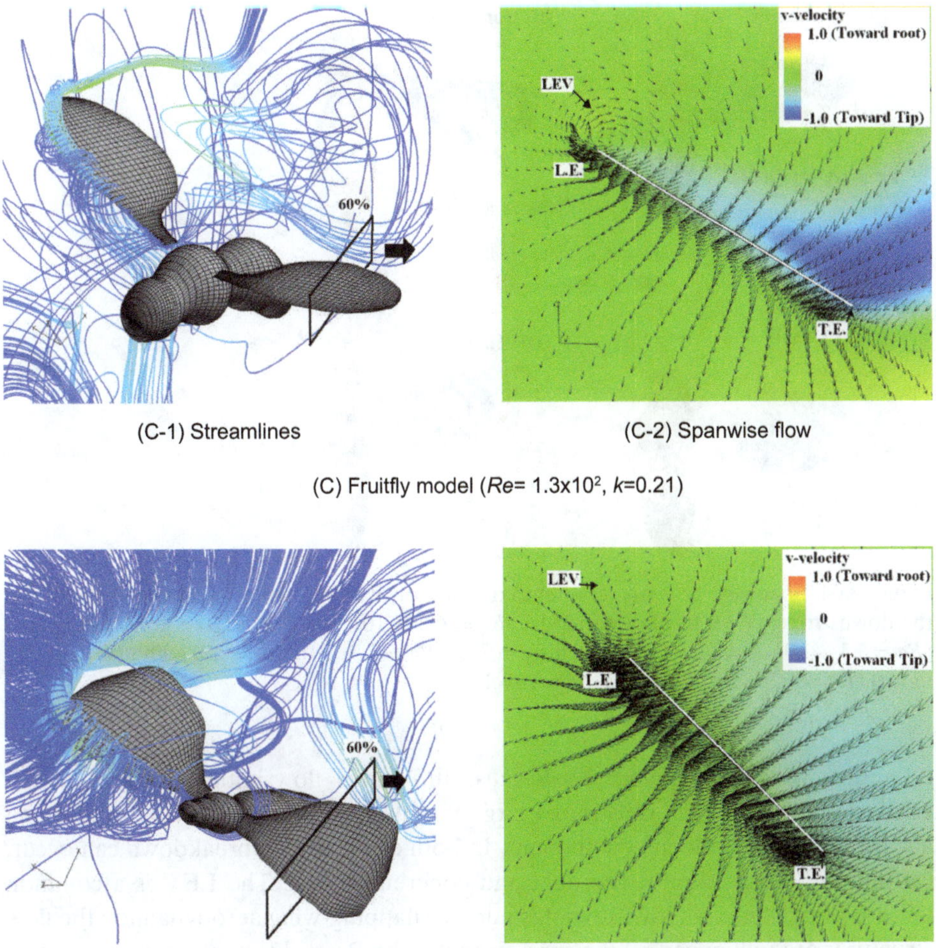

(C-1) Streamlines (C-2) Spanwise flow

(C) Fruitfly model (Re= 1.3x10^2, k=0.21)

(D-1) Streamlines (D-2) Spanwise flow

(D) Thrips model (Re= 1.2x10^1, k=0.25)

Figure 3.63 (*continued*)

the entire stroke, eventually breaking down during the subsequent supination or pronation [387].

In reality the thrips wings are extremely small – of the order of 1 mm – and are composed of a comblike planform with solidity ratios of 0.2 and less [398]. Consequently, the Reynolds numbers are so low that the viscous effects dominate and the unsteady viscous flow around their wings can be approximated as Stokes flow. Using slender-body theories the flow physics around an array of oscillating bodies was studied by Weihs and Barta [398] and Barta [399]. They showed that the comblike wings of thrips are able to produce forces similar to those of solid wings, while being able to save weight.

3.8 Concluding Remarks

This chapter highlighted the rigid flapping-wing aerodynamics at the low Reynolds number range between $O(10^2)$ and $O(10^4)$. Similar to a conventional delta wing,

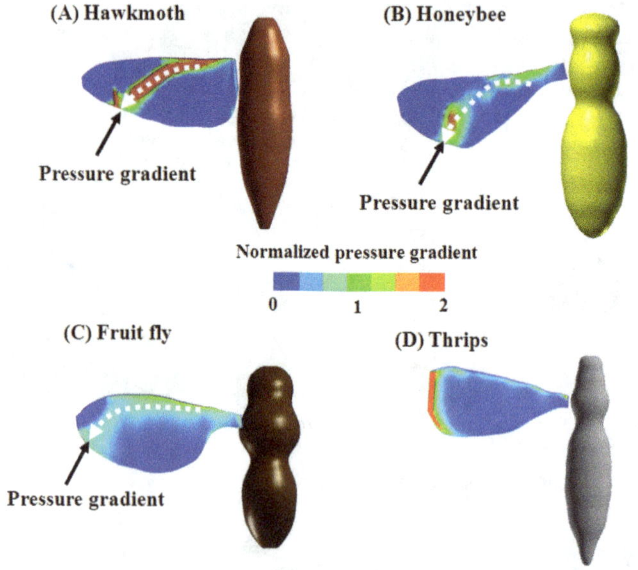

Figure 3.64. Spanwise pressure gradient contours on flapping wings around the middle of the downstroke: (A) a hawkmoth model ($Re = 6.3 \times 10^3$, $k = 0.30$); (B) a honeybee model ($Re = 1.1 \times 10^3$, $k = 0.24$); (C) a fruit fly model ($Re = 1.3 \times 10^2$, $k = 0.21$), and (D) a thrips model ($Re = 1.2 \times 10^1$, $k = 0.25$), respectively.

a flapping wing can enhance lift because the vortex flow initiates at the leading edge of the wing, which rolls into a large vortex over the leeward side, containing a substantial axial velocity component. In both cases, vortex breakdown can occur, causing the destruction of the tight and coherent vortex. The LEV is a common feature associated with low Reynolds number flapping-wing aerodynamics; the flow structures are influenced by the swirl strength, the Reynolds number, and the rotational The effectiveness of LEVs in promoting lift is correlated with a flyer's size. It seems that, at a fixed AoA, if the swirl is strengthened, then vortex breakdown occurs at lower Reynolds numbers. In contrast, a weaker swirling flow tends to break down at a higher Reynolds number. Hence, since the fruit fly, which operates at a lower Reynolds number range, exhibits a weaker LEV, it tends to better maintain the vortex structure than a hawkmoth and a honeybee, which create a stronger LEV. Furthermore TiVs associated with fixed finite wings are traditionally seen as phenomena that decrease lift and induce drag [297]. However, for flapping wings, the wing and vortex interactions, and consequently the aerodynamic outcome, can be far more complex. For a low-aspect-ratio hovering wing, with delayed rotation, TiVs can increase lift both by creating a low-pressure region near the wingtip and by anchoring the LEV to delay or even prevent it from shedding. Furthermore, for certain flapping kinematics such as synchronized rotation with modest AoA, the LEV remains attached along the spanwise direction and the tip effects are not prominent; in such situations, the aerodynamics is little affected by the AR of a wing. Appropriate combinations of advanced rotation and dynamic stall associated with large AoAs can produce more favorable lift. The combined effects of TiVs, LEV, and jet can be manipulated by the choice of kinematics to make a 3D wing aerodynamically better or worse than an infinitely long wing. Furthermore, the delayed stall of the LEV and interaction of TiVs and LEVs are significantly affected by the free-stream.

However, wing-wake interactions and TiVs can lead to a decrease in the aerodynamic performance when the wing orientation and the surrounding vortical flow structure are not well coordinated. Hence, the effectiveness of the unsteady flow mechanisms is strongly linked to the flapping wing kinematics, Reynolds number, and the free-stream environment.

This chapter also assessed the roles of *Re*, *St*, *k*, as well as kinematics at the Reynolds number of $O(10^4)$, for both flat plate and SD7003 airfoil in forward flight with a shallow stall (pitch/plunge) and a deep stall motion (plunge). Massive leading-edge separation is observed at the sharp leading edge of the flat plate; its geometric effect is seen to dominate over other viscosity effects and the *Re* dependence is limited. For the blunter SD7003 airfoil, the flow is mostly attached for the shallow stall motion at $Re = 6 \times 10^4$, whereas it experiences massive separation for the deep stall motion. The deep stall kinematics produces a more aggressive time history of effective AoA with a higher maximum. The flow is characterized by a stronger leading-edge separation with earlier generation of an LEV and formation of a secondary vortex near the surface of the flat plate. The maximum lift coefficients of the flat plate are substantially different from those for the SD7003 airfoil. Furthermore, there is a noticeable phase lag in the deep stall SD7003 airfoil and flat plate in the time history of the lift coefficient. The difference in curvature of the leading-edge shape causes the lag in the generation and evolution of the LEV. Overall, for the shallow stall motion, the force acting on the SD7003 airfoil has smaller lift and drag compared to the force on the flat plate due to the formation of an LEV in the flat plate case. In contrast, the shape effects become less significant for the deep stall motion with higher maximum effective AoA: the flow separates over the SD7003 airfoil during the latter half of the downstroke, and the resulting force is similar to that of the flat plate.

The studies based on the kinematics and geometries of biological flyers offer direct evidence of unsteady aerodynamic mechanisms. Compared to flow structures generated by simple flapping motion, those around insect flapping wings have unique features, such as ring-like vortex generation per stroke with a strong downward jet.

We also highlighted selected linearized aerodynamic models used for control applications or design optimizations because their computational cost is significantly smaller than that of Navier-Stokes equation solutions; however, their applicability is not comprehensively understood in the flapping wing regime. Whereas quasi-steady models tend to over-predict the lift generation and may not able to capture wing-wake interactions with varied Reynolds numbers, they can still provide reasonable time-averaged lift estimations.

The chapter also presented several studies focusing on biological flyers, aimed at offering case studies in nature, as well as snapshots of interesting flow characteristics discussed in "canonical" investigations based on simplified problem definitions.

The topics discussed and highlighted in this chapter are essential to understand flapping wing aerodynamics with rigid wings. In Chapter 4 we discuss the effects of wing flexibility that increases the degree of freedom and showcase the challenges and intriguing features of multidisciplinary physics.

4 Flexible Wing Aerodynamics

4.1 General Background of Flexible Wing Flyers

As discussed in Section 1.1, it is well known that flying animals typically have flexible wings to adapt to the flow environment. Birds have different layers of feathers, all of which are flexible and often connected to each other. Hence, they can adjust the wing planform for a particular flight mode. Furthermore, bats have more than two dozen independently controlled joints in the wing [400] and highly deforming bones [401] that enable them to fly at either positive or negative AoAs, to dynamically change wing camber, and to create complex 3D wing topology to achieve extraordinary flight performance. Bats have compliant thin-membrane surfaces, and their flight is characterized by highly unsteady and 3D wing motions (see Fig. 4.1).

As discussed in Chapter 1, birds and bats can also change their wingspan (flexing their wings) to decrease the wing area, increase the forward velocity, or reduce drag during an upstroke. In fast forward flight, birds and bats reduce their wing area slightly during the upstroke relative to the downstroke. At intermediate flight speeds, the flexion during the upstroke becomes more pronounced. However, bats and birds flex their wings in different manners. The wing surface area of a bird's wing consists mostly of feathers, which can slide over each other as the wing is flexed and still maintain a smooth surface. Bat wings, in contrast, are mostly a thin membrane supported by the arm bones and the enormously elongated finger bones. Given the stretchiness of the wing membrane, bats can flex their wings a little, reducing the span by about 20 percent, but they cannot flex their wings too much or the wing membrane will go slack. Slack membranes are inefficient, because drag goes up and the trailing edges are prone to flutter, making them harder for fast flight [22].

While making bending or twisting movements, biological flyers have natural capabilities of adjusting the camber of their wings in accordance with what the flow environment dictates, such as a wind gust, object avoidance, and target tracking. Bats are known for being able to change the shape of the wing passively, depending on the free-stream conditions. As shown in Figure 4.1 bats can change their wing shapes during each flapping cycle. In human-made devices, sails and parachutes operate according to similar concepts. This passive control of the wing surface can prevent flow separation and enhance lift-to-drag ratio. Birds adjust their wings based on different strategies. For example, some species have coverts that act as *self-activated*

Figure 4.1. A bat (*Cynopterus brachyotis*) in flight. (a) Beginning of downstroke, head forward, tail backward: the whole body is stretched and lined up in a straight line. (b) Middle of downstroke: the wing is highly cambered. (c) End of downstroke (also beginning of upstroke): the wing is still cambered. A large part of the wing is in front of the head and the wing is going to be withdrawn to its body. (d) Middle of upstroke: the wing is folded towards the body. From Tian et al. [231].

flaps to prevent flow separation. These features offer shape adaptation and help adjust the aerodynamic control surfaces; they can be especially helpful during landing and in an unsteady environment. In Figure 4.2 the coverts have popped up on a skua, and the flexible structure of the feathers is clearly shown. Figure 4.3 highlights three flyers in different flight modes: an egret fishing, a kingfisher trying to land, and a Black Kite during cruising.

Insect wings deform to a great extent during flight, and their wing properties are generally anisotropic because of the membrane-batten structures; the spanwise bending stiffness is about 1 to 2 orders of magnitude larger than the chordwise bending stiffness in a majority of insect species. Experimental efforts reported a scaling relationship between wing flexural stiffness and wing length scale [402] [403]. In Combes and Daniel's analysis each flexural stiffness is estimated based on a 1D beam model. The results showed that spanwise flexural stiffness scales with the third power of the wing chord, whereas the chordwise stiffness scales with the second power of the wing chord [402]. Figure 4.4 shows wings of cardinals, hummingbirds, bats, dragonflies, cicadas, and wasps, all exhibiting similar overall structural characteristics. They exhibit substantial variations in *ARs* and shapes, but share a common feature of a reinforced leading edge. A dragonfly wing has more local variations in its structural composition and is more corrugated than the wing of a cicada or a wasp. It was shown in the literature that wing corrugation increases both warping rigidity

Figure 4.2. The flexible covert feathers acting like self-activated flaps on the upper wing surface of a skua. Photo from Bechert et al. [416].

and flexibility [24]. The wing structure of a dragonfly helps prevent fatigue fracture [24].

Moreover, wing flexibility can allow for a passive pitching motion due to the inertial forces during the wing rotation as the stroke reverses [303]. There are three modes of passive pitching motion that are similar to those considered in the studies of rigid wing models (see also Fig. 3.9) [201] [245] [301]: (i) delayed pitching, (ii) synchronized pitching, and (iii) advanced pitching. The ratio of flapping frequency to the natural frequency of the wing is key to determining the modes of the passive pitching motion of the wing [404]. The thin nature of the insect wing skin structure makes it unsuitable for taking compressive loads, which may result in skin wrinkling and/or buckling. On the aerodynamics side, for example, wind-tunnel measurements show that corrugated fixed wings are aerodynamically insensitive to the Reynolds number variations, which is quite different from a typical low Reynolds number airfoil (see Section 3.5). For example, Figure 2.5 shows that a dragonfly wing is insensitive to the variation of the Reynolds number in its operating range, in contrast to the Eppler 374 airfoil, which displays a zigzag pattern at a certain Reynolds number range. Obviously, these features go beyond the fact that, as a flyer's size is reduced, the wing becomes thinner and tends to become more flexible. As discussed in detail, the flexibility of the wing has a profound impact on its performance in lift, drag, and thrust. Fundamentally, a passively compliant structure can help adjust the structures so that the resulting aerodynamics can remain desirable [405]–[408]. Realizing that the dimensionless scaling of both fluid dynamics and structural dynamics (and their interactions) cannot be invariant, instead of trying to map out all physical mechanisms versus different Reynolds numbers and sizes, frequencies, and the like, one needs to focus on identifying favorable combinations of materials (elasticity, anisotropy, spatially varying properties, and so on) and on controlling the shape deformation of combined flapping and flexible structural dynamics, with the goal of identifying the most optimized combination of the kinematics, structural behavior, and possible control strategies (including hovering and wind gust effects).

Figure 4.3. Flexible wing patterns of three flyers in different flight modes, including an egret as a predator, a kingfisher trying to land, and a Black Kite during cruising.

Overall, biological flyers have several outstanding features that may pose several design challenges in the design of MAVs. For example, (i) there is substantial anisotropy in the wing structural characteristics between the chordwise and spanwise directions; (ii) they employ shape control to accommodate spatial and temporal flow structures; (iii) they accommodate wind gust and accomplish station keeping with varying kinematics patterns; (iv) they use multiple unsteady aerodynamic mechanisms for lift and thrust enhancement; and (v) they combine sensing, control, and wing maneuvering to maintain not only lift but also flight stability. In principle, one might like to first understand these biological systems, abstract certain desirable features, and then apply them to MAV design. A challenge is that the scaling of fluid dynamics, structural dynamics, and flight dynamics between smaller natural flyers and practical flying hardware/lab experiments (larger dimension) is fundamentally difficult. Regardless, to develop a satisfactory flyer, one needs to meet the following objectives:

- Generate necessary lift, which scales with the vehicle/wing length scale as L_{ref}^3 (under geometric similitude); however, often, a flyer needs to increase or reduce lift to maneuver toward/avoid an object, resulting in the need for substantially more complicated considerations
- Minimize power consumption

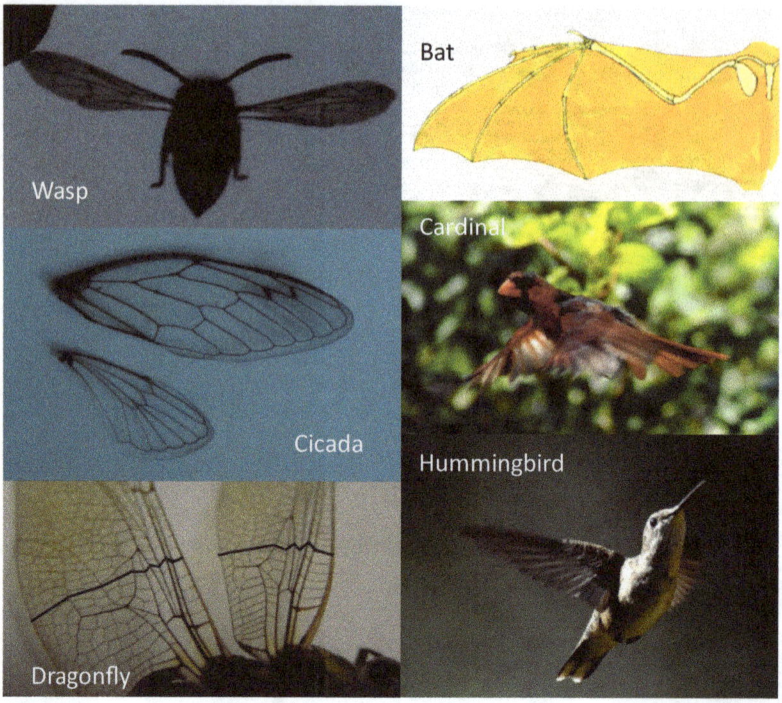

Figure 4.4. Flexible and selectively stiffened wings of a cardinal, hummingbird, bat, dragonfly, cicada, and wasp.

As illustrated in Section 1.1, when wind gust adjustment, object avoidance, or station keeping become major factors, highly deformed wing shapes and coordinated wing-tail movement are often observed (see Fig. 1.12). Understanding of the aerodynamic, structural, and control implications of these modes is essential for the development of high-performance and robust MAVs capable of performing desirable missions. The flexibility of animal wings leads to complex fluid-structure interactions, whereas the kinematics of flapping and the spectacular maneuvers performed by natural flyers result in highly coupled non-linearities in fluid mechanics, aeroelasticity, flight dynamics, and control systems. Can large flexible deformations provide a better interaction with the aerodynamics than those limited to the linear regimes? If torsion stiffness along the wingspan can be tailored, how does that affect the wing kinematics for optimum thrust generation? How do these geometrically non-linear effects and the anisotropy of the structure affect the aerodynamics characteristics of the flapping wing?

Interaction between air and a thin structure is not only observed in animal locomotion. For example, the lift enhancement due to LEV formation that is discussed in the context of unsteady aerodynamic mechanisms in Chapter 3 is observed for autorotating plant seeds [409]. Freely falling papers [410] and their characteristic rotations that are typical for business cards or leafs [411] [412] are also well modeled with the principles described in Chapter 3. More recently, deformations of a leaf and the flow field around it have been measured at different wind speeds in a wind tunnel [413] (see also Fig. 4.5). The leafs reconfigure themselves by streamlining their structure such that the resulting drag grows slower than the classical U^2

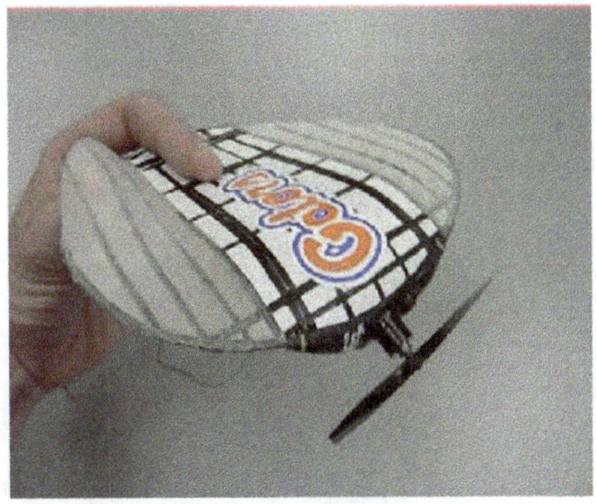

Figure 4.5. Six-inch (15-cm) MAV with flexible wing; from Ifju et al. [17].

relations [414]. The vibration of the trees due to strong wind is one of the main reasons why trees are damaged, and uncovering these daily life physics helps us meet our environmental challenges.

Nature's design of flexible structures can be put into practice for MAVs. Adopting a flexible wing design (see Figs. 4.6 and 4.7) similar to bat wings can improve the performance of MAVs, especially at high AoAs by passive shape adaptation, which results in delayed stall [147] [406]. Figure 4.8, adapted from Waszak et al. [406], compares the lift curves versus AoAs for rigid and membrane wings. The three different flexible wing arrangements are depicted in Figure 4.7. The one-batten design has the most flexibility characterized by large membrane stretch. The two-batten design is, by comparison, stiffer and exhibits less membrane stretch under aerodynamic load. The six-batten wing is covered with an inextensible plastic membrane that further increases the stiffness of the wing and exhibits less membrane deformation and vibration. The nominally rigid wing is constructed with a two-batten frame covered with a rigid graphite sheet. Under modest AoA, both rigid and membrane wings demonstrate similar lift characteristics, with the stiffer wings having a slightly higher lift coefficient. However, a membrane wing stalls at a substantially higher AoA than a rigid wing. This aspect is a key element in enhancing the stability and agility of MAVs.

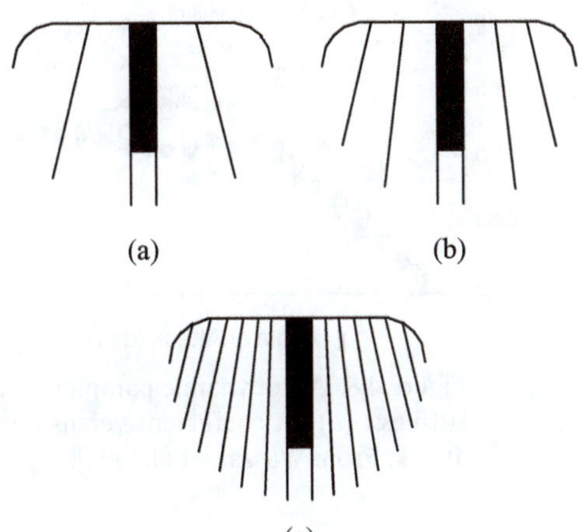

Figure 4.6. Three versions of the flexible wing were tested in the wind tunnel; from Waszak et al. [406]. (a) One-batten flexible wing; (b) two-batten flexible wing; and (c) six-batten flexible wing covered with an plastic inextensible membrane.

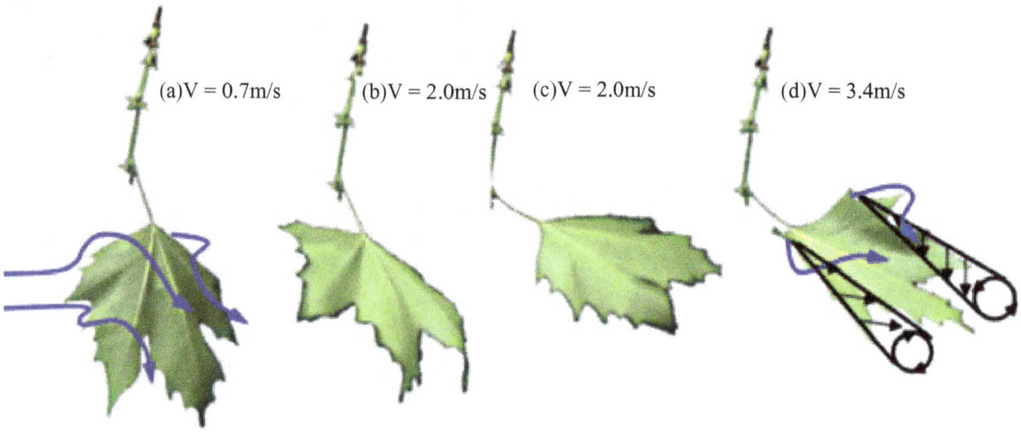

(a) Deformation of the leaf at various wind speeds

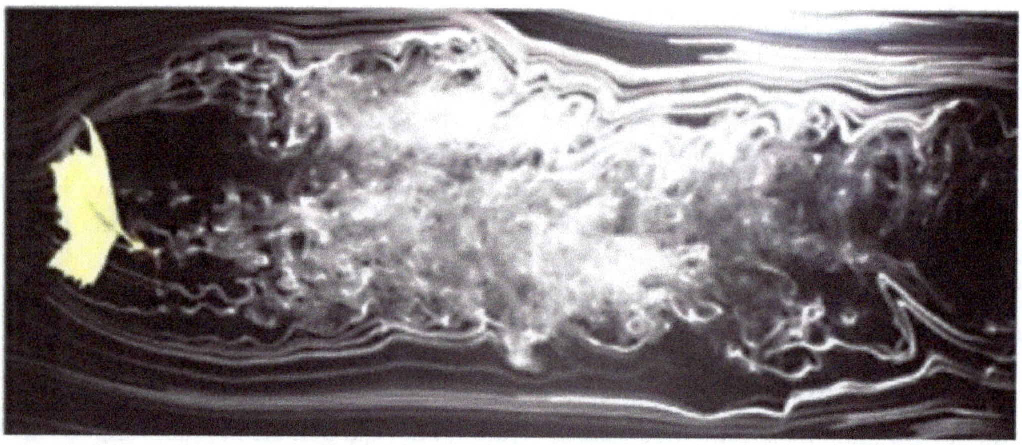

(b) Smoke-wire visualation of the flow behind a leaf at the AoA of 90°.

Figure 4.7. Aeroelastic study of a leaf in controlled laboratory setting. From Shao et al. [413].

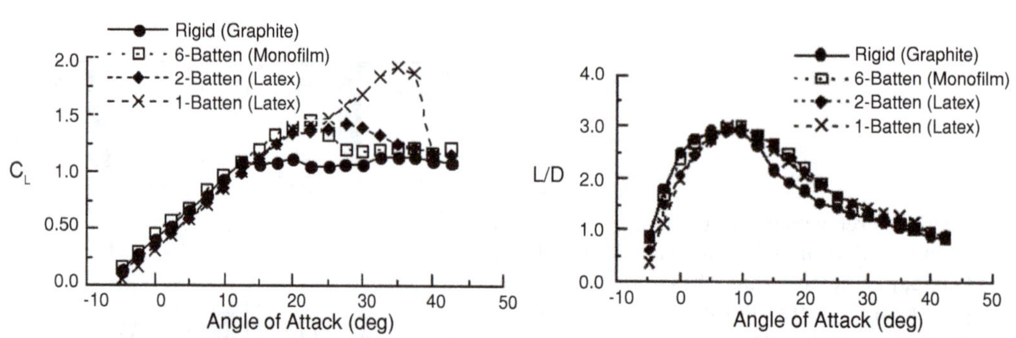

Figure 4.8. Aerodynamic parameters vs. angle of attack for configurations with varying wing stiffness: (a) lift coefficient versus angle of attack; (b) lift-to-drag ratio versus the angle of attack. From Waszak et al. [406].

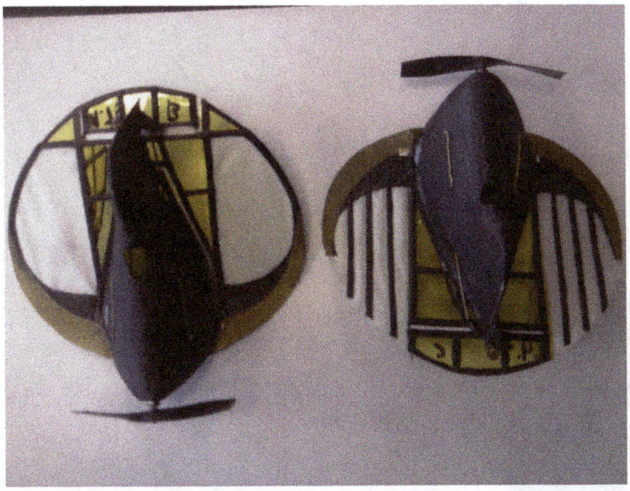

Figure 4.9. Representative MAVs with membrane wing developed by Peter Ifju and collaborators at the University of Florida. Left: the wing is framed along the entire peripherals. Right: the wing is flexible along the trailing edge while reinforced by battens.

The membrane concept has been successfully incorporated in MAVs designed by Ifju et al. [17] and Stanford et al. [407]. However, traditional materials such as balsawood, foam, and monocoat, are not appropriate for implementing the flexible wing concept on these small vehicles. In their design illustrated in Figure 4.6 and Figure 4.9, unidirectional carbon fiber and cloth prepreg materials are used for the skeleton (leading-edge spar and chordwise battens). These are the same materials used for structures that require fully elastic behavior yet undergo large deflections. The fishing rod is a classic example of such a structure. For the membrane, extensible material is chosen to allow for deformations even under very small loads, such as for lightly loaded wings. Latex rubber sheet material has been used in this case. The stiffness of the whole structure can be controlled by the number of battens and the membrane material used.

As presented earlier, the experimental data for rigid and flexible wings (Fig. 4.8), with configurations similar to that shown in Figure 4.9, show that a membrane wing stalls at a substantially higher AoA than a rigid wing. Some aspects of low AR, low Reynolds number, and rigid wing aerodynamics have been presented by Torres and Mueller [162] as well as in Chapter 2. The lift-curve slope in Figure 4.8 is approximately 2.9 with the prop pinned. The lift-curve slopes of similar rigid wings reported by Torres and Mueller [162] at a comparable Reynolds number and aspect ratio ($Re = 7 \times 10^4$, $AR = 2$) are approximately 2.9 as well. However, these wings have stall angles between 12° and 15°. The stall angles of the flexible wings are between 30° and 45° and are similar to those of much lower aspect ratio rigid wings ($AR = 0.5$ to 1.0) [159]. However, low-aspect-ratio rigid wings exhibit noticeably lower lift-curve slopes, typically between 1.3 and 1.7 [159]. Hence, flexible wings can effectively maintain the desirable lift characteristics with better stall margins [406]. Figure 4.8 shows fixed, flexible-wing MAVs designed by Ifju and co-workers of the University of Florida [17]. The general specifications of the design are presented in Table 4.1. Stanford et al. [407] and Shyy et al. [408] reviewed the development of membrane-based fixed wing MAVs.

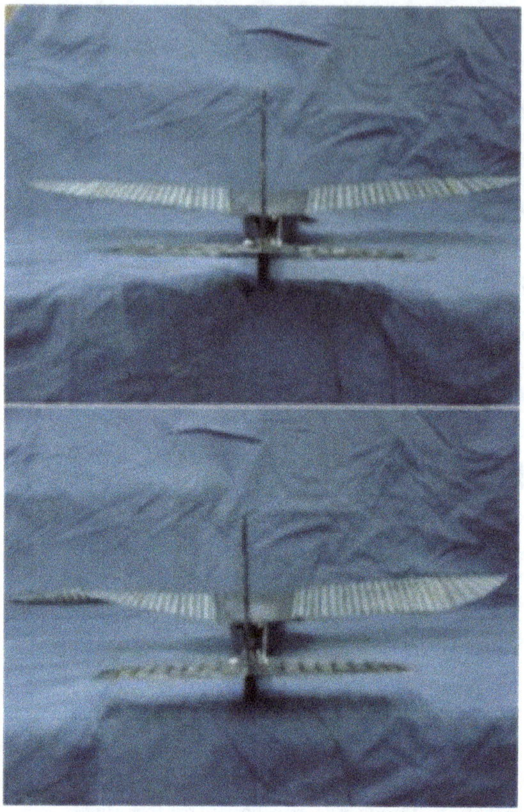

Figure 4.10. The flexible wing allows for wing warping to enhance vehicle agility [415].

From an engineering point of view, flexibility can be used for purposes other than flight quality improvement. These objectives include shape manipulation and reconfiguration for both improved maneuvering and storage. Traditional control surfaces such as rudders, elevators, and ailerons have been used almost exclusively for flight control. By morphing or reshaping the wing using distributed actuation such as piezoelectric and shape memory material, preferred wing shapes can be developed for specific flight regimes. Such reconfiguration, however, would require substantial authority and power if the wings are nominally rigid. The flexible nature of the wing allows for such distributed actuation with orders of magnitude less authority. For example, the individual battens on the wing can be made from shape memory alloys, piezoelectric materials, or traditional actuators, such as servos, which can be used to manipulate the shape and properties of the wing.

Figure 4.10 illustrates a model with morphing technology. It employs a thread connecting the wingtips to a servo in the fuselage of the airplane. As the thread is tightened on one side of the aircraft, it acts like an aileron and causes the AoA of the wing to increase. The roll rate developed by such an actuation mechanism is considerably higher than that from a rudder. Additionally, it produces nearly pure roll with little yaw interaction. Detailed information on the related technical approaches has been presented by Garcia et al. [415].

In some applications, it is desirable to store MAVs in small containers before releasing them. Flexible wing MAVs can be easily reconfigured for storage purposes. Figure 4.11 illustrates a 28 cm (11 inch) wingspan foldable wing MAV that can be

Figure 4.11. A foldable wing to enhance MAV portability and storage (Courtesy Peter Ifju).

stored in a 7.6 cm (3 inch) diameter canister. The wing uses a curved shell structural element on the leading edge, which enables the wing to readily collapse downward for storage, yet maintain rigidity in the upward direction to react to the aerodynamic loads. The effect is similar to that of a common tape measure, where the curvature in the metallic tape is used to retain the shape after it has unspooled from the casing, yet it can be rolled back into the casing to accommodate the small-diameter spool. The curvature ensures that the positive (straight) shape is developed after it is unwound from the case and can actually be cantilevered for some distance. The curvature of the leading edge of the wing acts as the curvature in the tape measure.

4.2 Governing Equations for Wing Structures

As an illustration, consider the hummingbird wing shown in Figure 3.5a. A major research interest is to probe the coupled dynamics between the fluid flow and the flexible wing. The fluid flow creates pressure and viscous stresses, which cause the wing to deform. The wing, in turn, affects the fluid flow structure via the shape change, resulting in a moving boundary problem [417].

Table 4.1. *General specification for the university of Florida MAV*

Wingspan	4.5′
Fuselage length	4.5′
Takeoff weight	45 g
Engine	Maxon Re10
Propeller	U-80 (62 mm)
RC receiver	PENTA with customized half-wave antenna
Maximum mission radius	0.9 miles
Video transmitter	SDX-22 70 mw
Camera	CMOS camera (350 lines resolution)

Partitioned analyses have been very popular in the area of computational fluid-structure interactions/computational aeroelasticity. A main motivating factor in adopting this approach is that one can develop and use state-of-the-art fluid and structure solvers and recombine them with minor modifications to allow for the coupling of the individual solvers. The accuracy and stability of the resulting coupled scheme depend on the selection of the appropriate interface strategy, which depends on the type of application. The key requirements for any dynamic coupling scheme are (i) kinematic continuity of the fluid-structure boundary, which leads to the mass conservation of the wetted surface, and (ii) dynamic continuity of the fluid-structure boundary, which accounts for the equilibrium of tractions on either sides of the boundary. This leads to the conservation of linear momentum of the wetted surface. Energy conservation at the fluid-structure interface requires both of these two continuity conditions to be satisfied simultaneously. The subject of fluid and structural interactions is vast. Recent reviews by Friedmann [418], Livne [419], and Chimakurthi et al. [420] offer substantial information and references of interest.

To take wing deformation into account, it is necessary to solve governing equations of wing structures. There are various wing structural models, and the choice of models depends on the problems of interest. In this section we consider the wing structure as a beam, a membrane, an isotropic flat plate, and a shell. Governing equations of each structural model are presented in each subsection.

4.2.1 Linear Beam Model

Linear beam theory is a well-known simplification of the linear theory of elasticity for the structures that can be regarded as largely one-dimensional whose widths are small compared with their lengths. It relates the deflections of a beam, which are assumed to be small, to the transverse loads. For a dynamic linear beam, the transverse deflection $w(x, t)$ is given by the Euler-Bernoulli beam equation as

$$\rho_s h_s \frac{\partial^2 w}{\partial t^2} + \frac{E h_s^3}{12} \frac{\partial^4 w}{\partial x^4} = f_{\text{ext}}, \tag{4–1}$$

where ρ_s is the density of the wing, h_s the thickness of the wing, E the modulus of elasticity, and f_{ext} the distributed external force per unit length acting in vertical direction. Uniform wing properties are assumed in Eq. (4–1). The linear beam theory has been widely used in the aeroelastic community to model and study the dynamic structural response and stability of the wing of an airplane [333].

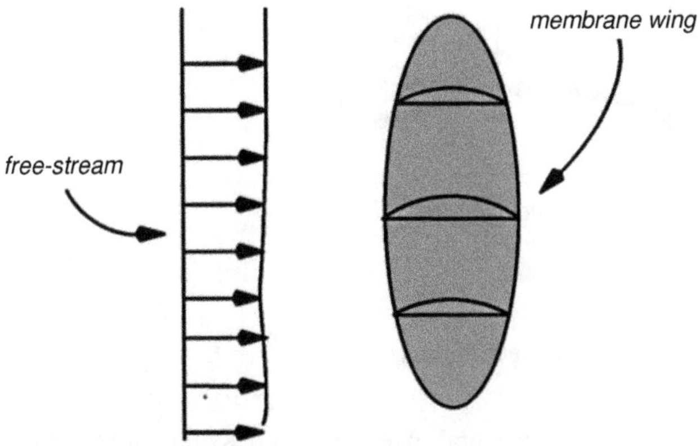

Figure 4.12. Schematic of membrane wing of finite span operating in a free-stream [406].

4.2.2 Linear Membrane Model

As an illustration, consider the membrane wing of Figure 4.12, which is shown operating in a free-stream. The analysis of membrane wings begins with the historical works of Voelz [421], Thwaites [422], and Nielsen [423]. These works consider the steady, 2D, irrotational flow over an inextensible membrane with slack. As a consequence of the inextensible assumption and the additional assumptions of small camber and incidence angle, the membrane wing boundary value problem is linearized and may be expressed compactly in non-dimensional integral equation form as

$$1 - \frac{C_T}{2}\int_0^2 \frac{\frac{d^2(y/\alpha)}{d\xi^2}}{2\pi(\xi - x)}d\xi = \frac{dy/\alpha}{dx}, \tag{4–2}$$

where $y(x)$ defines the membrane profile as a function of the x coordinate, α is the flow incidence angle, C_T is the tension coefficient, and ξ is the arc length along the membrane wing surface. Equation (4–2) was referred to as the "Thwaites sail equation" by Chambers [424] and simply as the "sail equation" by Greenhalgh et al. [425] and Newman [426]. This equation, together with a dimensionless geometric parameter ε, completely defines the linearized theory of an inextensible membrane wing in a steady, inviscid flow field. Parameter ε specifies the excess length of an initially flat and taut membrane and is defined as follows:

$$\varepsilon = \frac{L_0 - c}{c}, \tag{4–3}$$

where L_0 is the unstrained length of the membrane and c is the chord length. The meaning of these symbols can be better understood from Figure 4.13.

Different analytical and numerical procedures have been applied to the basic equation set to determine the membrane shape, aerodynamic properties, and membrane tension in terms of the AoA and excess length. In particular, Thwaites [422] obtained the eigensolutions of the sail equation that are associated with the wing at an ideal angle of incidence. Nielsen [423] obtained solutions to the same equation using a Fourier series approach that is valid for wings at angles of incidence other than the ideal angle. Other more recent but similar works are those by Greenhalgh

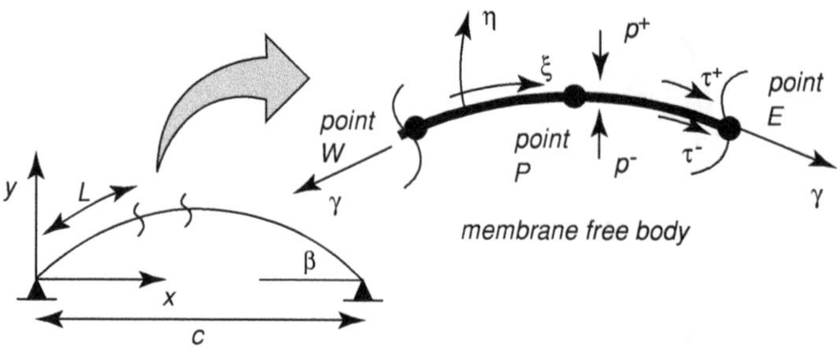

Figure 4.13. Leading edge constrained elastic membrane.

et al. [425], Sugimoto and Sato [427], and Vanden-Broeck and Keller [428]. Various extensions of the linear theory have appeared in the literature over the years. Vanden-Broeck [429] and Murai and Maruyama [430] developed non-linear theories valid for large camber and incidence angle. The effect of elasticity has been included in the membrane wing theories of Jackson [431] and Sneyd [432], and Murata and Tanaka [433] investigated the effects of membrane porosity. In a paper by de Matteis and de Socio [434], experimentally determined separation points were used to modify the lifting potential flow problem in an attempt to model flow separation near the trailing edge. A comprehensive review of the work published before 1987 related to membrane wing aerodynamics is given by Newman [426].

The extent of agreement between the various potential flow-based membrane wing theories and experimental data has been reported by several authors, including Greenhalgh et al. [425], Sugimoto and Sato [427], and Newman and Low [141]. In general, there has been considerable discrepancy between the measurements made by different authors [431], which have all been in the turbulent flow regime at Reynolds numbers between 10^5 and 10^6. As a result of the discrepancies in the reported data, primarily due to differences in Reynolds number and experimental procedure, the agreement between the potential-based membrane theories and the data has been mixed. In particular, the measured lift is in fair agreement with the predicted value when the excess length ratio is less than 0.01 and the AoA is less than 5°. However, even for this restricted range of values, the measured tension is significantly less than that predicted by theory. Furthermore, for larger excess lengths and incidence angles the theory poorly predicts lift and tension.

The main reason for the disagreement is the existence of the viscous effect, which significantly affects the force distribution on the wing and therefore the effective shape of the wing. To illustrate the flexible structural dynamics in response to aerodynamic forces, consider equilibrium equations for a 2D elastic membrane subjected to both normal and shearing stresses. As discussed by Shyy et al. [417], the membrane is considered to be massless, and the equilibrium conditions are stated in terms of the instantaneous spatial Cartesian coordinates and the body-fitted curvilinear coordinates. The basic formulation is essentially identical to many previously published works such as de Matteis and de Socio [434] and Sneyd [432].

Figure 4.13 illustrates an elastic membrane restrained at the leading and trailing edges and subjected to both normal and tangential surface tractions p and τ, respectively. Imposing equilibrium in the normal and tangential directions requires that

$$\frac{d^2y}{dx^2}\left(1+\left(\frac{dy}{dx}\right)^2\right)^{-3/2} = -\frac{\Delta p}{\gamma}, \tag{4–4}$$

$$\frac{d\gamma}{d\xi} = -\tau, \tag{4–5}$$

where γ is the membrane tension. Equation (4–4) is the Young-Laplace equation cast in Cartesian coordinates. The net pressure and shear stress acting on a segment of the membrane are given, respectively, by

$$\Delta p = p^- - p^+, \tag{4–6}$$

$$\tau = \tau^- - \tau^+, \tag{4–7}$$

where the superscripts + and – indicate the values at the upper and lower surfaces of the membrane, as shown in the figure. If the membrane material is assumed to be linearly elastic, the nominal membrane tension $\overline{\gamma}$ may be written in terms of the nominal membrane strain $\overline{\delta}$ as

$$\overline{\gamma} = (S^0 + E\overline{\delta})h_s, \tag{4–8}$$

where S^0 is the membrane prestress, E is the elastic modulus, and h_s is the membrane thickness. The nominal membrane strain is given by

$$\overline{\delta} = \frac{L - L_0}{L_0}, \tag{4–9}$$

where L_0 is the unstrained length of the membrane and L is the length of the membrane after deformation, which may be expressed in terms of the spatial Cartesian coordinates as

$$L = \int_0^c \sqrt{1+\left(\frac{dy}{dx}\right)^2}\, dx, \tag{4–10}$$

where c is the chord length.

The aeroelastic boundary value problem can be written in non-dimensional form after introducing the following dimensionless variables:

$$x^* = \frac{x}{c}, \tag{4–11}$$

$$y^* = \frac{y}{c}, \tag{4–12}$$

$$P = \frac{p}{\rho U_{\text{ref}}^2} = \frac{p}{q_\infty}, \tag{4–13}$$

$$\gamma^* = \frac{\gamma}{S^0 h_s} \text{ or } \gamma^* = \frac{\gamma}{Eh_s}, \tag{4–14}$$

where Eq. (4–14) is used to non-dimensionalize the membrane tension depending on whether the tension is dominated by pretension or by elastic strain. The resulting dimensionless equilibrium equation when membrane tension is dominated by elastic strain is

$$\frac{d^2y^*}{dx^{*2}}\left(1+\left(\frac{dy^*}{dx^*}\right)^2\right)^{-3/2} = -\frac{1}{\Pi_1}\frac{\Delta P}{\gamma^*}, \tag{4–15}$$

with Π_1 the effective stiffness parameter, defined to be

$$\Pi_1 = \frac{Eh_s}{q_\infty c}. \tag{4–16}$$

When membrane tension is dominated by pretension, Eq. (4-4) leads to the following dimensionless equation:

$$\frac{d^2y^*}{dx^{*2}}\left(1+\left(\frac{dy^*}{dx^*}\right)^2\right)^{-3/2} = -\frac{1}{\Pi_{1,\text{pret}}}\frac{\Delta P}{\gamma^*}, \tag{4–17}$$

with the effective pretension, $\Pi_{1,\text{pret}}$ defined to be

$$\Pi_{1,\text{pret}} = \frac{S^0 h_s}{q_\infty c}. \tag{4–18}$$

If the two ends of a 2D membrane are fixed, the boundary conditions in dimensionless form are

$$y^* = 0 \text{ at } x^* = 0 \text{ and } 1. \tag{4–19}$$

Regarding the physical significance of the aeroelastic parameters Π_1 and $\Pi_{1,\text{pret}}$ we note that the dimensionless deformation of an initially flat elastic membrane is inversely proportional to Π_1 in the absence of pretension. The dimensionless deformation of a membrane is inversely proportional to $\Pi_{1,\text{pret}}$ in the presence of large initial pretension. Consequently, the steady-state, inviscid aeroelastic response of an initially flat membrane wing at a specified AoA is controlled exclusively by Π_1 in the limit of vanishing pretension, and exclusively by $\Pi_{1,\text{pret}}$ in the limit of vanishing material stiffness. Similarly Song et al. [435] also presented a theoretical model for membrane camber due to aerodynamic loading and showed an expression for non-dimensional aerodynamic loading, expressed as a Weber number, *We*:

$$We = \frac{\Delta pc}{Eh_s} = 2\left(\frac{S^0}{Eh_s} + \frac{\bar{\delta}}{c}\right)\sin(\beta), \tag{4–20}$$

where β is the contact angles at the leading and trailing edges (Fig. 4.13).

This scaling analysis is based on a massless structure. If the airfoil mass is considered, then the inertia scaling needs to be considered. Between the elastic and inertia scaling, one can also deduce the structural natural frequency.

4.2.3 Hyperelastic Membrane Model

A rubberlike material can be used to cover the rigid skeleton of a MAV design to obtain flexibility of the wing (see Fig. 4.6). The large deformations observed for this kind of material in the Reynolds number range of operation indicate that the linear elasticity assumption may not be valid. To address this issue, a hyperelastic model to describe the 3D membrane material behavior is employed [162]. The stress-strain curve of a hyperelastic material is non-linear, but follows the same path in loading and unloading (below the plastic limit, which is significantly higher than in metals). Compared with the previously discussed 2D linear model, a 3D membrane model introduces several complicated factors. First, for three-dimensional membranes, the tension is defined as a biaxial tension along the lines of principal stress [436]. Second, the geometric and material properties may vary along the spanwise direction and need to be described in detail. A third factor is membrane compression, leading to wrinkles when one of the principal tensions vanishes. In addition, it is desirable to account for the membrane mass when solving for the dynamic equations of the membrane movement.

A finite element analysis of the static equilibrium of an inflated membrane undergoing large deformations is presented by Oden and Sato [437]. A review of the earlier work on the dynamic analysis of membranes can be found in Jenkins and Leonard [438]. An update of the state-of-the-art models in membrane dynamics is presented by Jenkins [439]. Verron et al. [440] studied, both numerically and experimentally, the dynamic inflation of a rubberlike membrane. Ding et al. [441] numerically studied partially wrinkled membranes.

In a recent effort, Stanford et al. [442] proposed an accurate linear model for 3D membranes used in MAV design. Their experimental measurements showed that the maximum strain value is quite small; therefore they constructed a linear approximation of the stress-strain curve, centered about the membrane wing's prestrain value. The linear constitutive equation used for membrane modeling is Poisson's equation:

$$\frac{\partial^2 W}{\partial x^2} + \frac{\partial^2 W}{\partial y^2} = -\frac{p(x, y)}{T}, \tag{4–21}$$

where W is the out-of-plane membrane displacement, $p(x, y)$ is the applied pressure (wing loading, in this case), and T is the membrane tension per unit length. The aerodynamic loads are computed on a rigid wing and fed into the structural model, assuming that the change in shape of the membrane wing did not overtly redistribute the pressure field. They obtained good agreement between the experimental data and computations.

A 3D membrane model was developed by Lian et al. [163]. The model gives good results for membrane dynamics with large deformations, but has limited capability to handle the wrinkle phenomenon that occurs when the membrane is compressed. The membrane material considered obeys the hyperelastic Mooney-Rivlin model [443]. A brief review of their membrane model is given next.

The Mooney-Rivlin model is one of the most frequently employed hyperelastic models because of its mathematical simplicity and relatively good accuracy for reasonably large strains (less than 150 percent) [443]. For an initially isotropic membrane, Green and Adkins [444] defined a strain energy function, W, as

$$W = W(I_1, I_2, I_3), \tag{4–22}$$

where I_1, I_2, and I_3 are the first, second, and third invariants of the Green deformation tensor, respectively. More details about the model, its validation, and its numerical implementation can be found in the literature (e.g., [154] [160] [440]).

4.2.4 Flat Plate and Shell Models

A flat 3D wing can be modeled as a plate that allows for spanwise and chordwise bending and twist. For a thin isotropic plate, the small transverse displacement is governed by the classical plate equation,

$$\rho_s h_s \frac{\partial^2 W}{\partial t^2} + \frac{E h_s^3}{12(1-\nu^2)}\left(\frac{\partial^4}{\partial x^4} + 2\frac{\partial^4}{\partial x^2 y^2} + \frac{\partial^4}{\partial y^4}\right) W = f_{\text{ext}}, \tag{4–23}$$

where W is transverse displacement, ρ_s the density, h_s the thickness, E the modulus of elasticity, ν the Poisson's ratio of the wing, and f_{ext} the distributed external force per unit length acting in the vertical direction.

The most general structural class, which includes the previously introduced model, is the shell model. A thin shell is modeled as a curved 2D structure with the in-plane and out-of-plane displacements coupled via its curvature. A shell finite element (FE) can carry bending and membrane forces. With a thin shell FE model, Nakata and Liu [445] investigated the non-linear dynamic response of anisotropic flexible hawkmoth wings flapping in air, and Chimakurthi, Cesnik, and Stanford [446] simulated flapping plate/shell-like wing structures undergoing small strains and large displacements/rotations.

More research is needed to further refine the structural model for the anisotropic, batten-enforced mechanical properties of the highly flexible structures such as wings and to better understand the dynamics of the resulting fluid-structure interactions.

4.3 Scaling Parameters for the Flexible Wing Framework

As previously discussed in Section 1.2, scaling parameters resulting from dimensional analysis help identify key characteristics of the model, via Buckingham's Π-theorem, and also reduces the number of involved parameters to the sufficient number of combinations [333], [447]–[449]. Under certain circumstances, the result obtained from the dimensional analysis can be reduced to a simpler relationship, with a reduced number of arguments, as a property of the special problem under consideration. The non-dimensional parameters arising from such a scaling analysis can identify similarity variables, which can be of critical value even if a complete mathematical solution is missing [447].

Generally for a flapping rigid wing framework (as discussed in Chapter 3), two non-dimensional parameters, such as Reynolds numbers and reduced frequency or the Strouhal number, are manifest in the governing equations of the fluid. For the field of flexible flapping wing aerodynamics, numerous efforts using scaling arguments have increased our knowledge of the complex interplay between flexibility and the resulting aerodynamics. However, depending on the type of model and on the governing equations, the resulting set of scaling parameters may vary. For instance, for flexible flapping wings, Shyy et al. [450] considered the flapping wing aeroelastic system based on the Navier-Stokes equation coupled with out-of-plane motion of an isotropic flat plate. Ishihara et al. [451] [452] introduced the Cauchy number that describes the ratio between the fluid dynamic pressure and elastic reaction force from the scaling argument (the Navier-Stokes equation along with the linear isotropic elasticity equations) and presented the correlation between time-averaged lift and the Cauchy number. Thiria and Godoy-Diana [453] and Ramananarivo, Godoy-Diana, and Thiria [454] introduced the elastoinertial number, using scaling arguments to define the ratio between the inertial forces and the elastic restoring forces, and showed the correlation between time-averaged thrust and the elastoinertial number, based on flight velocity measurement using a self-propelled flapping flyer with flexible wings in air. Since the density ratio between the air and the wing is high ($\sim O(10^3)$), the elastic deformation of the wing was mostly balanced by the wing inertia. Furthermore, Ramananarivo et al. [454] linked the cubic non-linear damping term due to the aerodynamics to the effects of flexibility in the aerodynamic performance. However, it is very difficult for any one study to cover all of the parameter-space of the flapping flexible wing system in which we are interested. For example, the effects of density ratios on the force generation of flexible flapping wings have not been adequately addressed. Ideally the parameter-space involving the scaling parameters for the fluid-structure interactions should be mapped out in a systematic fashion to understand the impact of flexibility and the density ratio on the force generation and propulsive efficiency of coupled systems. However, as is shown next, the number of dimensionless (scaling) parameters involved is large, making it practically impossible to examine all combinations. Consequently, insight is needed so that efforts can be directed toward creating a suitable combination of these scaling parameters.

We considered the relevant physical quantities related to the system of flexible flapping wing fluid dynamics proposed in Shyy et al. [450] [455] and Kang et al. [351]. There are 13 variables (see Figure 3.5a): the density, ρ, and the viscosity, μ, of the fluid; the reference velocity, U_{ref}, of the fluid flow field; the half-span, R, the mean chord, c_{m}, and the thickness, h_s, of the wing geometry; the density, ρ_s, the Young's modulus, E, and the Poisson's ratio, v, of the wing structure; the flapping (plunging) amplitude, ϕ_a (h_a), the flapping frequency, f, and the geometric AoA, α; and finally the resulting aerodynamic force, F. Three fundamental dimensions lead to 10 non-dimensional parameters. With ρ, U_{ref}, and c_{m} as the basis variables to independently span the fundamental dimensions, the dimensional analysis leads to the non-dimensional parameters shown in Table 4.2. The resulting set of non-dimensional parameters includes most of the well-known parameters in the flapping wing aerodynamics community; however, there are other sets of non-dimensional parameters. More importantly, some of these dimensionless parameters scale with a flyer's physical dimensions in different ways. As pointed out in Section 1.2, the different scaling relationships between the various dimensionless parameters and size, speed, and other parameters of flapping wings pose fundamental difficulties in utilizing a laboratory model of different sizes: as the physical dimension and speed of a wing change, the scaling parameters vary differently, making it virtually impossible to conduct experiments capable of maintaining dynamic similarity.

Although Table 4.2 offers a complete set of dimensionless parameters based on the dimensional quantities selected, alternative parameters can be derived by combining these dimensionless parameters. For example, the advance ratio J given

Table 4.2. *List of Non-dimensional parameters for flexible wing aerodynamics*

Non-dimensional parameter	Symbol	Definition	Note
Reynolds number	Re	$\rho U_{\mathrm{ref}} c_{\mathrm{m}}/\mu$	
Aspect ratio	AR	R/c_{m}	
Thickness ratio	h_s^*	h_s/c_{m}	
Density ratio	ρ^*	ρ_s/ρ	
Poisson's ratio	v	v	
	Π_1	$Eh_s^{*3}/\{12(1-v^2)\rho U_{\mathrm{ref}}^2\}$	plate
Effective stiffness		$Eh_s^{*3}/\{12\rho U_{\mathrm{ref}}^2\}$	beam
		$Eh_s^*/\{\rho U_{\mathrm{ref}}^2\}$	membrane wing
Effective pretension	$\Pi_{1,\mathrm{pret}}$	$S^0 h_s^*/(\rho U_{\mathrm{ref}}^2)$	membrane wing
Effective rotational inertia	Π_2	$I_B/(\rho U_{\mathrm{ref}}^5)$	
Reduced frequency	k	$\pi f c_{\mathrm{m}}/(U_{\mathrm{ref}})$	
Strouhal number	St	$\phi_a AR\, k/\pi$	flapping
		$h_a k/(c_{\mathrm{m}}\pi)$	plunging
Effective angle of attack	α_e	$a + \mathrm{atan}(2\pi\, St)$	plunging
Force coefficient	C_F	$F/\left(\frac{1}{2}\rho U_{\mathrm{ref}}^2 c_{\mathrm{m}}^2 AR\right)$	

by Eq. (3–14) results from the Strouhal number. Another example is the effective inertia Π_0 (Eq. 4–24),

$$\Pi_0 = \rho^* h_s^* (k/\pi)^2, \tag{4–24}$$

which is used frequently later to discuss the scaling of aerodynamic forces of a flexible wing, is a combination of ρ^*, h_s^*, and k. Two other non-dimensional parameters that prove to be important are the frequency ratio, f/f_1, which is the ratio between the motion frequency and the first natural frequency of the wing, and the non-dimensional tip deformation parameter, γ, which scales with the aerodynamic performance. The relation between the non-dimensional parameters shown in Table 4.2 and these two parameters is derived in Section 4.5. Furthermore, additional dimensionless parameters arise as the scope of the mechanical system broadens. As discussed by Shyy et al. [450], if an anisotropic shear deformable plate is considered for the wing, an additional dimensionless parameter appears that describes the ratio between rotational inertia forces and aerodynamic forces. This effective rotational inertia parameter Π_2 is defined as

$$\Pi_2 = I_B/\left(\rho U_{\mathrm{ref}}^5\right), \tag{4–25}$$

where I_B is the mass moment of inertia. The effects of the twist on the aerodynamic performance have been considered many times in the literature [24].

The flexible wing structure can be modeled locally as a beam (see Eq. (4–1)) that oscillates in time due to its flexibility under the aerodynamic loading. Following the same non-dimensionalization process as in Chapter 3, Eq. (4–1) becomes

$$\rho^* h_s^* \left(\frac{k}{\pi}\right)^2 \left(\frac{l}{c_{\mathrm{m}}}\right) \frac{\partial^2 w^*}{\partial t^{*2}} + \Pi_1 \left(\frac{c_{\mathrm{m}}}{l}\right)^3 \Delta^{*2} w^* = f_{\mathrm{ext}}^*, \tag{4–26}$$

where special care is given in the direction of the wing bending, because the correct length scale for the spanwise bending is the half-span R and not the chord c_{m}. The correction factor that arises is expressed as (l/c_{m}), where $(l/c_{\mathrm{m}}) = 1$ for the chordwise flexible airfoil case (Section 4.4.2.1) and $(l/c_{\mathrm{m}}) = AR$ for the spanwise flexible wing case (Section 4.4.2.2) or isotropic Zimmerman wing case (Section 4.4.2.3) where AR is the aspect ratio of the wing: for the 3D wings the bending motion is aligned with R, so that a factor of AR is required to renormalize the transverse displacement. All non-dimensional parameters appearing in Eq. (4–26) are consistent with the parameters listed in Table 4.2. The effective stiffness $\Pi_1 = Eh_s^{*3}/\{12\,\rho U_{\mathrm{ref}}^2\}$ [450] gives the ratio between the elastic bending forces and the fluid dynamic forces. The equivalent effective stiffness for the plate and membrane can be found easily; that is, $Eh_s^{*3}/\{12(1-\nu^2)\rho U_{\mathrm{ref}}^2\}$ and $Eh_s^{*}/(12\rho U_{\mathrm{ref}}^2)$, respectively. The coefficient of the inertial term in Eq. (4-26), abbreviated as the effective inertia, is Π_0. Finally, the force coefficient is then given by a to-be-determined relation

$$C_F = \tilde{\Psi}(Re, AR, h_s^*, \rho^*, \Pi_1, k, St). \tag{4–27}$$

4.4 Interactions between Elastic Structural Dynamics and Aerodynamics

4.4.1 Fixed Membrane Wing

Much of the earlier efforts in membrane wing studies focused on fixed-wing-based vehicles [207]. Of particular importance to the membrane wing MAVs is the passive ability of the wing skin to change shape to reflect flight conditions. Shyy et al. [408] provided a comprehensive review of MAV performance characteristics, with particular emphasis on the effects of low AR and AoA on TiVs, laminar boundary-layer separation, stall characteristics, and vibrations in MAVs. Ifju et al. [17] presented methods for design and construction of a variety of flexible wing MAVs and demonstrated that the adaptive nature of the flexible wing design greatly improved stability and reduced drag.

Shyy et al. [456] compared the performance between a low Reynolds number membrane and rigid airfoil in terms of the lift-to-drag ratio. They investigated three airfoils – one rigid, one membrane based, and the third a hybrid of both – of the same nominal camber at 6 percent in a fluctuating free-stream. To mimic the effect of wind gust the fluctuations in the free-stream were modulated for 25 percent or more, and the sinusoidal modulation frequency of the free-stream was 1.7 Hz. The linear membrane model was used to account for the airfoil's flexibility. The hybrid airfoil was built with a curved wire screen beneath the membrane, so that it could achieve a camber deformation greater than 6 percent, but not less because the wire construction prevented a decrease in the camber. The size of the wing chords and the average wind tunnel speed gave a Reynolds number of 7.5×10^4. The experiments for all three airfoils were conducted at an AoA of 7° and the results are shown in Figure 4.14. Detailed numerical simulations based on the Navier-Stokes equations and two-equation turbulence closure, along with a moving grid technique to track the shape variations of the membrane and hybrid airfoils, were conducted for various configurations.

Shyy et al. [456] found that, at modest AoA, the flow over the rigid airfoil is attached to the surface at all time and the lift-to-drag ratio follows the free-stream fluctuations. When the AoA is increased to 7° substantial flow separations occur, causing a modification to the effective shape of the rigid airfoil. As the AoA is

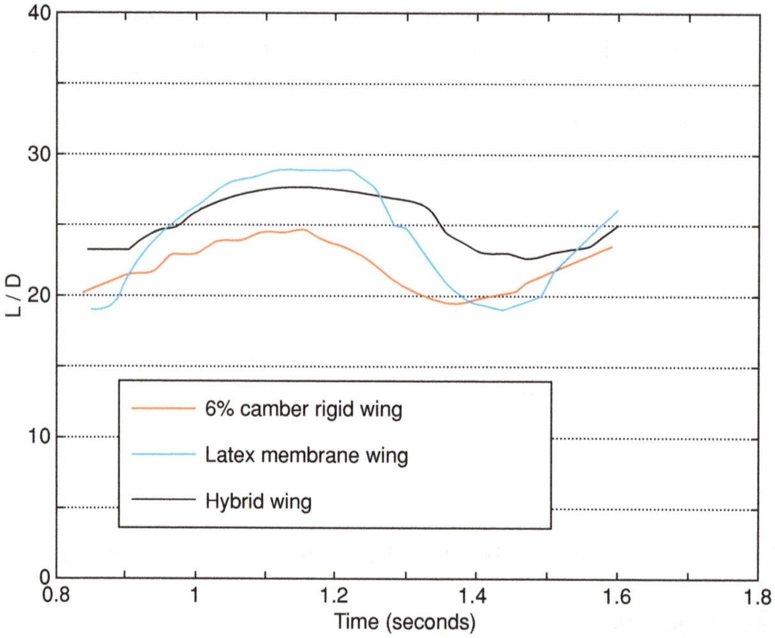

Figure 4.14. Experimental L/D results for rigid, flexible, and hybrid wings at $Re = 7.5 \times 10^4$ and AoA $= 7°$. The latex membrane wing exhibits about 6% camber at 35.4 fps. The hybrid wing has a curved-wire screen camber stop. From Shyy et al. [456].

increased, C_L tends to increase as well, but the lift-to-drag ratio decreases due to flow separation. When the flow is separated at higher AoAs, the airfoil is less sensitive to an unsteady free-stream. At both AoAs, the lift coefficients between rigid and flexible airfoils are comparable, but the lift-to-drag ratio is higher for the flexible airfoil. For the membrane airfoil, at AoA$=7°$ the flow separation is confined to the leading edge, resulting in a better aerodynamic performance. However, some negative effects occur with the flexible membrane. When the free-stream velocity reached its lower values during a fluctuating cycle, the camber of the flexible membrane tends to collapse, and a massive separation over the whole surface occurs. This phenomenon is due to the smaller pressure differences between the upper and lower surface of the membrane, and hence a degraded performance is obvious.

The hybrid airfoil, which has a curved wire screen stop to prevent the camber from becoming too low, shows interesting results. For a lower AoA the aerodynamic characteristics associated with a hybrid wing are essentially the same as for those with a flexible airfoil because the flow field near the wing is non-separating. However, when the AoA is increased, the characteristics are considerably better for the hybrid airfoil compared to the flexible profile. The separation zone is smaller, compared to the rigid airfoil, and the sensitivity to fluctuations in the free-stream is reduced when compared to the flexible configuration.

A computational study [469] of unsteady structural response of the membrane airfoil is shown in Figure 4.15i. This figure displays the membrane oscillations by means of *x-t* diagrams showing the difference between the instantaneous and mean membrane deflection. The membrane is nominally stationary at AoA $= 4°$ because the flow is steady over the major portion of the airfoil (see Fig. 4.15ii). At

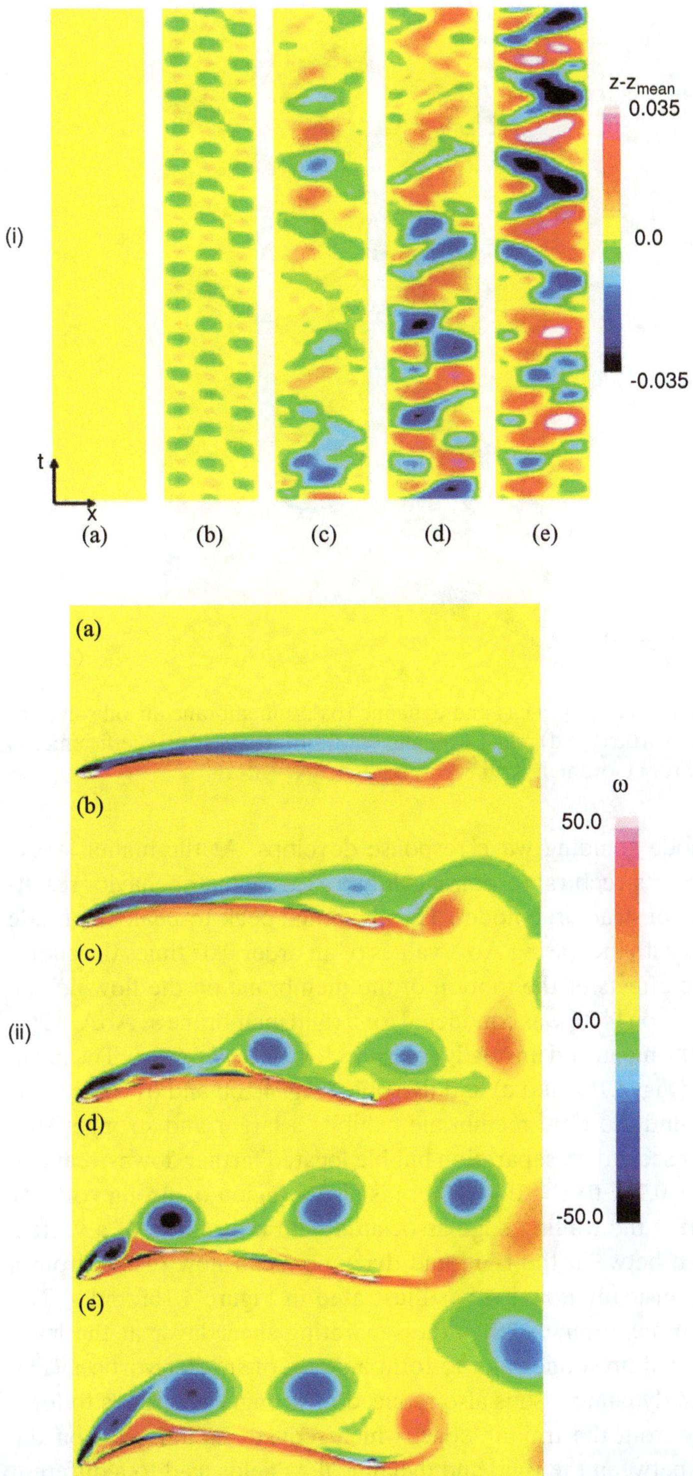

Figure 4.15. (i): *x-t* diagram for perturbation from the mean membrane deflection and (ii) instantaneous vorticity for various AoAs: (a) AoA = 4°, (b) AoA = 8°, (c) AoA = 12°, (d) AoA = 16°, and (e) AoA = 20°. From Gordnier [469].

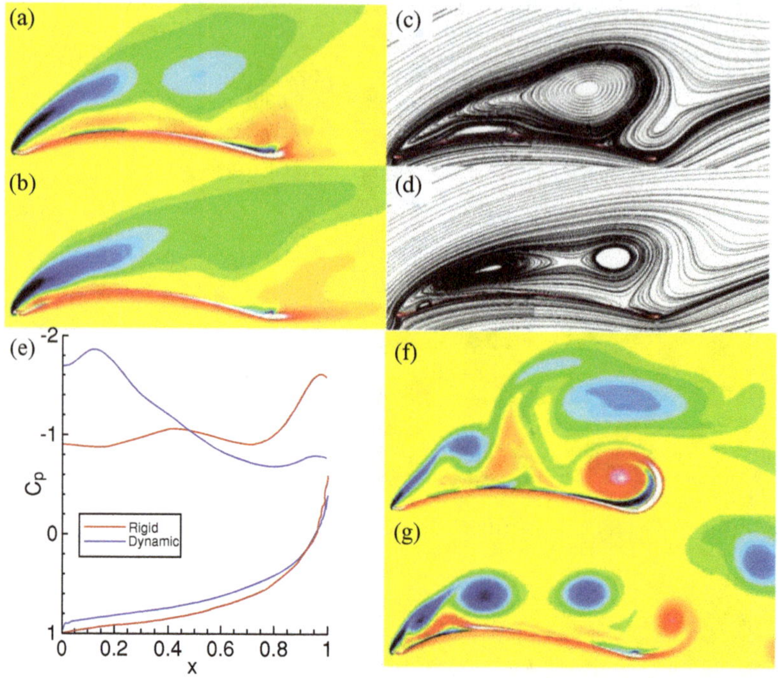

Figure 4.16. Comparison of the rigid (a,c,f) and dynamic (b,d,g) membrane airfoil solutions for AoA = 20°: a,b) mean vorticity, c,d) mean streamlines, e) surface pressure coefficient, f,g) instantaneous vorticity. From Gordnier [469].

AoA = 8°, a third mode standing wave response develops. At the higher AoAs (12°~20°), the membrane structural response exhibits a less regular behavior resulting from a combination of structural modes. The maximum peak-to-peak amplitude of the deflection grows with increasing AoA values of an order 0.07 times the chord.

To understand the effects of the motion of the membrane on the flow field, a computational investigation [470] was conducted for a rigid membrane at AoA = 20°. Figure 4.16 compares the mean and unsteady solutions for these two cases. The mean solutions for the rigid (Fig. 4.16a and c) and dynamic (Fig. 4.16b and d) cases show that the flow field around the rigid membrane exhibits a larger and stronger stall vortex and also a larger secondary separation bubble located farther downstream on the airfoil between $x = 0.171$ to $x = 0.44$. There is also a region of strong vorticity of the opposite sign near the trailing edge associated with the roll-up of a vortex. This notable difference between the rigid and dynamic mean flow results from a distinct change in the unsteady flow field as illustrated in Figure 4.16f and g. The dynamic motion of the membrane excites the separating shear layer at the leading edge, causing it to roll up sooner and to form a series of smaller vortices (Fig. 4.16g). The TEV in the dynamic case is also reduced in strength and tends to form downstream and away from the trailing edge, which reduces its influence on the airfoil. These changes between the rigid and dynamic flow fields lead to a different mean pressure distribution around the airfoil. This difference is greater on the upper surface where the rigid case exhibits a flat pressure distribution characteristic of a fully stalled flow, except near the trailing edge where the influence of the TEV is observed. In contrast, the dynamic case shows a strong suction region over the front portion of the wing but higher values of pressure downstream. The overall impact of the dynamic motion therefore appears to be a delay in the airfoil stall with an 8 percent increase in lift and a 15 percent reduction in drag compared to the rigid wing. Further investigation of the impact of the dynamic structural response on the overall aerodynamics is warranted, because it may provide a means for passively controlling the flow to provide improved airfoil performance through judicious aeroelastic tailoring.

Termed *adaptive washout* by Albertani et al. [457], this effect has been shown to reduce drag, give improved stall behavior, and allow for passive gust rejection. Several factors dictate the in-flight performance of a membrane wing MAV, including skeletal layout, stiffness, and membrane prestrain. The effect of batten thickness and spacing on MAVs during flight was investigated by Abudaram et al. [458], who showed that thin (less stiff) battens led to reduced wing bending and adaptive washout. Rojratsirikul et al. [459] investigated experimentally the effect of membrane prestrain and excess length on the fluid-structure interaction of 2D membrane airfoils. They tested the airfoils with prestrains of 0, 2.5, and 5 percent at AoAs varying from 9–30° and flow velocities of 5, 10, and 15 m/s ($Re = 5 \times 10^4 \sim 1.5 \times 10^5$). Time-averaged membrane tension was dominated by membrane pretension for low-flow velocities, gradually increasing to a uniform value for all cases of pretension. This study also indicated that models with pretension exhibited large flow-separation regions.

Hui et al. [460] studied the effect of structural stiffness on the aerodynamic response of MAVs ($Re = 7 \times 10^4$). They constructed five MAVs using identical airfoils; the first was made entirely of unidirectional carbon fiber to serve as a rigid basis, whereas the remaining four were constructed to be batten-reinforced membrane MAVs. The number of battens varied between the models: 1, 2, 3, and 10. They analyzed the aerodynamic performance of the various structures in conjunction with PIV measurements. The results showed that batten-reinforced membrane MAVs of extremely low stiffness, such as the one-batten model, are too flexible to maintain the airfoil shape during flight and provide the lowest L/D of any of the models. The remaining batten-reinforced models provide delayed stall, reduced drag, and higher L/D compared to the rigid airfoil wing models. It was also shown that a significant increase in battens, and therefore stiffness, as in the case of the 10-batten model, provides very similar aerodynamic performance to the rigid airfoil model. PIV measurements showed that at AoA > 10°, the rigid airfoil undergoes flow separation, whereas the batten-reinforced models deforms, reducing the effective AoA and allowing flow to stay attached longer. Vibrations of the membrane skins of BR models are noted for low AoA (<10°).

Lian and Shyy [154] compared the aerodynamic performance of a rigid wing to that of a batten-reinforced MAV using computations and showed that they have a comparable prestall performance, whereas the membrane MAV demonstrates delayed stall and increased lift after stall. Stanford et al. [461] characterized the deflection of a membrane wing using a steady-state computation. They compared the computed membrane MAV pressure distributions and static structural deflections to the visual image correlation (VIC) results. They identified Hooke's law as an accurate means of approximating the prestress of the stretched membrane, despite latex rubber's hyperelastic material properties.

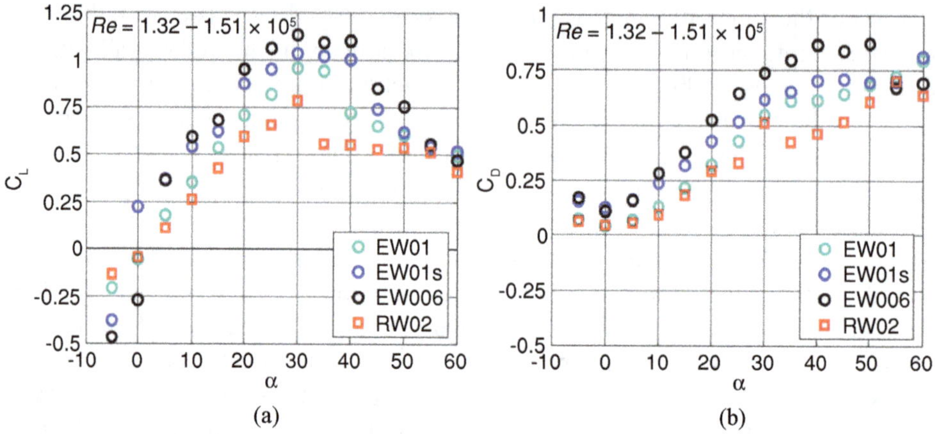

Figure 4.17. Lift and drag coefficients over rectangular wings with different flexibilities: RW02, a rigid steel-plate wing; EW006 and Eq01, thin and thick latex membranes; EW01s, a latex membrane with 6% slack [462].

Galvao et al. [462] measured the lift, drag, and deflection of a compliant rectangular membrane wing at a Reynolds number range of 7×10^4 to 2×10^5 and a AoA range from -5° to 60°. The wing was composed of a compliant latex membrane held between two stainless steel posts located at the leading and trailing edges. They tested four wing models in total: a thin non-compliant wing composed of steel shim stock (denoted RW02 in Figs. 4.17 and 4.18), two compliant membrane wings using latex rubber sheets of thickness 0.25 mm and 0.15 mm (denoted EW01 and EW006 in Figs. 4.17 and 4.18, respectively), and a latex membrane wing (0.25 mm thick) in which the membrane is given 6 percent slack (denoted EW01s in Figs. 4.17 and 4.18). Figure 4.17 depicts the lift coefficient for the test. The compliant wings have greater lift slope than the rigid wing, and the thinner compliant wing has greater lift slope than the thick compliant wing. Wing deflection measurements show that this greater lift slope is due to the increased camber for the compliant wing; this is consistent with the numerical results [158]. The thinner compliant wing stretches to a greater degree

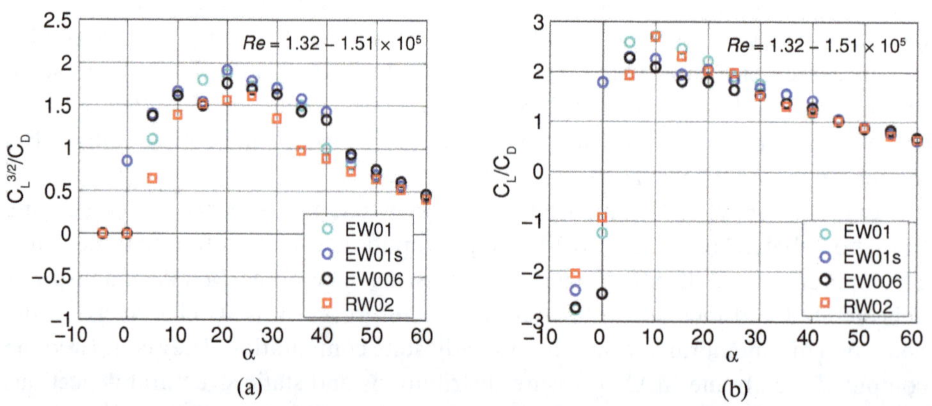

Figure 4.18. Measured power efficiency and lift-to-drag ratio over four wings with different flexibilities: RW02, a rigid steel-plate wing; EW006 and EW01, thin and thick latex membranes; EW01s, a latex membrane with 6% slack [462].

than the thicker membrane and therefore has a larger camber at the same AoA, resulting in a larger lift coefficient. Whereas Figure 4.8 shows that membrane wings have a similar lift slope to the rigid wing, Figure 4.17a indicates that flexible wings have a greater lift slope than a rigid wing. The seemingly contradictory conclusion is due to different experimental setups. Figure 4.8 is based on measurement of a MAV with a free trailing edge, which can be tilted up under forces [406]. As pointed out by Lian and Shyy [154], this trailing-edge deflection reduces the effective AoA.

The interplay among the camber, effective AoA, and the lift can be complicated. Before stall, the flexible structure in the experiments by Waszak et al. exhibits smaller effective AoA because the trailing edge is not fixed and the effective camber is reduced accordingly. In contrast, in the experiment of Galvao et al. [462], the trailing edge is fixed, resulting in a higher camber. They reported that a compliant wing can delay stall by 2 to 8 degrees of AoA, which is qualitatively consistent with the observation of Waszak et al. [406]. After stall, the lift coefficients for the compliant wings decrease in a more attenuated manner compared to the rigid wing. Close to the stall, the camber of the wing is observed to decrease. The de-cambering decreases the severity of the separation, thus delaying the sharp drop in lift force. This behavior enables the wing to sustain high lift at high AoAs. Furthermore, the compliant wings generate more lift at AoAs from 5° to 55°. The compliant wings are also found to yield more drag (Fig. 4.17) for two possible reasons. First, the enlarged camber increases the form drag. Second, the high-frequency fluctuation and vibration heighten the drag. This vibration becomes more noticeable when the trailing edge is not fixed, possibly leading to flutter. As we have explained before, during slow-forward flight, bats can only flex their wings slightly to avoid flutter. The compliant wings also demonstrate their superiority in terms of power efficiency ($C_L^{3/2}/C_D$) over a wide range of AoAs (Fig. 4.18a) [462]. This becomes more evident at higher AoAs. However, in terms of flight range efficiency (C_L/C_D), compliant wings have comparable performance to the rigid wing (Fig. 4.18b).

Lian and Shyy [463] numerically investigated the flexible airfoil aerodynamics. In their test, the upper surface of the airfoil is covered with a membrane that extends from 33–52 percent of the chord. No pretension is applied to the membrane. The membrane has a uniform thickness of 0.2 mm with a density of 1200 kg/m^3 and is considered as hyperelastic material. A computational test is performed at AoA = 4° and $Re = 6 \times 10^4$. It is observed that when flow passes the flexible surface, the surface experiences self-excited oscillation and the airfoil displays varied shape over time (Fig. 4.19). Analysis shows that the transverse velocity magnitude can reach as much as 10 percent of the free-stream speed (0.3 m/s). During the vibration, energy is transferred from the wall to the flow, and the separated flow is energized. Compared with corresponding rigid airfoil simulation, the surface vibration causes both the separation and transition positions to exhibit a standard variation of 6 percent of the chord length.

Figure 4.20 presents the time history of the lift coefficient for the membrane wing. Even though the time-averaged lift coefficient 0.60 of a flexible wing is comparable to that of the corresponding rigid wing, the membrane wing lift coefficient displays a time-dependent variation, with a maximum magnitude as much as 10 percent of its mean. The drag coefficient shows a similar pattern, but the time-averaged value closely matches that of the rigid wing. These observations are consistent with

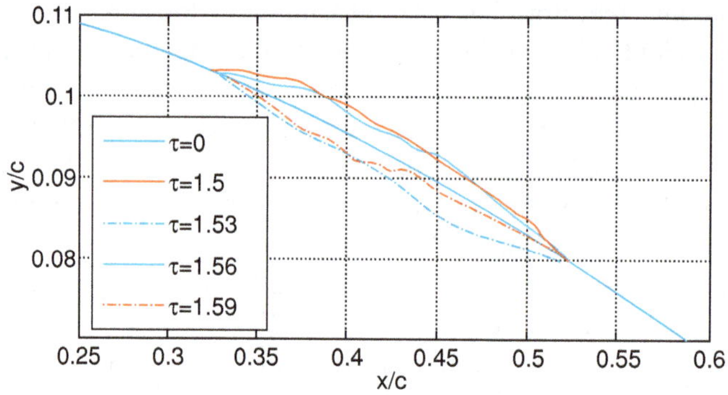

Figure 4.19. Membrane airfoil shapes in a steady free-stream at different time instants. The vibration changes the effective wing camber, where τ is the non-dimensional time, defined by tc/U [463].

previous efforts in 3D MAV wing simulations, without transitional flow models [154]. Furthermore, the experimental evidence also supports that, until the stall condition is reached, the membrane and rigid wings exhibit comparable aerodynamic performance. The flexible wing, in contrast, can delay the stall margin substantially [406] [462]. Using discrete Fourier transformation analysis the primary frequency of this flexible airfoil is found to be 167 Hz (Fig. 4.21). Given the airfoil chord (0.2 m) and free-stream speed (0.3 m/s), this high vibration frequency is unlikely to affect the vehicle stability. Figure 4.20 indicates that a low-frequency cycle exists in the high-frequency behavior in the lift coefficient history. This cycle, with a frequency of about 14 Hz, seems to be associated with the vortex shedding (Fig. 4.22). In a different simulation with laminar flow over a six-inch membrane wing (i.e., the entire wing surface is flexible), Lian and Shyy observed a self-excited structural vibration with a frequency around 120 Hz [154]; the experimental measurement of similar wings records a primary frequency around 140 Hz [406].

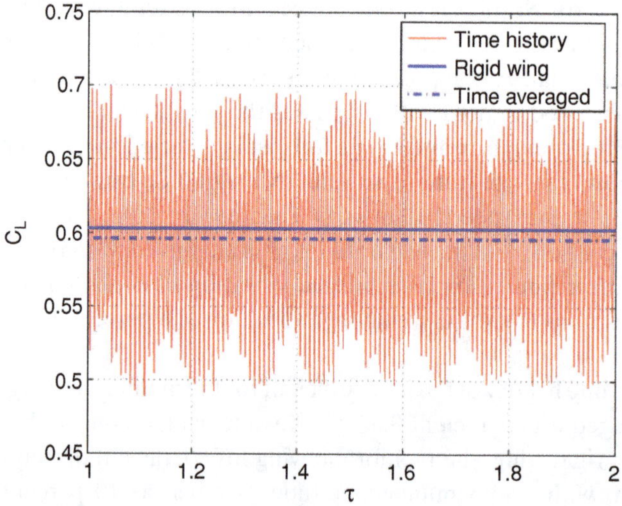

Figure 4.20. Time history of lift coefficient for membrane wing, showing both a high- and a low-frequency oscillation [463].

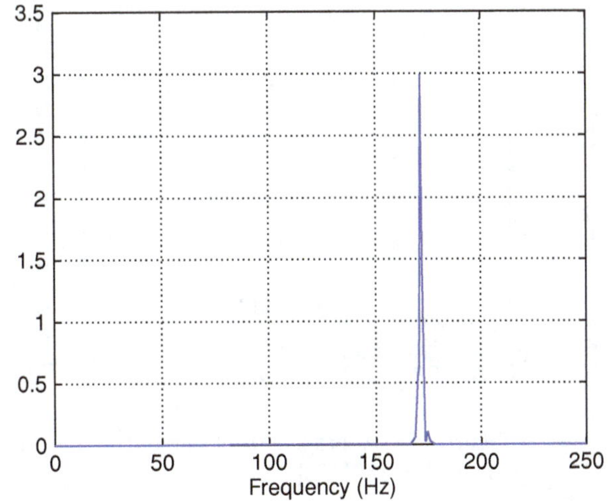

Figure 4.21. Power spectrum of the lift force. The dominated frequency is 167 Hz [463].

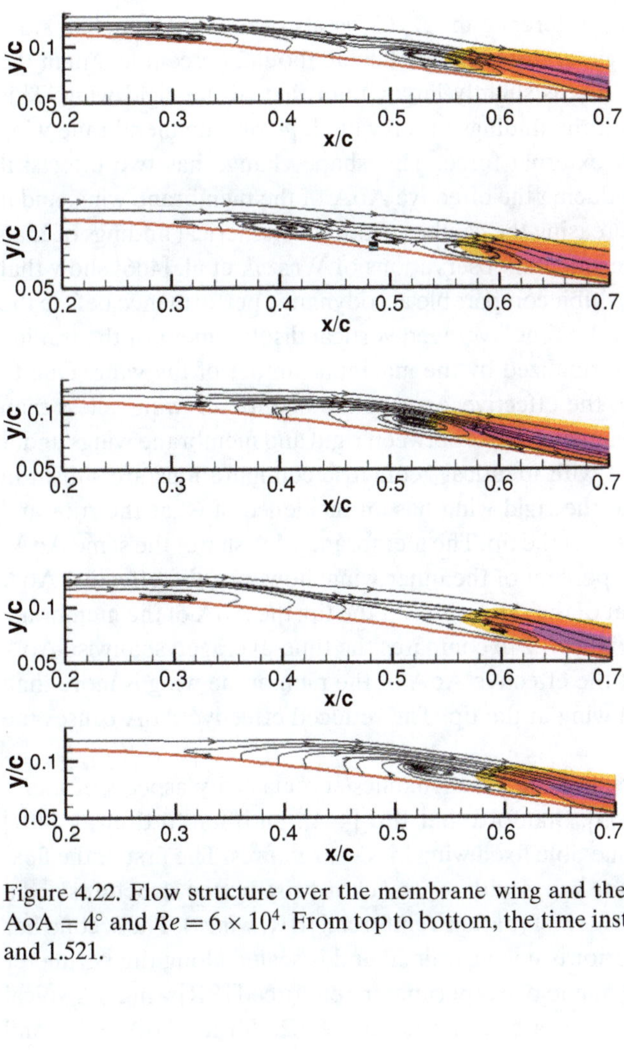

Figure 4.22. Flow structure over the membrane wing and the associated vortex shedding at AoA = 4° and $Re = 6 \times 10^4$. From top to bottom, the time instant $\tau = 1.5, 1.506, 1.512, 1.515$, and 1.521.

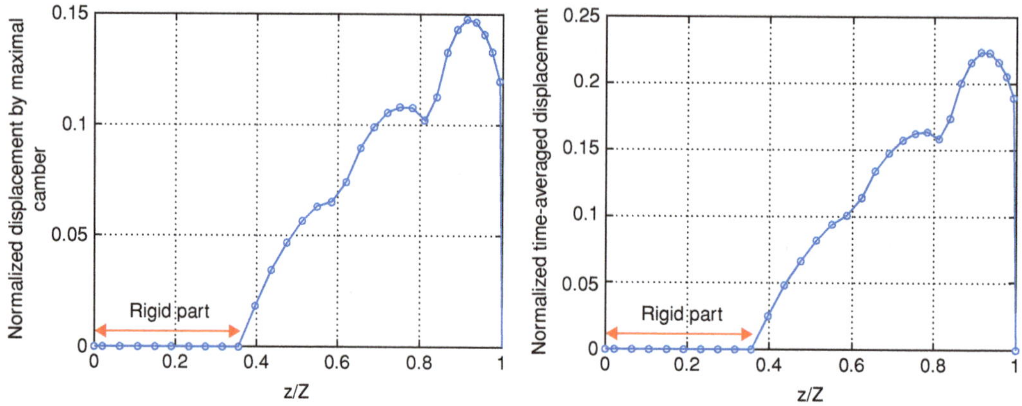

Figure 4.23. Averaged displacement of the membrane-wing trailing edge: (a) AoA = 6°; (b) AoA = 15° [158].

Lian et al. [160] compared the aerodynamics between membrane and rigid wings for MAV applications. The flexible wing exhibits slightly less lift coefficient than the rigid one at AoA = 6°. The difference in C_L/C_D is also small. At a higher AoA of 15°, the membrane wing generates a lift coefficient about 2 percent less than the rigid wing; however, its C_L/C_D is slightly larger than that of the rigid wing. This observation is consistent with the findings of Shyy et al. [456]. The membrane wing changes its shape under an external force. This shape change has two effects: it decreases the lift force by reducing the effective AoA of the membrane wing, and it increases the lift force by increasing the camber. Both the numerical findings of Lian and Shyy [158] and the experimental observations of Waszak et al. [406] show that membrane and rigid wings exhibit comparable aerodynamic performance before the stall limit. Figure 4.23 shows the time-averaged vertical displacement of the trailing edge. The displacement is normalized by the maximal camber of the wing. Due to the membrane deformation, the effective AoA of the membrane wing is less than that of the rigid wing. The spanwise AoAs between rigid and membrane wings under the same flow condition and with identical geometric configurations are shown in Figure 4.24. In Figure 4.24a, the rigid wing has an incidence of 6° at the root and monotonically increases to 9.5° at the tip. The membrane wing shares the same AoAs with the rigid wing in the 36 percent of the inner wing; however, the effective AoA toward the tip is less than that of the rigid wing. At the tip, the AoA of the membrane wing lowers by about 0.8°. Figure 4.24b compares the time-averaged spanwise AoA at AoA = 15°, showing that the effective AoA of the membrane wing is more than 1° less than that of the rigid wing at the tip. The reduced effective AoA causes the decrease in the lift force.

In an attempt to understand the aerodynamics/aeroelasticity aspects of membrane wings, Shyy et al. [455], Stanford and Ifju [464], and Stanford et al. [407] studied a rigid wing and two flexible fixed wing MAV structures. The first of the flexible wing MAV has membrane wings with several chordwise batten structures and a free trailing edge for geometric twist (batten-reinforced [BR] wings). The second has membrane wings whose interior is unconstrained and is sealed along the perimeter to a stiff laminate for aerodynamic twist (perimeter-reinforced [PR] wings). Typical flow structures for all three wings are shown in Figure 4.25 for an AoA = 15° and

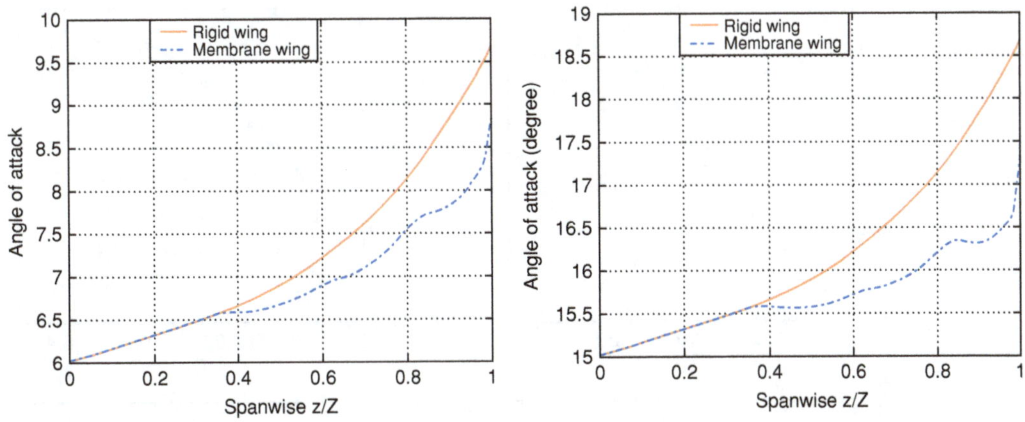

Figure 4.24. Time-averaged spanwise AoA for membrane wing: (a) AoA = 6°; (b) AoA = 15° [463].

$U_\infty = 15$ m/s. The two hallmarks of MAV aerodynamics can be seen from the flow over the rigid wing: the low Reynolds number (10^5) causes the laminar boundary layer to separate against the adverse pressure gradient at the wing root, and the low aspect ratio ($AR = 1.2$) forces a strong TiV swirling system, leaving a low-pressure region at the wingtip. Flow over the flexible BR wing is characterized by pressure undulations over the surface [465], where the membrane inflation between each batten redirects the flow. The shape adaptation decreases the strength of the adverse pressure gradient and thus the size of the separation bubble. A large pressure spike develops over the PR wing at the leading edge of the membrane skin. The pressure recovery over the wing is shifted aftward, and the flow separates as it travels down the inflated shape, where it is then entrained into the low-pressure core of the TiV. This interaction between the TiVs and the longitudinal flow separation is known to lead to unsteady vortex destabilization at high AoA [167]; no such relationship is obvious for the BR and rigid wings. The low-pressure regions at the wingtips of the two membrane wings are weaker than those observed on the rigid wing, presumably due to energy considerations: strain energy in the membrane may remove energy from the lateral swirling system. Furthermore, the inflated membrane shape may act as a barrier to the tip vortex formation.

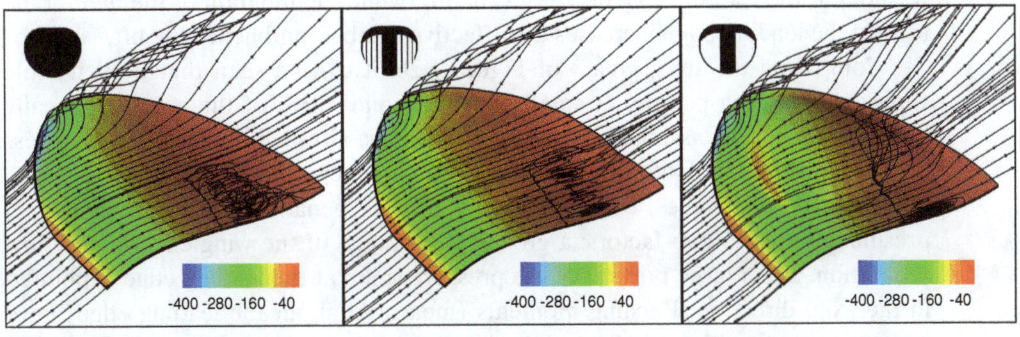

Figure 4.25. Streamlines and pressure distributions (Pa) over the top wing surface: AoA = 15°, $U_\infty = 15$ m/s [407].

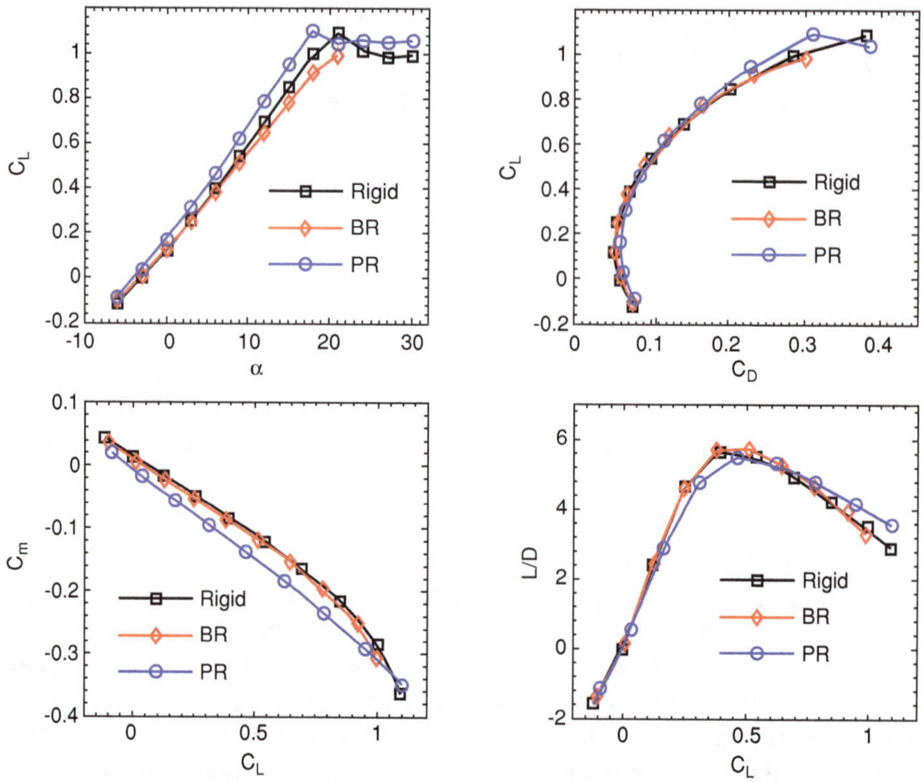

Figure 4.26. Computed aerodynamic performance: AoA = 15°, U_∞ = 15 m/s [407].

The lift, drag, and pitching moment coefficients for these three wings through an α-sweep are shown in Figure 4.26. The C_L-α relationships are mildly non-linear (20–25 percent increase in C_L between 0° and 15°) due to growth of the low-pressure cells at the wingtip. Further characteristics of a low *AR* are given by the high stall angle, computed as being 21° for the rigid case. The aerodynamic twist of the PR wing increases C_L (by as much as 8 percent), making the MAV more susceptible to gusty conditions. $C_{L,max}$ is slightly higher as well, subsequently lowering the stall angle to 18°. The adaptive washout of the BR wing decreases C_L (by as much as 15 percent over that of the rigid wing), though the change is negligible at lower AoAs. This decrease is thought to be a result of two offsetting factors: the adaptive washout at the trailing edge decreases the lift, while the inflation of the membrane toward the leading edge increases the effective camber, and hence the lift.

Comparing the drag polars of Figure 4.26, it can be seen that both flexible wings incur a drag penalty at small lift values, indicative of the aerodynamically non-optimal shapes assumed by the flexible wings, although the BR wing has less drag at a given AoA [466]. The drag difference between the rigid and the BR wing is very small, while the PR wing displays a larger penalty. This larger penalty is presumably due to two factors: a greater percentage of the wing experiences flow separation, and a large portion of the pressure spike at the leading edge is pointed in the axial direction. Pitching moments (measured about the leading edge) have a negative slope with both C_L and AoA, as necessitated by stability requirements. Non-linear trends due to low *AR* effects are again evident. Both the BR and the PR

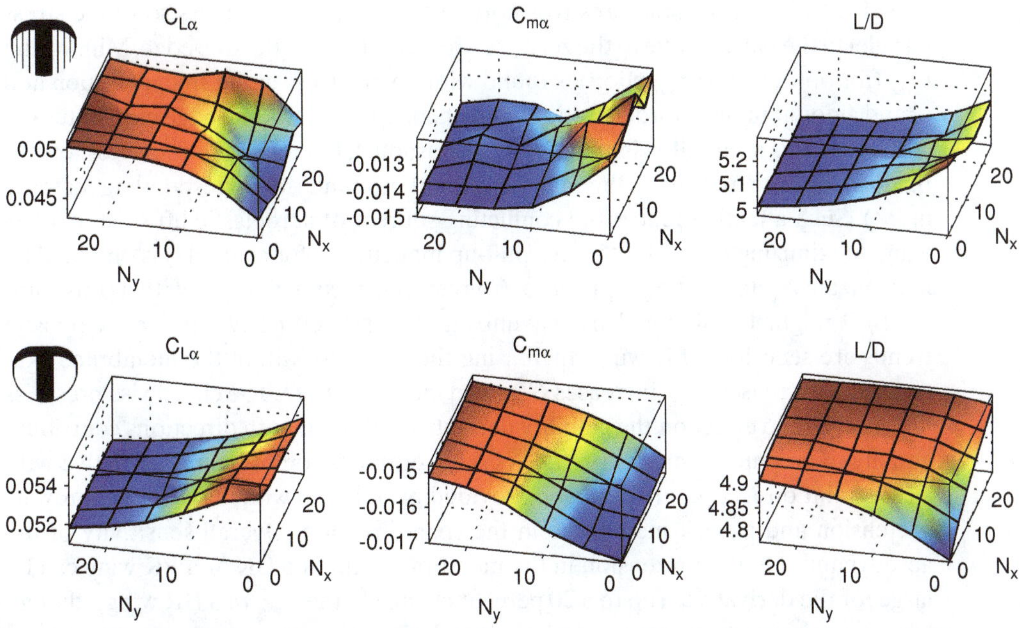

Figure 4.27. Aeroelastic tailoring of chordwise (N_x) and spanwise (N_y) membrane prestress resultants (N/m): contour represents z-axis values [407].

wings have a lower $\partial C_m/\partial C_L$ than the rigid wing, though only the PR wing shows a drastic change (by as much as 15 percent) as a result of the membrane inflation, which shifts the pressure recovery toward the trailing edge, adaptively increasing the strength of the restoring pitching moment with increases in lift/AoA [461]. Steeper C_m slopes indicate larger static margins: stability concerns are a primary target of design improvement from one generation of MAVs to the next. The range of flyable center of gravity (CG) locations is generally only a few millimeters long; meeting this requirement represents a strenuous weight management challenge. Furthermore, the PR wing displays a greater range of linear C_m behavior, possibly because the adaptive membrane inflation quells the strength of the low-pressure cells, as discussed earlier. No major differences appear between the L/D characteristics of the three wings for low AoAs. At moderate angles, the large drag penalty of the PR wing decreases the efficiency, while the BR wing slightly outperforms the rigid wing. At higher angles, both the lift and drag characteristics of the PR wing are superior to the other two, resulting in the best L/D ratios.

Aeroelastic tailoring conventionally uses unbalanced laminates for bend/twist coupling, but the pretension within the membrane skin has an enormous impact on the aerodynamics: for the 2D case, higher pretension generally pushes flexible wing performance to that of a rigid wing. For a 3D wing, the response can be considerably more complex, depending on the nature of the membrane reinforcement. Effects of increasing the membrane pretension may include a decrease in drag, decrease in C_L, linearized lift behavior, increase in the zero-lift AoA, and more abrupt stalling patterns. Furthermore, aeroelastic instabilities pertaining to shape hysteresis at low AoAs can be avoided with specific ratios of spanwise-to-chordwise pretensions [467].

Increasing the prestress within the membrane skin of a BR wing (Fig. 4.27) generally increases $C_{L\alpha}$, decreases $C_{m\alpha}$, and decreases L/D. The system is very sensitive to changes in the prestress normal to the battens and less so to the stress parallel to the battens, due to the zero prestress condition at the free edge. Minimizing $C_{L\alpha}$ (for optimal gust rejection) is found with no prestress in the span direction and a mild amount in the chord direction. The unconstrained trailing edge eliminates the stiffness in this area (allowing for adaptive washout), but retains the stiffness toward the leading edge, removing the inflation seen there (and the corresponding increase in lift). Such a tactic reduces the conflicting sources of aeroelastic lift seen in a BR wing. Maximizing $C_{L\alpha}$ (for effective pull-up maneuvers, for example) is obtained by maximizing N_y and setting N_x to zero. Conversely, maximizing $C_{L\alpha}$ with a constraint on L/D might be obtained by maximizing N_x and setting N_y to zero. Opposite trends are seen for a PR wing. Increasing the prestress within the membrane skin generally decreases $C_{L\alpha}$, increases $C_{m\alpha}$, and increases L/D. The chordwise prestress has a negligible effect on the stability derivatives, though both directions contribute equally to an improvement in L/D. As such, optimization of either derivative with a constraint on L/D could easily be provided by a design with maximum chordwise pretension and a slack membrane in the span direction. Overall sensitivity of the aerodynamics to the pretension in the membrane skin of a BR or a PR wing can be large for the derivatives (up to a 20 percent change in the $C_{m\alpha}$ of a BR wing), though less so for the wing efficiency. Variations in L/D are never more than 5 percent.

More recently, membrane wing MAVs have been developed and tested to determine the structural response of BR membrane wings for varying conditions: small AoAs, number of battens, and membrane pretension [468]. A self-excited instability (flutter) was noted for each model with limit cycle oscillations occurring post-flutter flow velocities. These experiments showed that increasing the membrane pretension and the number of structural battens for the membrane wing MAVs delays the flutter velocity and reduces the magnitude of limit cycle oscillations at a given flow velocity.

4.4.2 Flapping Flexible Wings

Research into aeroelasticity of flapping wings has recently increased though a full picture of the basic aeroelastic phenomena has not yet been obtained [451]. One important question that needs to be answered to understand the key fundamentals of flapping flexible wings is, Are aerodynamic loads essential to determine the wing deformation during the flights of biological flyers? Daniel and Combes [470] analytically addressed this question and suggested that aerodynamic loads are relatively unimportant in determining the bending patterns in oscillating wings, when the ratio between the wing structure and the surrounding air is high. Subsequently, experimental investigations by Combes [471] and Combes and Daniel [472] found that the overall bending patterns of a Hawkmoth wing when flapped (single degree-of-freedom flap rotation) in air and helium are quite similar, despite a 85 percent reduction in fluid density in the latter, suggesting that the contribution of aerodynamic forces is relatively small compared to that of inertial-elastic forces during flapping motion. However, they also mentioned that realistic wing kinematics may include rapid rotation at the stroke reversals that may lead to increased aerodynamic forces due to unsteady aerodynamic mechanisms (see also Section 3.3). Furthermore, static bending tests done by Combes and Daniel [402] showed anisotropy of wing structures in a variety of insect species. Mountcastle and Daniel [473] investigated the influence of wing compliance on the mean advective flows (indicative of induced flow velocity) using PIV measurement. Their results demonstrate that flexible wings yield mean advective flows with substantially greater magnitudes and orientations that are more beneficial to lift than do stiff wings.

For simpler flapping wing configurations, Zhu [474] numerically investigated the unsteady oscillation of a flexible wing and found that, when the wing is immersed in air, the chordwise flexibility reduces both the thrust and the propulsion efficiency, whereas spanwise flexibility (through equivalent plunge and pitch flexibility) increases the thrust without efficiency reduction within a small range of structural parameters. However, when the wing is immersed in water, the chordwise flexibility increases the efficiency and the spanwise flexibility reduces both the thrust and the efficiency. Shkarayev et al. [475] investigated the aerodynamics of cambered membrane flapping wings. Specifically, they introduced a cambered airfoil into the wing by shaping metal ribs attached to the membrane skin of the 25 cm wingspan model. They found that the thrust force generated by a 9 percent camber wing is 30 percent higher than that of a flat wing of the same size. Adding a dihedral angle to the wings and keeping the flapping amplitude constant improves the cambered wing's performance even further. The aerodynamic coefficients are defined using a reference velocity as a sum of two components: a free-stream velocity and a stroke-averaged wingtip flapping velocity. The lift, drag, and pitching moment coefficients obtained using this procedure collapse well for studied advance ratios, especially at lower AoAs.

Hui et al. [476] examined various flexible wing structures to evaluate their implications on flapping wing aerodynamics. They showed that the flexible membrane wings have better overall aerodynamic performance (i.e., lift-to-drag ratio) over the rigid wing for soaring flight, especially for high-speed soaring flight or at a relatively high AoA. The rigid wing has better lift production performance for flapping flight in general. The latex wing, which is the most flexible among the three tested wings, has the best thrust generation performance for flapping flight. The less flexible nylon wing, which has the best overall aerodynamic performance for soaring flight, is the worst for flapping flight applications.

Kim et al. [477] developed a bio-mimetic flexible flapping wing using micro-fiber composite actuators and experimentally investigated the aerodynamic performance of the wing under flapping and non-flapping motion in a wind tunnel. Results showed that the camber due to wing flexibility could produce positive effects (i.e. stall delay, drag reduction, and stabilization of the LEV) on flapping wing aerodynamics in quasi-steady and unsteady regions. Mueller et al. [478] presented a versatile experimental test for measuring the thrust and lift of a flapping wing MAV. They showed an increase in average thrust due to increased wing compliance and the detrimental influence of excessive compliance on drag forces during high-frequency operation. Also they observed the useful effect of compliance on the generation of extra thrust at the beginning and end of flapping motions.

Watman and Furukawa [479] investigated the effects of passive pitching motions of flapping wings on aerodynamic performance using robotic wing models. They considered two types of passive flapping wing models. The first model used a rigid connection between all parts of the structure. This design was utilized in several

MAVs [480] and served as a common design used for the analysis of flapping wings. The second model was designed to allow the free rotation of ribs and membrane over a limited angle. This design was used recently in a small MAV prototype [481]. They showed that the former passive flapping wing (constrained) has better performance compared to the latter design because of favorable variation in the passive pitching angle of the wing.

Wu et al. [482] conducted a multidisciplinary exercise correlating flapping wing MAVs' aeroelasticity and thrust production by quantifying and comparing the elasticity, dynamic responses, and air flow patterns of six different pairs of MAV wings (in each one, the membrane skin was reinforced with different leading-edge and batten configurations) of the Zimmerman planform (two ellipses meeting at the quarter chord) with varying elastic properties. In their experiments, single degree-of-freedom flapping motion was prescribed to the wings in both air and vacuum. Among many conclusions, they found that, within the range of flexibility considered, more flexible wings are more thrust-effective at lower frequencies, whereas stiffer wings are more effective at higher frequencies. They hypothesized that flexible wings may have a certain actuation frequency for peak thrust production and that the performance would degrade once that frequency is passed. A rapidly growing number of studies on flexible flapping wings have been reported recently [351] [451]. Because the complicated problems arise from the anisotropy of the wing structure and the interaction among aerodynamic loadings, inertia, and elastic forces, it is worth understanding the aerodynamics and wing deformation of the simplified flexible wing model before tackling the anisotropic wing structure. This approach helps us to make the link between MAVs and biological flyers. In the following section, we highlight the studies focusing on simplified flexible wing models. namely chordwise and spanwise flexibility and isotropy.

4.4.2.1 Chordwise Flexible Wing

Katz and Weihs [483] analyzed the generation of hydrodynamic forces by the motion of a uniform and massless flexible foil in a large-amplitude curved motion in an inviscid incompressible flow. They found that the chordwise flexibility increases the propulsive efficiency by up to 20 percent while causing small decreases in the overall thrust, compared with similar motion with rigid foils.

Pederzani and Haj-Hariri [484] performed computational analyses on a rigid wing from which a portion was cut out and covered with a very thin and flexible material (latex). They showed that due to a snapping motion (i.e., non-zero velocity in the direction opposite to that of the following stroke of the latex at the beginning of each stroke), the strength of the vortices that are shed is higher in lighter wing structures, leading to the generation of more thrust. Furthermore, snapping such structures requires less input power than snapping heavier ones. Using inviscid flow theory and beam equations, Chaithanya and Venkatraman [485] investigated the influence of inertial effects due to prescribed motion on the thrust coefficient and propulsive efficiency of a plunging/pitching thin plate. Their results demonstrated that flexible airfoils with inertial effects yield more thrust than those without inertial effects. This is due to the increase in the fluid loading in the former that subsequently leads to an increase in the deformation. Due to their shape, deformed airfoils produce a force component along the forward velocity direction [485].

Gopalakrishnan [486] analyzed the effects of elastic cambering of a rectangular membrane flapping wing on aerodynamics in forward flight using a linear elastic membrane solver coupled with an unsteady LES method. They investigated different membrane prestresses to give a desired camber in response to the aerodynamic loading. The results showed that the camber introduced by the wing flexibility increases the thrust and lift production considerably. Analysis of flow structures revealed that the LEV stays attached on the top surface of the wing, follows the camber, and covers a major part of the wing, which results in high force production. Attar et al. [487] examined the effect of Strouhal number, reduced frequency, and static AoAs on the structural and fluid response of the plunging membrane airfoil. They showed that, at a low AoA and a low Strouhal number, increasing reduced frequency results in a decrease in the mean sectional lift and an increase in the drag coefficients; increasing the Strouhal number significantly affects the lift generation at a low AoA and an intermediate value of reduced frequency. They also observed that, when the effective AoA is studied for fixed values of the Strouhal number and reduced frequency, the act of plunging gives improved mean sectional lift when compared with the case of a fixed flexible airfoil (see Section 4.4.1). In contrast, for rigid wings, which they also considered, the LEV lifts off from the surface, resulting in low force production.

To evaluate the role played by the LEV for a flexibly cambered airfoil, Gulcat [488] investigated (i) a thin rigid plate in a plunging motion, (ii) a flexibly cambered airfoil whose camber was changed periodically, and (iii) the plunging motion of a flexibly cambered airfoil. The leading-edge suction force for all cases was predicted by means of the Blasius theorem, and the time-dependent surface velocity distribution of the airfoil was determined by unsteady aerodynamic considerations. Gulcat [488] reported that the viscous effects obtained by the unsteady boundary-layer solution produce very little alteration to the oscillatory behavior of the net propulsive force. These forces only reduce the amplitude of the leading-edge suction force obtained by the unsteady aerodynamic theory. Heaving plunging makes the major contribution to the thrust; therefore, it is possible to get high propulsion efficiency with limited camber flexibility.

Miao and Ho [489] prescribed a time-dependent flexible deformation profile for an airfoil in pure plunge and investigated the effect of flexure amplitude on the unsteady aerodynamic characteristics for various combinations of Reynolds numbers and reduced frequency. For the specific combination of Reynolds number, reduced frequency, and plunge amplitude, the results showed that thrust-indicative wake structures are observed behind the trailing edge of those airfoils with flexure amplitudes of 0.0–0.5 of the chord length. This wake structure evolves into a drag-indicative form as the flexure amplitude of the airfoil is increased to 0.6 and 0.7 of the chord length. Studies conducted under various combinations of Reynolds numbers and reduced frequency showed that the propulsive efficiency of a chordwise flexible airfoil in pure plunge is influenced primarily by the value of the reduced frequency rather than by the Reynolds number.

Toomey and Eldredge [490] performed numerical and experimental investigations to understand the role of flexibility in flapping wing flight using two rigid elliptical sections connected by a hinge with a torsion spring. The section at the leading edge was prescribed with fruit-fly-like hovering wing kinematics [201], whereas the trailing-edge section responded passively due to the fluid dynamic and inertial/elastic forces. It was found that the lift force and wing deflection are primarily controlled by the nature of the wing rotation. Faster wing rotation, for example, leads to larger peak deflection and lift generation. Advanced rotation also leads to a shift in the instant of peak wing deflection, which increases the mean lift. In contrast to the rotational kinematics, the translational kinematics has very little impact on spring deflection or force. And although the rotational kinematics is nearly independent of the Reynolds number, the translational kinematics increases with increasing Reynolds number.

Poirel et al. [491] conducted a wind-tunnel experimental investigation of self-sustained oscillations of an aeroelastic NACA 0012 airfoil occurring in the transitional *Re* regime; in particular, aeroelastic limit cycle oscillations for the airfoil constrained to rotate in pure pitch. The structural stiffness and the position of the elastic axis were varied. Their investigation suggested that laminar separation plays a role in the oscillations, either in the form of trailing-edge separation or due to the presence of a laminar separation bubble.

Vanella et al. [492] conducted numerical investigations on a similar structure and found that the best performance (up to approximately 30 percent increase in lift) is realized when the wing is excited by a non-linear resonance at one third of its natural frequency. For all Reynolds numbers considered, the wake-capture mechanism is enhanced by a stronger flow around the wing at stroke reversal, resulting from a stronger vortex at the trailing edge.

Heathcote et al. [493] investigated the effect of chordwise flexibility on aerodynamic performance of an airfoil in pure plunge under hovering conditions. Because the trailing edge is a major source of shedding of vorticity at zero free-stream velocity, they showed that the amplitude and phase angle of the motion of the trailing edge affect the strength and spacing of the vortices, as well as the time-averaged velocity of the induced jet. Direct force measurements confirmed that, at high plunge frequencies, the thrust coefficient of the airfoil with intermediate stiffness is highest, although the least stiff airfoil could generate larger thrust at low frequencies. It was suggested that there is an optimum airfoil stiffness for a given plunge frequency and amplitude. Similar conclusions were made in another study [494] that analyzed the influence of resonance on the performance of a chordwise flexible airfoil prescribed with pure plunge motion at its leading edge. It was shown that although the mean thrust could increase with an increase in flexibility, below a certain threshold the wing is too flexible to communicate momentum to the flow. Yet, too much flexibility leads to a net drag, and hence, only a suitable amount of flexibility is desirable for thrust generation.

Although most of the recent computational and experimental studies have explored the role of wing flexibility in augmenting aerodynamic performance while focusing on single wings at relatively higher Reynolds numbers, Miller and Peskin [495] numerically investigated the effect of wing flexibility on the forces produced during clap-and-fling/peel motion [68] of a small insect ($Re \approx 10$) focusing on wing–wing interactions. They prescribed both clap-and-fling kinematics separately to a rigid and a chordwise flexible wing and showed that, although lift coefficients produced during the rigid and flexible clap strokes are comparable, the peak lift forces are higher in the flexible cases than in the corresponding rigid cases. This is due to the peel motion that delays the formation of the TEVs, thereby maintaining vortical asymmetry and augmenting lift for longer periods [496].

Zhao et al. [497] investigated the chordwise flexibility effects on the LEVs and aerodynamic force generation using 16 different dynamically scaled mechanical flexible model wings in quiescent fluid at a Reynolds number of 2×10^3 [498]. Findings from their experiments are that the magnitude of the LEV correlates with the aerodynamic forces generated by the flapping wing, and the camber influences instantaneous aerodynamic forces through the modulation of the LEV. Du and Sun [499] numerically investigated the effect of prescribed time-varying twist and chordwise deformation on the aerodynamic force production of a fruit-fly-like model wing in hover. They showed that aerodynamic forces on the flapping wing are not affected much by the twist, but by the camber deformation. The effect of combined camber and twist deformation is similar to that of camber deformation alone. With a deformation of 6 percent camber and 20° twist (typical values observed for wings of many insects), the lift increases by 10–20 percent compared to the case of a rigid flat plate wing. They therefore showed that chordwise deformation could increase the maximum lift coefficient of a fruit fly model wing and reduce its power requirement for flight.

Lee et al. [500] numerically investigated a tadpole-type wing that consists of a rigid leading edge and a flexible plate at *Re* of 2.5×10^3. They considered two elastic property distributions: homogeneous and linear. Their findings showed that the structural deformation changes the effective AoA, the vortex intensity, and the direction of the net force vector, resulting in a 13 percent increase in lift for the homogeneous type and a 33 percent increase in propulsive efficiency and force for the linear type distribution. Mountcastle and Daniel [501] numerically studied how aerodynamic force production and control potential are affected by pitch/elevation phase and variations in wing flexural stiffness. They showed that lift and thrust forces are highly sensitive to flexural stiffness distributions with performance optima that lie in different phase regions.

In another study, Heathcote and Gursul [336] performed water-tunnel studies to examine the thrust and efficiency of a flexible 2D airfoil plunging in forward flight. Their experimental setup is shown in Figure 4.28. Their airfoil was made up of a rigid teardrop 30 percent chord and a flexible flat plate 70 percent chord (see Fig. 4.28b). They changed the chordwise flexibility of the airfoil by changing the thickness of the flat plate. Following the experiment of Heathcote and Gursul [336], Kang et al. [351] explored the thrust enhancement induced by chordwise flexibility; they computed the thrust of a purely plunging chordwise flexible airfoil for different thickness ratios ($h_s^* = 4.23 \times 10^{-3}$, 1.41×10^{-3}, 1.13×10^{-3}, 0.85×10^{-3}, and 0.56×10^{-3}) and motion frequencies that produce Strouhal numbers between $St = 0.085$ and 0.3 with 0.025 increments, with the plunge amplitude kept fixed to $h_a/c_{\mathrm{m}} = 0.194$. The reduced frequency k then varies between 1.4 and 4.86. The airfoil consisted of a rigid teardrop leading edge and an elastic plate that plunged sinusoidally in the free-stream. As shown in Figure 4.29 variation in the thickness changes Π_1, whereas the motion frequency affects both k and St. Detailed experimental setup and discussion of fluid physics are in Heathcote and Gursul [336].

Shyy et al. [450] obtained a numerical solution for $St = 0.17$ for different thickness ratios. They used an Euler-Bernoulli beam solver to solve Eq. (4–1) for the

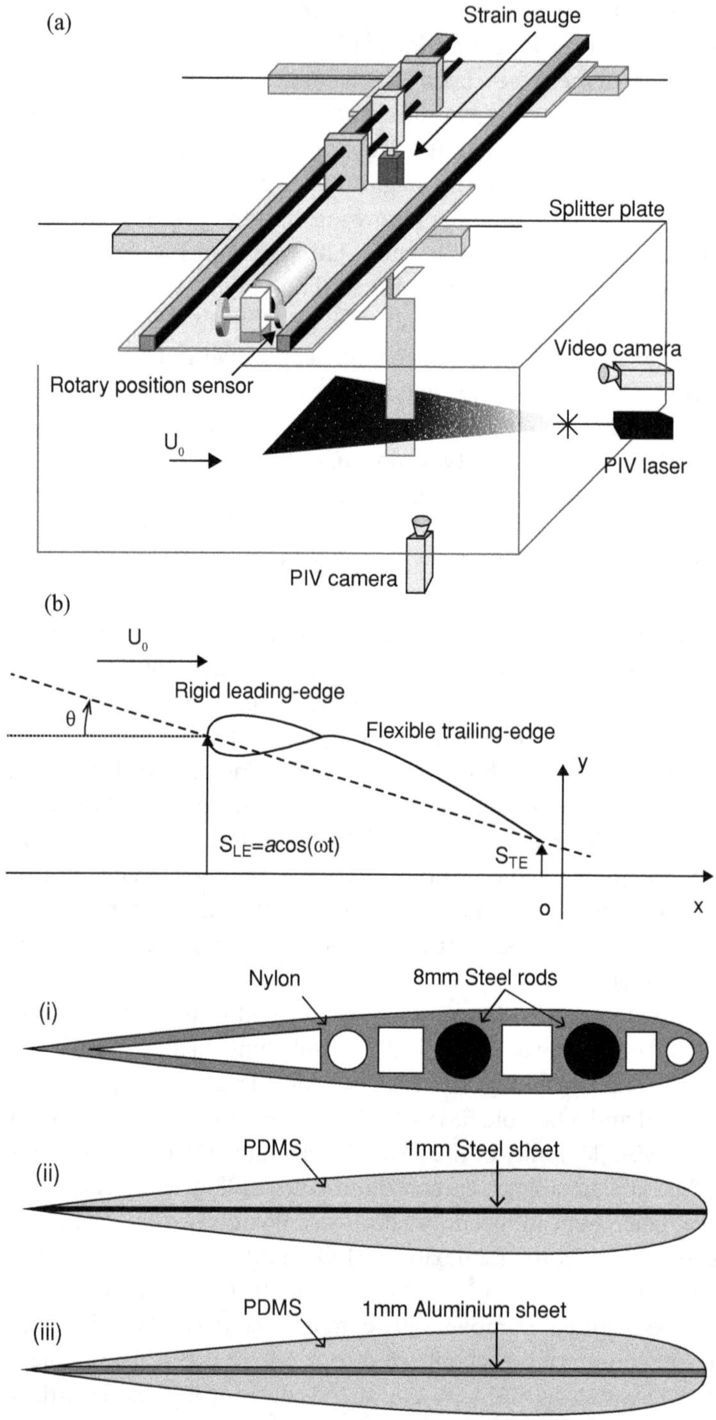

Figure 4.28. (a) Water-tunnel experimental setup for force measurements, wing deformation measurements, and PIV measurements; (b) chordwise flexible airfoil configuration; (c) cross-sections of spanwise flexible wing configuration. From Heathcote and Gursul [336] and Heathcote, Wang, and Gursul [502].

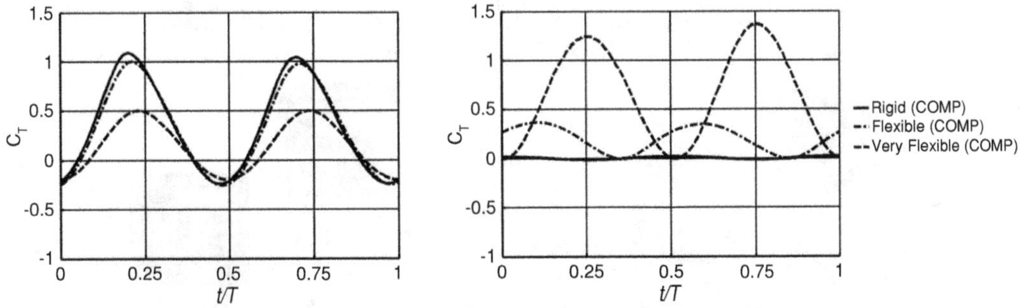

(a) C_T generated by the rigid teardrop (b) C_T generated by the flexible flat plate

Figure 4.29. Time histories of thrust coefficient contribution due to the teardrop and the flexible plate separately at $St = 0.17$: (a) response of the teardrop; (b) response of the flexible plate. From Shyy et al. [450].

deformation of the elastic flat plate, while the rigid teardrop moved with the imposed kinematics. Furthermore, the Reynolds number $Re = 9.0 \times 10^3$ and the density ratio $\rho^* = 7.8$ were held constant in all cases. One of the mechanisms found by Shyy et al. [450] is that the chordwise deformation of the rear flexible plate in both flexible and very flexible cases results in an effective projected area for the thrust forces to develop. For $St = 0.17$, the thrust coefficient as a function of normalized time for the rigid, flexible, and very flexible wings is shown in Shyy et al. [450] and in Figure 4.29. To estimate the individual contribution of the teardrop and the flexible plate to force generation, the time histories of the thrust coefficient are shown separately for each element. The thrust response to variation in flexibility is different for each element: with increasing chordwise flexibility of the plate, the instantaneous thrust contributed by the flexible plate increases.

The interplay between the motion frequency indicated by St and the resulting thrust and wingtip displacement is further illustrated in Figure 4.30. For the flexible airfoil the resulting thrust generation increases with the increased motion frequency (k and St); the maximum wingtip displacement also shows monotonic increase with the motion frequency. A striking observation is that the vorticity field looks similar for all Strouhal numbers shown, but the pressure contours and the resulting thrust time histories differ in value. This could be related to the scaling proposed in Section 4.5 that the force acting on a moving body is largely dominated by the motion of the airfoil and less by the vorticity in the flow field at high reduced frequencies. A similar trend is shown for the very flexible airfoil, whose thickness ratio is about 2.5 times smaller than for the flexible airfoils. The thrust increases with higher k and St, but the maximum tip amplitude saturates for $St = 0.15, 0.25$, and 0.4. Instead of resulting in a larger tip amplitude motion, higher motion frequency leads to a larger phase lag of the wingtip relative to the wing root. Increasing motion frequency leads to higher acceleration of the wing, and hence greater force generation. However, eventually the fluid dynamics time scale and reponse become limiting factors, as discussed in Section 4.5.

Figure 4.31 shows the time-averaged thrust coefficient for a range of motion frequencies from the numerical computation [351] and the experimental measurements [336]. For the thickest flat plate ($h_s^* = 4.23 \times 10^{-3}$) the computed thrust compares well with the experimental measurements. At higher motion frequencies, $St = 0.28$ and

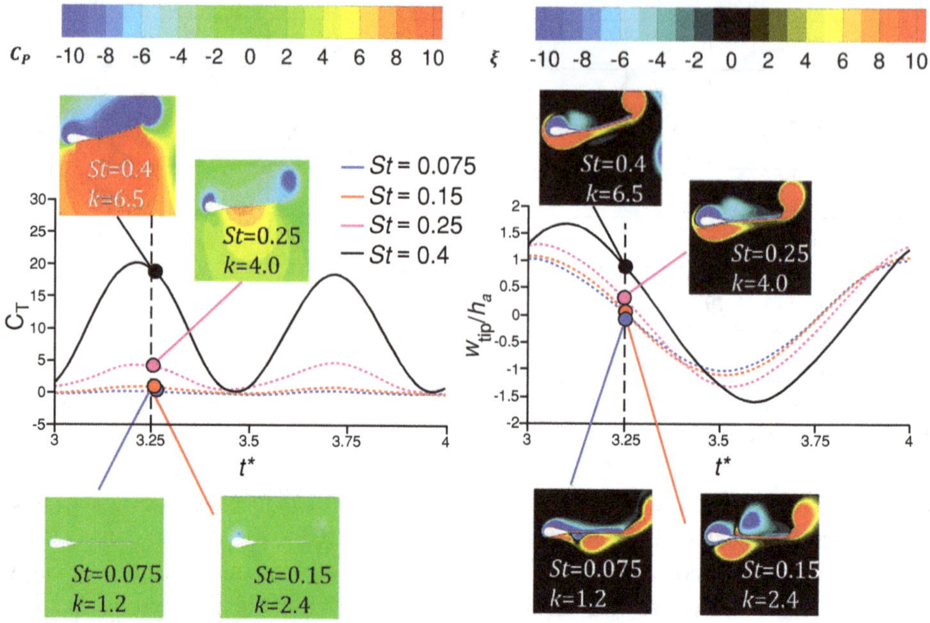

(a) *Flexible* ($h_s^* = 1.41 \times 10^{-3}$)

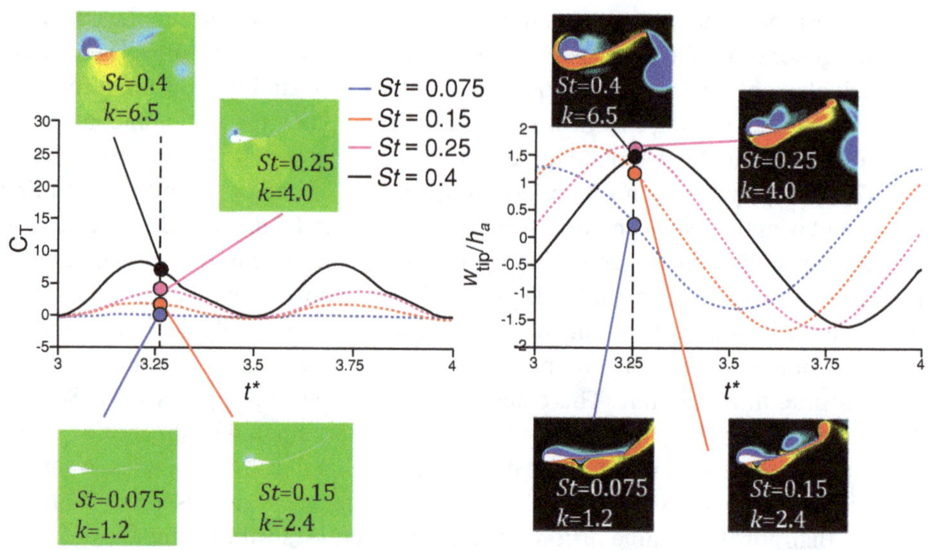

(b) *Very Flexible* ($h_s^* = 0.56 \times 10^{-3}$)

Figure 4.30. Time histories of thrust and wingtip displacement normalized by the plunge amplitude as a function of non-dimensional time. Pressure coefficient and vorticity contours at $t^* = 0.25$ for each Strouhal number are shown as well ($Re = 9.0 \times 10^3$ and $\rho^* = 7.8$). From Kang et al. [351].

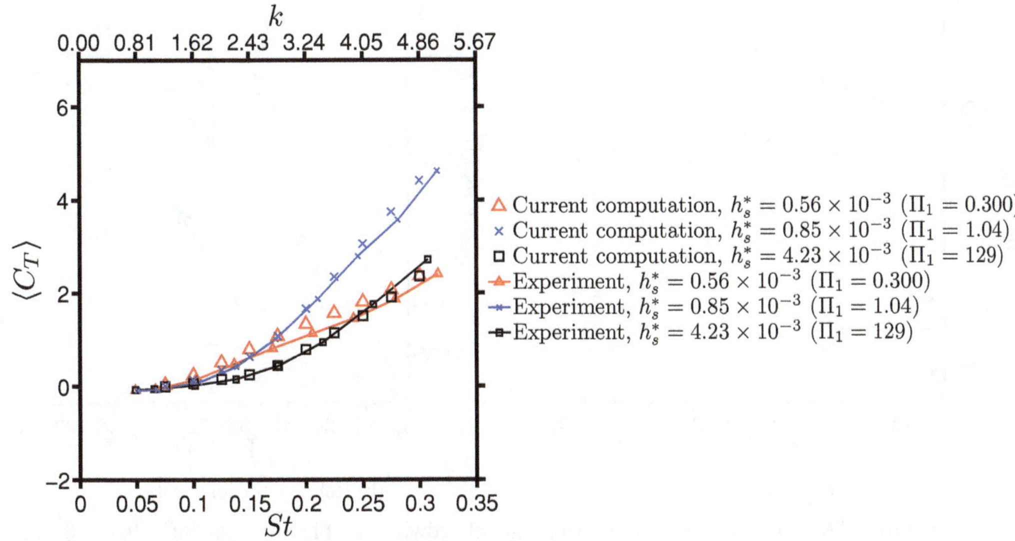

Figure 4.31. Time-averaged thrust coefficient for a plunging chordwise flexible airfoil at $Re = 9.0 \times 103$ and $\rho^* = 7.8$ for different flat plate thickness and motion frequencies. The experimental data are extracted from Heathcote and Gursul [336]. From Kang et al. [351].

0.3, the computed thrust starts to deviate. A similar trend is observed for the other thicknesses: at $h_s^* = 0.85 \times 10^{-3}$ the correlation between the numerical result and the experimental measurement is good until $St = 0.23$, and at $h_s^* = 0.56 \times 10^{-3}$ it is good only at the lowest frequencies. Modeling uncertainties, such as laminar-to-turbulent transitions, non-linearities in the structural modeling, or non-negligible twist or spanwise bending in the experimental setup, which are not accounted for in the numerical computations, may be attributed to the observed differences. The thrust for the thickest airfoil ($h_s^* = 4.23 \times 10^{-3}$) can be enhanced by increasing the motion frequency, which results in higher St and k. Increased St leads to greater fluid-dynamic force, and as shown in Section 3.6.4 increasing St and k further results in a greater acceleration-reaction force. Figure 4.31 also shows that the thrust generation depends on the thickness of the wing: At $St = 0.125$, $\langle C_T \rangle$ for $h_s^* = 0.56 \times 10^{-3}$ is the maximum, but for higher Strouhal numbers the thrust generated by the thinnest airfoil is the lowest. At $St = 0.3$, $h_s^* = 0.85 \times 10^{-3}$ generates the highest thrust, while the thinnest wing, $h_s^* = 0.56 \times 10^{-3}$ deteriorates in thrust.

To characterize the structural response, the tip displacement normalized to the plunge amplitude, w_{tip}/h_a, is plotted in Figure 4.32 as a function of the phase lag relative to the leading edge for the thicknesses and frequencies considered. The phase lag Φ is calculated by determining the time instant at which the trailing-edge displacement is a maximum. For the thickest airfoil, $h_s^* = 4.23 \times 10^{-3}$, both the deformations and the phase lag are small. As we decrease the airfoil thickness, both w_{tip}/h_a and Φ increase with increasing frequency. Eventually, w_{tip}/h_a saturates when Φ approaches 90°: when $\Phi > 90°$ the motion of the deformed trailing edge is out of phase with the imposed leading edge. Relative to the leading-edge displacement, $(w_{\text{tip}} - W_{\text{root}})/h_a$ shows that, by decreasing the stiffness and increasing the motion frequency, not only does the tip deformation increase monotonically but also the phase lag does, so that the resulting wingtip displacement reduces in magnitude

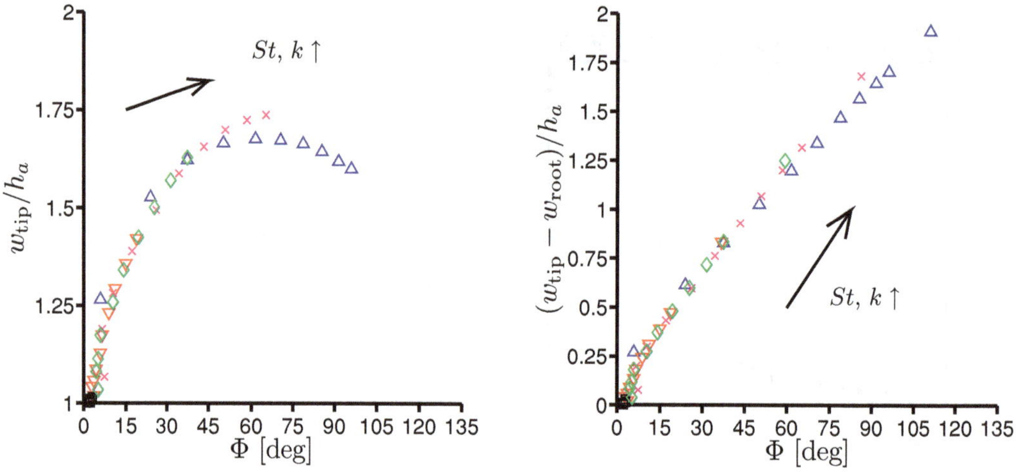

(a) Tip displacement (b) Relative tip deformation

Figure 4.32. Tip deformations of a plunging chordwise flexible airfoil at $Re = 9.0 \times 10^3$ and $\rho^* = 7.8$ for different flat plate thickness and motion frequencies. △ $h_s^* = 0.56 \times 10^{-3}$; × $h_s^* = 0.85 \times 10^{-3}$; ◇ $h_s^* = 1.13 \times 10^{-3}$; ▽ $h_s^* = 1.41 \times 10^{-3}$; □ $h_s^* = 4.23 \times 10^{-3}$. From Kang et al. [351].

when the motion is out of phase (Fig. 4.31). The phase lag is known to be a critical parameter for flight efficiency [223] [454]. Due to the monotonic relation observed between $(w_{tip} - w_{root})/h_a$ and Φ, for the cases considered in this study, the role of the phase lag may be related to the relative maximum deformation, which is shown later to be the main parameter that describes the force and the propulsive efficiency scalings (Section 4.5.2). In Section 4.5 a relationship between the mean thrust and the structural response is established. Furtheremore, Spagnolie et al. [223] showed that when $\Phi < 180°$ the freely moving wing moves forward, and when $\Phi > 180°$ it moves backward. The phase differences observed in this study are smaller than 180°, producing thrust consistent with their observation. It is also remarkable to observe that the critical phase lag of 180° pointed out by Spagnolie et al. [223] corresponds to the maximum obtainable relative tip deformation, which occurs when the tip motion and the root motion are 180° apart.

In summary, considering the outcome from the studies conducted on chordwise flexible plunging/pitching structures discussed so far, three factors may play a vital role in aerodynamic force generation in hover/forward flight: (i) airfoil plunge motion modifies the effective AoA and aerodynamic forces; (ii) the relative motion of the leading edge and trailing edge creates the pitch angle that dictates the direction of the net force; and (iii) airfoil shape deformation modifies effective geometry such as camber.

4.4.2.2 Spanwise Flexible Wing

Flexibility in spanwise direction of the wing affects the resulting flapping velocity along the spanwise direction. Liu and Bose [503] considered a 3D pitching and plunging wing in forward flight. Their results showed that the phase of the flexing motion of the wing relative to the prescribed heave motion plays a key role in determining thrust and efficiency characteristics of the fin.

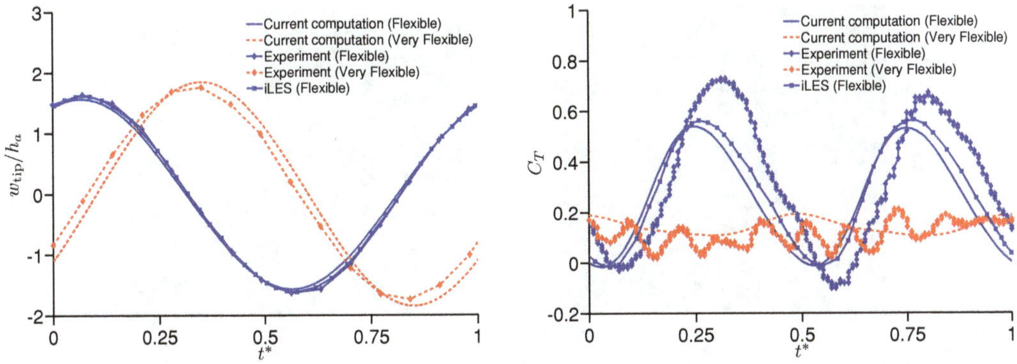

(a) tip displacement (b) thrust coefficient

Figure 4.33. Time history of (a) tip displacements and (b) thrust coefficient of a plunging spanwise flexible wing at $Re = 3.0 \times 10^4$ for different wing stiffness, wing density, and motion frequencies. The experimental data are extracted from Heathcote, Wang, and Gursul [502] and the implicit LES from Gordnier et al. [504]. From Kang et al. [351].

Heathcote et al. [502] conducted water-tunnel studies to study the effect of spanwise flexibility on the thrust, lift, and propulsive efficiency of a plunging flexible-wing configuration in forward flight. The leading edge at the wing root was actuated with a prescribed sinusoidal plunge displacement profile. Wing shape was recorded with a 50-frames-per-second, high-shutter-speed, digital video camera. They measured the overall wing thrust coefficient and tip displacement response. Whereas the "inflexible" wing cross-section was built up from nylon and reinforced by steel rods, the "flexible" and "highly flexible" cross-sections were built up from PDMS (rubber) and stiffened with a thin metallic sheet made out of steel and aluminum, respectively (see Fig. 4.28i-iii). Subsequently, computations were conducted on these wing configurations by Chimakurthi et al. [420], Gordnier et al. [504], and Aono et al. [505] at the chord-based Reynolds number of 3×10^4 using an aeroelastic framework. Recently, Kang et al. [351] examined the effect of spanwise flexibility on the thrust generation for same cases.

In the following discussion we highlight the thrust enhancement mechanism associated with spanwise flexible plunging wings in forward flight for two combinations of density ratios and effective stiffnesses and several motion frequencies at $Re = 3.0 \times 10^4$ based on the results reported by Kang et al. [351] and Shyy et al. [450]. The vertical displacements of the wingtip from the computations and the experiments for the flexible and very flexible wings are shown in Figure 4.33a. The displacement is normalized with respect to the amplitude of prescribed wing root movement (see Eq. (3–12)). For the flexible wing, in comparison to the tip response presented in previous studies (experiment [502]; implicit LES (iLES) computation [504]), the tip response of the current computation shows good correlation. For the very flexible wing, however, the tip response of the current computation exhibits slight larger amplitude and phase advance compared to the measurements [502].

Time histories of thrust coefficient for the flexible and very flexible wings are shown in Figure 4.33b. For the flexible wing the thrust in the current computation is under-predicted and shows a phase advance compared to the measurements [502]. However, the magnitude and the timing of thrust peaks compare well with

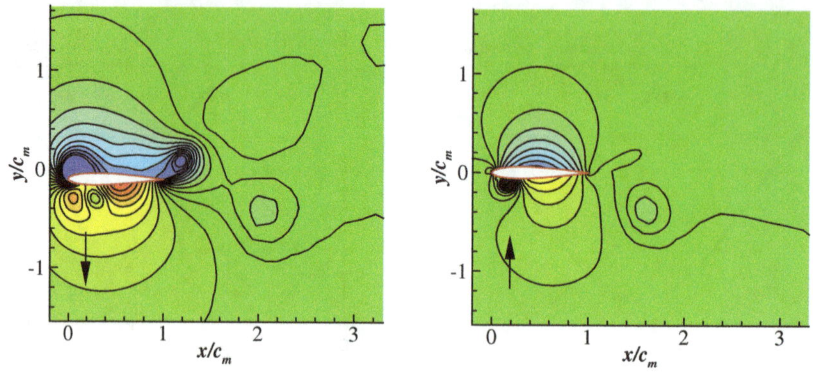

(a) pressure coefficient contour levels: 20; range: -2.5 to 2.5

$h_s^* = 0.01, St = 0.1, \ k = 1.82, \ t^* = 2.25$

(left) $\Pi_1 = 212, \ \rho^* = 7.8$; (right) $\Pi_1 = 38, \ \rho^* = 2.7$

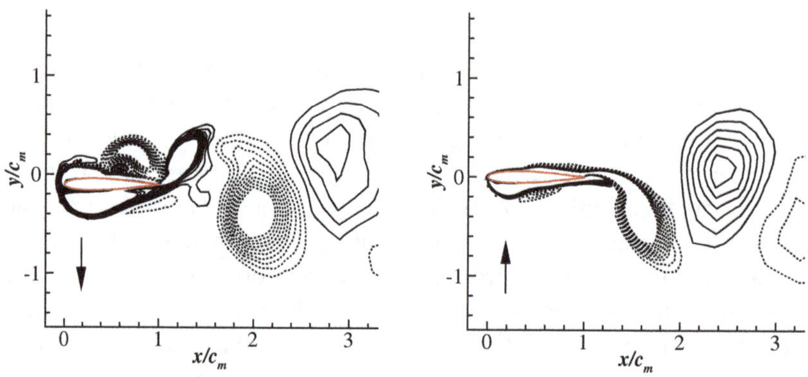

(b) vorticity contour levels: 20; range: -3 to 3

$h_s^* = 0.01, St = 0.1, \ k = 1.82, \ t^* = 2.25$

(left) $\Pi_1 = 212, \ \rho^* = 7.8$; (right) $\Pi_1 = 38, \ \rho^* = 2.7$

Figure 4.34. Pressure coefficient $(p - p_\infty)/(\frac{1}{2}\rho U_{\text{ref}}^2)$ and vorticity contours at 75 percent span location for flexible and very flexible wing configurations. The arrow indicates the direction of the airfoil motion. From Kang et al. [351].

the thrust prediction using the iLES [504] coupled with a geometrically non-linear beam solver. Furthermore, the measured thrust is asymmetric in the down- and upstroke, whereas in both computations the thrust has symmetric behavior. For the very flexible wing, the computed thrust history is in a reasonable agreement with the experimental measurements in terms of the amplitude and the trend of thrust. It is worth pointing out that the measurements include higher frequency components, whereas the waveforms of the computed thrust are smooth for all cases [351] [504]. As shown in Figure 4.34, there is no evidence of high-frequency behavior in the tip response. The experimental flow field measurements [502], however, show that the vortex fragments into a collection of weak vortices, and Heathcote et al. [502] suggested that this fragmentation is likely to be responsible for the multiple peaks

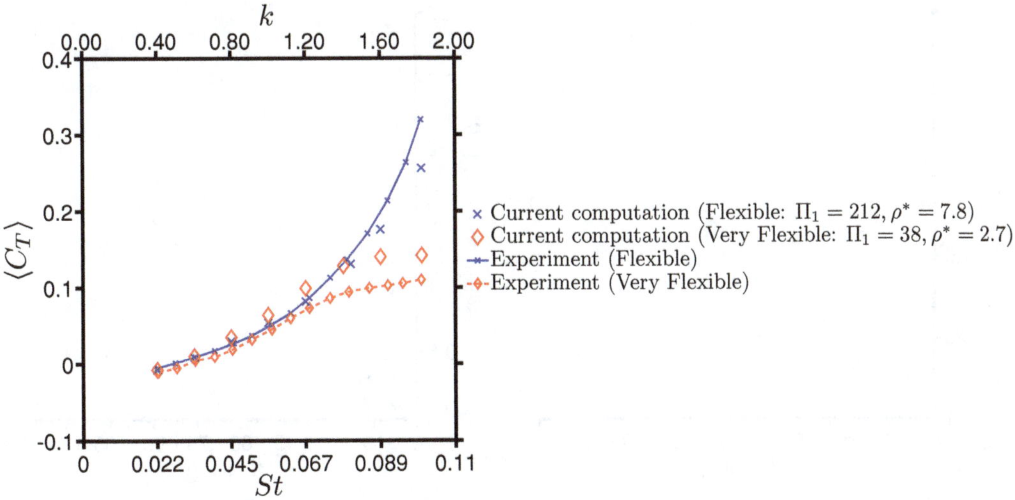

Figure 4.35. Time-averaged thrust coefficient of a plunging spanwise flexible wing at $Re = 3.0 \times 10^4$ for different wing stiffness, wing density, and motion frequencies. The experimental data are extracted from Heathcote, Wang, and Gursul [502]. From Kang et al. [351].

observed in the thrust coefficient time histories. Figure 4.34 shows vorticity and pressure contours for the flexible and the very flexible wing configurations at the mid-span section at time instant $t^* = 0.25$, when the wing is at the center of the downstroke. The dominance of leading-edge suction in the flexible case and the reduction of it in the very flexible case are visible in Figure 4.34. The phase lag between the prescribed motion and the deformation of the wing could be used to explain the thrust generation in flexible flapping wings [450]. For the very flexible case the cross-sectional motion is in the opposite direction to the imposed kinematics at the wing root. The phase lags at the wingtip with respect to the prescribed motion for the flexible and very flexible cases are 126°, and –26°, respectively. As a result of the substantial phase lag in the very flexible case, the wingtip and root move in opposite directions during most of the stroke, resulting in lower effective AoAs and consequently lesser aerodynamic force generation; in the contour plots, see also the direction of the arrow denoting the direction of the wing movement.

Figure 4.35 shows the time-averaged thrust coefficients for the two different materials for various frequencies of k from 0.4 to 1.82. At higher motion frequencies the time-averaged thrust of the flexible wing is under-predicted by the current computational results, whereas the thrust of the very flexible wing is over-predicted, which again may be ascribed to the uncertainties in the computational modeling or experimental setup. However, the qualitative trend of the thrust response to the variation of the motion frequency is well captured. When the plunging motion is slow (i.e., $k < 1.2$), the thrust generation is similar for both materials. For higher motion frequencies, the flexible wing benefits more from the flexibility than does the very flexible wing: the thrust saturates for the very flexible wing with increasing *St*. Furthermore, similar trends are observed: increasing motion frequency enhances thrust, and decreasing the effective stiffness does not necessarily lead to higher thrust.

The structural response is depicted in Figure 4.36. For the flexible wing, the phase lag between the prescribed motion and the tip response for $k > 1.4$ is from

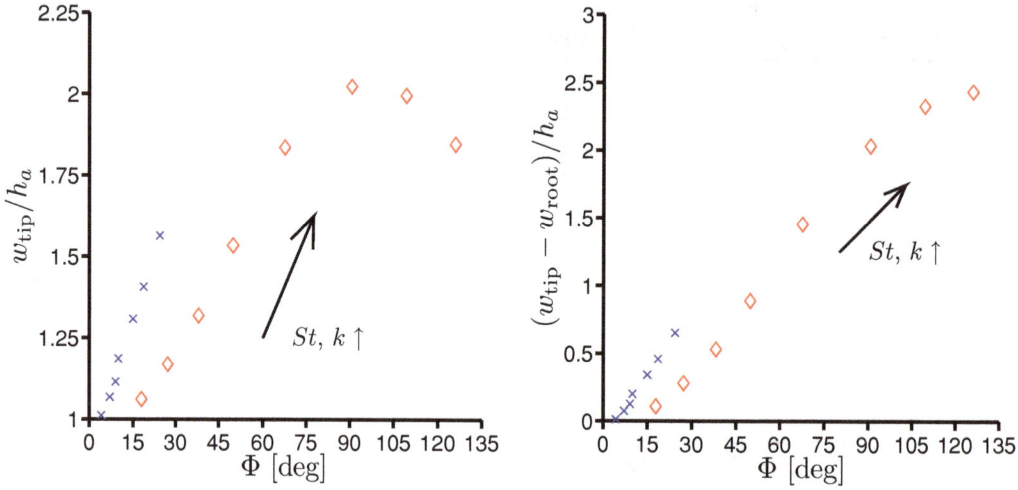

(a) Tip displacement (b) Relative tip deformation

Figure 4.36. Tip deformations of a plunging spanwise flexible wing at $Re = 3.0 \times 10^4$ for different wing stiffness, wing density, and motion frequencies. $\times\ \Pi_1 = 121$, $\rho^* = 7.8$; $\diamond \Pi_1 = 38$, $\rho^* = 2.7$. From Kang et al. [351].

17.6°–23.8°, whereas for the very flexible wing, Φ varies from 108.8° to 125.9°. The wingtip of the very flexible wing moves in the opposite direction from the root for most of the stroke for higher motion frequencies, whereas for the flexible wing, the wing root and the tip are in phase. This is confirmed in Figure 4.36, where all flexible wing cases show a phase lag of the wingtip relative to the wing root, Φ, less than 90°, and for $k = 1.6$ and $k = 1.82$, $\Phi > 90°$ for the very flexible wing. Again, the correlations of the dynamics from the root to tip play a key role in the tip displacement, as shown in Figure 4.36, where the relative tip displacement is shown to be monotonic to Φ. The results discussed so far focused on a single Reynolds number 3×10^4. In a parametric study Heathcote et al. [502] found that the mean thrust coefficient is a function of the Strouhal number and is only very weakly dependent on the Reynolds number.

4.4.2.2 Isotropic Wing

When the properties of a material are the same in all directions, the material is called *isotropic*. Aono et al. [506] reported on a combined computational and experimental study of a well-characterized flapping wing structure. An aluminum wing was prescribed with single degree-of-freedom flapping at 10 Hz and ±21° amplitude. Flow velocities and deformation were measured using digital image correlation and digital PIV techniques, respectively. In the most flexible flapping wing case, the elastic twisting of the wing produces substantially larger mean and instantaneous thrust due to shape-deformation-induced changes in effective AoA. Relevant fluid physics were documented including the counter-rotating vortices at the leading and the trailing edge, which interact with the tip vortex during the wing motion.

In this section we focus on a 3D hovering flapping isotropic wing in air. The wing is a flat plate wing of $c_{\mathrm{m}} = 0.0196$ m and $AR = 3.825$, with a thickness ratio of $h_s^* = 2.0 \times 10^{-2}$ having a Zimmerman planform (see Fig. 4.37, hovering in air at

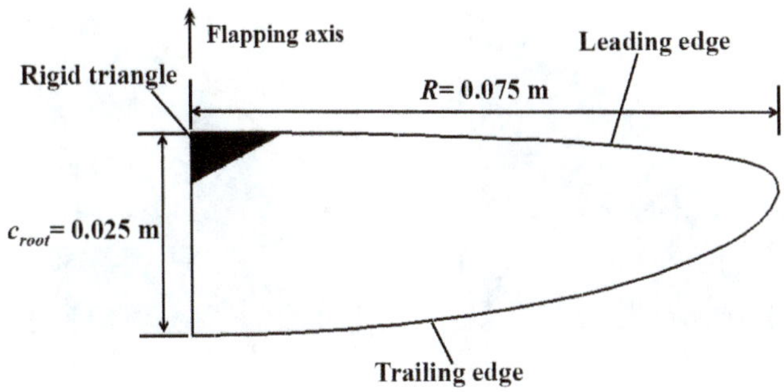

Figure 4.37. Geometry of the Zimmerman planform.

$Re = 1.5 \times 10^3$. A sinusoidal flapping motion is introduced at the rigid triangle at the leading edge at the wing root following Eq. (3–15) with $St = 0.25$ and $k = 0.56$. The flapping axis is parallel to the wing root. Note that in the axes definition by Wu et al. [507] the wing flaps up and down to generate thrust due to wing flexibility; however, in the study considered here, the flapping wing axis has been rotated so that the flapping axis is parallel to the lift direction, such that any flexibility in the wing leads to lift generation. The triangular rigid region near the root at the leading edge undergoes prescribed motion and is constrained in all degrees of freedom in the structural solver, since the flapping mechanism in the experiment [506] is actuated at this region on the wing.

To assess the effects of wing properties – the effective stiffness Π_1 and the density ratio ρ^* – on the resulting lift and wing deformations, surrogate models are constructed to qualitatively explore their implications. The range for these variables in the design space is chosen to cover a wide range of applications, as shown in Figure 4.38. To effectively assess the order of magnitude of the design variables, a logarithmic scaled design space is populated.

The objective functions are (i) the lift coefficient averaged over one motion cycle between the second and the third cycle and (ii) the twist angle θ given as

$$\theta = \max\{\mathrm{acos}(c_3 \cdot e_1)\}, \tag{4–28}$$

where c_3 is the unit vector in the direction from the leading edge to the trailing edge at near the mid-span of the wing and e_1 is the unit vector in the lift direction.

Figure 4.38. Design of experiment in logarithmic scale for the design variables Π_1 and ρ^*. The training points are indicated by circles and the training points by crosses.

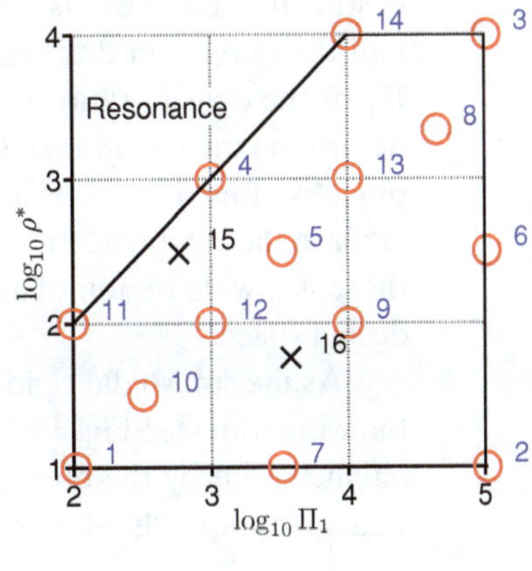

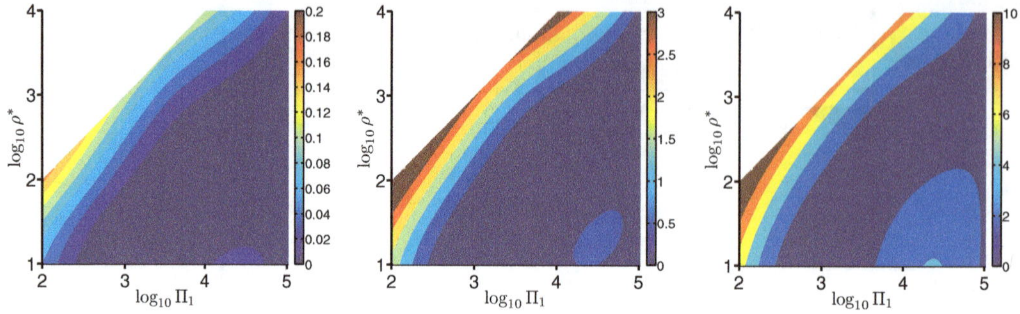

(a) $\langle C_L \rangle$ (b) θ (c) ψ

Figure 4.39. Surrogate model responses for (a) lift, (b) twist, and (c) bending angles for a flapping isotropic Zimmerman wing, hovering at $Re = 1.5 \times 10^3$ and $k = 0.56$.

So θ gives the degree of the lift-favorable projectional area of the wing due to the deformation, and (iii) the bending angle ψ that is defined as

$$\psi = \max\left\{\tan^{-1}\left(\frac{w}{R}\right) - \phi\right\}, \tag{4–29}$$

which measures the wing deformation in the spanwise direction x as the maximum tip displacement angle relative to the imposed flap angle ϕ. For simplicity we refer to lift as the time-averaged lift coefficient in this section.

The experiments use a face-centered cubic design (FCCD). Then the remainder of the design space is filled evenly in the design space, with the cases 8 and 10 together with the testing points generated by the latin hypercube algorithm. In total 14 training points are selected. The design space with a logarithmic bias toward the softer Π_1 and lighter ρ^* structures is shown in Figure 4.38. The region where $\log_{10}\rho^* > \log_{10}\Pi_1 + 2$ is out of the scope of the current study as this region showed largely unstable behavior of the wing motion because the imposed frequency of 10 Hz is close to the natural frequencies [351].

The resulting surrogate models are shown in Figure 4.39 for the lift, twist, and bending angle. Notice that the time-averaged lift for the rigid wing would be zero due to the symmetry in the hovering kinematics without pitching motion. The lift, twist, and bending are at the maximum at case 4 (shown in Fig. 4.38). However, these three objective functions follow qualitatively similar trends in the design space, suggesting that there exists a correlation between the resulting time-averaged lift force and the maximum deformations. Furthermore, it is not only the effective stiffness Π_1 or the density ratio ρ^* but also the balance between these two parameters that determine the resulting deformation and the lift generation. The region of increased objective functions between $\log_{10}\Pi_1 = 4$ and 5 and $\log_{10}\rho^* = 1$ to 2 is caused by the error in the surrogate model due to the high gradient near the resonance region, yet there is a wide region of almost zero values at the stiffer and lighter portion of the design space.

As the sinusoidal rigid-body motion is imposed at the triangular rigid part near the wing root (see Fig. 4.37), the wing inertia and the resulting aerodynamic load are balanced out by the elastic force. Since the wing is made of isotropic material the structure responds with both spanwise bending and twisting.

For the chordwise flexible airfoil in Section 4.4.2.1 and the spanwise flexible wing in Section 4.4.2.2, the thrust generation in forward flight was shown to be mostly dependent on the resulting tip motion relative to the imposed kinematics at the wing root. For the flapping Zimmerman wing in hover in air, maximum horizontal tip displacement w, normalized by the prescribed amplitude $h_a = R \sin \phi_a$, is plotted against the phase lag with respect to the top of the stroke of the rigid body motion. The higher tip amplitude corresponds with the larger phase lag compared to the imposed kinematics, while the tip motion is in phase. The cases with the lowest Π_1/ρ^* ratio have larger deformation consistent with the surrogate model responses shown in Figure 4.39.

4.5 A Scaling Parameter for Force Generation for Flexible Wings

The maximum propulsive force, such as thrust in forward flight or lift in hovering motion, was generated at a frequency that was slightly lower than the natural frequency of the system [222, 367, 453, 454, 494, 508] as shown in Section 4.4.2. Zhang, Liu, and Lu [222] using the lattice Boltzmann method, numerically studied a flexible flat plate modeled as a rigid plate with a torsional spring at the pivot point on the leading edge of the wing. They concluded that the flat plate would move forward and hence generate thrust when the leading edge plunges at a motion frequency that is lower than the natural frequency of the system; the flat plate would move backward if the frequency ratio, the ratio between the motion frequency and the natural frequency, is greater than one. Similarly, Masoud and Alexeev [508] used the lattice Boltzmann method to show that the maximal propulsive force is obtained at the frequency ratio of 0.95. The magnitude of the maximal force would increase when the inertial effects became more important than the fluid inertia. Michelin and Llewellyn Smith [494] used potential flow theory to describe the flow over a plunging flexible wing. They showed that the trailing-edge flapping amplitude and the propulsive force are maximal at resonance conditions. In a series of experiments using a self-propelled simplified insect model, Thiria and Godoy-Diana [453] and Ramananarivo, Godoy-Diana, and Thiria [454] also showed that the maximum thrust force os around a frequency ratio of 0.7. More recently, Gogulapati and Friedmann [367] coupled an approximate aerodynamic model, which was extended to forward flight including the effects of fluid viscosity, to a non-linear structural dynamic model. For various setups of composite anisotropic Zimmerman wings [507], they investigated the propulsive force generation in forward flight. They found that the maximum propulsive force is also obtained at a frequency ratio slightly lower than one. These observations are consistent with the general perception of the resonance phenomena, in which even small external forces can induce large-amplitude deformations and potentially be efficient as well.

However, it has been reported for insects that the flapping frequency is below the natural frequencies of the wing and is only a fraction of the resonance frequency [509, 510] . Sunada, Zeng, and Kawachi [509] measured the natural frequencies of vibration in air and the wing-beat frequencies for four different dragonflies. The wing-beat frequency ratios were in the range of 0.30–0.46. Chen, Chen, and Chou [510] also measured the wing-beat frequencies and natural frequencies of dragonfly wings. In their measurements the average flapping frequency was 27 Hz, whereas the natural frequency, calculated using a spectrum analyzer, was 170 Hz when clamped at the wing base, resulting in a frequency ratio of about 0.16.

The propulsive efficiency was also investigated numerically [492, 508] and experimentally using a self-propelled model [453] [455]. Vanella et al. [492] conducted numerical investigations on a two-link model and found that the optimal performance is realized when the wing is excited at a frequency ratio of 0.33. For all Reynolds numbers considered in the range of $7.5\times10^1-1\times10^3$, the wake-capture mechanism is enhanced due to a stronger flow around the wing at stroke reversal, resulting from a stronger vortex at the trailing edge. Using the experimental setup described earlier, Thiria and Godoy-Diana [453] and Ramananarivo, Godoy-Diana, and Thiria [454] also showed that the maximum efficiency is obtained at a frequency ratio of 0.7, lower than that of the maximum propulsion. They concluded that the performance optimization is obtained not by looking at the resonance but by adjusting the temporal evolution of the wing shape. In contrast, Masoud and Alexeev [508] showed that the optimal efficiency for a hovering flat plate at $Re = 100$ occurs when the motion is excited at a frequency ratio of 1.25. In their setup a flexible flat plate had a geometric AoA of 40° in contrast to the previously mentioned studies where the plunging motion was symmetric.

4.5.1 Propulsive Force and Non-Dimensional Wingtip Deformation Parameters

From the results presented in Sections 4.4.2.1–4.4.2.3 for the three different cases Kang et al. [351] made the following observations: (i) the time-averaged force increases with increasing motion frequency; (ii) changes in structural properties, such as the thickness ratio, Young's modulus, or wing density (mass), lead to a non-monotonic response in the force generation; and (iii) for the hovering isotropic Zimmerman wing, the ratio between the density ratio and the effective stiffness is monotonic with the time-averaged lift generation. To explain the observed trends Kang et al. [351] mainly analyzed the physics based on Eq. (4–1) with simplifying approximations for the fluid dynamic force, f^*_{ext} (see also Section 3.6.4), based on scaling arguments. The flow field and the structural displacement field should simultaneously satisfy Eq. (3–18) and Eq. (4–1); of these two equations Eq. (4–1) was considered, which has the advantage that it is linear, except for the fluid dynamic force term; in contrast, the Navier-Stokes equation is non-linear in the convection term. Subsequently, a relation between the time-averaged force and the maximum relative tip displacement was established by considering the energy balance.

To capture the essence of the mechanism involved in the force enhancement due to flexibility, it is necessary to analyze the interplay among the imposed kinematics, the structural response of the wing, and the fluid force acting on the wing. The derivation leading to the relation between the time-averaged force acting perpendicular to the wing motion, $\langle C_F\rangle$ and the maximum relative tip deformation v_{max} is lengthy, where $v^*(x^*,t^*) = w^*(x^*,t^*) - h(t^*)$ is the displacement of the wing relative to the imposed kinematics motion; see Kang et al. [351] for the detailed steps. A treatment using simplifying approximations is helpful in enabling the analysis, but mainly serves to elucidate the scaling analysis and is not meant to offer a complete solution. Consider Eq. (4–26) in one dimension in space with $0 \le x^* \le 1$ and time

$t^* \geq 0$, for the vertical displacement w^*, with the wing approximated as a linear beam. A plunge motion, Eq. (3-12), is imposed at the leading edge at $x^* = 0$. The trailing edge $x^* = 1$ is considered as a free end (i.e., with the boundary conditions,

$$w^*(0, t^*) = h(t^*) = St\frac{\pi}{k}\cos(2\pi t^*),$$

$$\frac{\partial w^*(0, t^*)}{\partial x^*} = \frac{\partial^2 w^*(1, t^*)}{\partial x^{*2}} = \frac{\partial^3 w^*(1, t^*)}{\partial x^{*3}} = 0$$

and the initial conditions

$$w^*(x^*, 0) = St\frac{\pi}{k}, \quad \frac{\partial w^*(x^*, 0)}{\partial t^*} = 0,$$

where the factors involving l/c_{m} become unity for the chordwise flexible airfoil case). For the spanwise flexible wing and the isotropic Zimmerman wing cases that are discussed in Section 4.4.2.2 and Section 4.4.2.3, respectively, Π_0 and Π_1 need to be corrected as $l/c_{\mathrm{m}} = AR$. Following the procedure described in Mindlin and Goodman [511], a partial differential equation with homogeneous boundary conditions can be found by superimposing the plunge motion on the displacement $v(x^*, t^*) = w(x^*, t^*) - h(t^*)$. The consequence of having a sinusoidal displacement at the root is that the vibrational response of the wing is equivalent to a sinusoidal excitation force, which is the inertial force. The dynamic motion given by Eq. (3-12) is coupled to the fluid motion via the fluid force term f^*_{ext}, which cannot be solved in a closed form due to its non-linearities.

For high-density-ratio FSI systems, Daniel and Combes [470] and Combes and Daniel [472] showed that the inertial force arising from the wing motion is larger than the fluid dynamic forces. To cover a wider range of density ratios, Kang et al. [351] included the fluid dynamic forces by considering the added mass effects. The motivation stems from the scaling discussed in Section 3.6.4, where for high k the added mass terms due to an accelerating body (see also Combes and Daniel [472]) contribute more on the wing than do the fluid dynamic forces from the hydrodynamic impulse; see Table 4.2 for a summary of the non-dimensional numbers considered in this study. Figure 3.55 illustrates the domain of importance of the various effects of the unsteady fluid force. Hence, the wing dynamics is modeled with external forces depending on the imposed wing acceleration as

$$f^*_{ext}(t^*) \approx 2\pi^2 St\, k\, \cos(2\pi t^*), \tag{4–30}$$

and the external force on the structural dynamics does not have spatial distribution explicitly accounted for and is being simplified in temporal form.

When dombined with the inertial force, the total external force $g(t^*)$ becomes

$$g(t^*) = f^*_{ext}(t^*) - \Pi_0\frac{d^2h(t^*)}{dt^{*2}} \approx 2\pi^2\left(1 + \frac{4}{\pi}\rho^* h^*_s\right) St\, k\, \cos(2\pi t^*). \tag{4–31}$$

The solution for the tip displacement is given by Kang et al. [351] who used the method of separation; that is, $v^*(x^*, t^*) = X(x^*)T(t^*)$ as

$$T_n(t^*) = \frac{2\pi\left(1 + \frac{4}{\pi}\rho^* h^*_s\right)\cdot St\cdot k\cdot Q_n}{\Pi_0(f_n^2/f^2 - 1)}\{\cos(2\pi t^*) - \cos(\omega_n t^*)\}, \tag{4–32}$$

for the temporal component, with Q_n, the Fourier coefficient of a unit function in the spatial modes, satisfying

$$Q_n = \frac{\int_0^1 X_n dx^*}{\int_0^1 X_n^2 dx^*},$$

and the natural frequency

$$\omega_n^2 = \left(\frac{k_n 1}{1}\right)^4 \frac{\Pi_1}{\Pi_0} = \left(\frac{2\pi f_n}{f}\right)^2, \tag{4–33}$$

with k_n as the eigenvalue belonging to the spatial mode X_n with $k_1 L \approx 1.875$. The spatial component X_n is the same as for a cantilevered oscillating beam. The temporal evolution of the wing given by Eq. (4–32) indicates there is an amplification factor of $1/(f_n^2/f^2 - 1)$, depending on the ratio between the natural frequency f_n of the beam and the excitation frequency f.

The full solution is $w^*(x^*, t^*) = h(t^*) + \sum_{n=1}^{\infty} X_n(x^*) T_n(t^*)$. The amplitude of the tip deformation, γ, for the first mode ($n = 1$) is given as

$$\gamma = \frac{\left(1 + \frac{4}{\pi}\rho^* h_s^*\right) \cdot St \cdot k}{\Pi_0 (f_1^2/f^2 - 1)}, \tag{4–34}$$

relative to the imposed rigid body motion normalized by the chord. The parameter γ can be rewritten as

$$\frac{\gamma}{h_a/c_{\mathrm{m}}} = \left(\frac{\rho c_{\mathrm{m}}}{\rho_s h_s}\frac{\pi}{4} + 1\right) \frac{4}{\left(\frac{f_1}{f}\right)^2 - 1} \sim \frac{A+1}{\left(\frac{f_1}{f}\right)^2 - 1} \tag{4–35}$$

where $f_1/f = \omega_1/(2\pi)$ is the inverse frequency ratio and $A = \pi \rho c_{\mathrm{m}}/(4\rho_s h_s)$ is the ratio between the acceleration-reaction force (added mass) and the wing inertia. Depending on the order of this ratio, either the acceleration-reaction force term or the wing inertia force can be neglected. Equation (4–35) gives the relative wingtip deformation normalized by the plunge amplitude, which can be related to the Strouhal number based on the deformed tip displacement. Note that when A is sufficiently large, the inertia force term can be neglected and γ is then proportional to $Ah_a^* \sim \rho h_a/(\rho_s h_s)$.

The proposed scaling parameter to estimate the resulting force on the flapping wing follows from the observation that there exists a correlation between the dynamic deformation of the wing at the tip, γ, given by Eq. (4–34), and the static tip deflection, which is $\langle C_F \rangle / \Pi_1$. By considering the non-dimensional energy equation, Kang et al. [351] derived the following relation between the time-averaged force $\langle C_F \rangle$ and the maximum relative tip displacement represented with the scaling factor γ, under the assumption of $f_1/f \ll 1$, resulting in

$$\frac{< C_F >}{\Pi_1} \sim \gamma. \tag{4–36}$$

The resulting scaling – Eq. (4–36) for the three canonical cases – is shown in Figure 4.40. When plotted in the log-scale, see Figure 4.40, the scaling for all cases

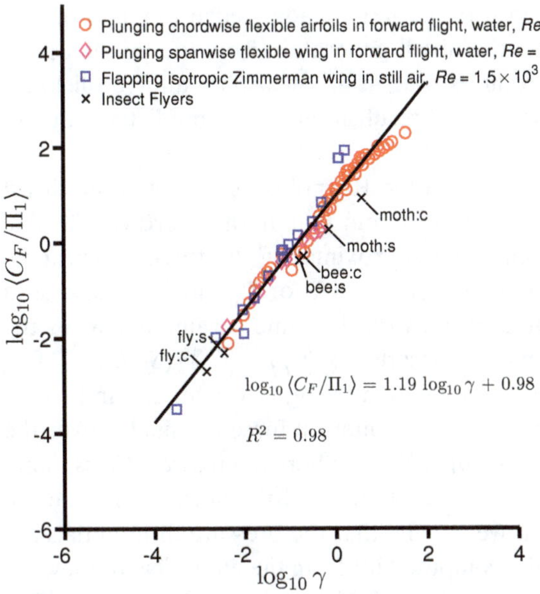

Figure 4.40. Normalized time-averaged force coefficients as function of Y, following Eq. (4–36). For the insect flyers the letters c and s correspond to chordwise and spanwise flexibility directions, respectively. From Kang et al. [351].

considered becomes more evident. A linear fit on the data set with the coefficient of determination of $R^2 = 0.98$ indicates that the relation between the normalized force and γ is a power law with the exponent of 1.19. The relation originating from the dimensional analysis, Eq. (4–36), then simplifies to

$$\langle C_F \rangle = \Pi_1 \Psi(\gamma), \tag{4–37}$$

with $\Psi(\gamma) = 10^{0.98}\gamma^{1.19}$. The elastoinertial number, $\mathcal{N}_{\text{ei}}$ proposed by Thiria and Godoy-Diana [454] as the thrust scaling parameter in air is a special case of γ; that is,

$$\gamma \xrightarrow{\rho^* h_s^* \gg 1 \text{ and } f/f_1 \gg 1} \mathcal{N}_{\text{ei}}. \tag{4–38}$$

It is important to note that the y-axis in Figure 4.40 shows $\langle C_F \rangle$. Recall that $\langle C_F \rangle$ was defined as the force acting normal to the wing that is responsible for the wing deformation; hence $\langle C_F \rangle$ is normal to $\langle C_T \rangle$ or $\langle C_L \rangle$ depending on the direction of the wing deformation. For the purely plunging chordwise flexible airfoil cases in forward flight in water, $\langle C_F \rangle = \langle C_T \rangle / (St \cdot k)$ where the factor $St \cdot k$ is the ratio between $C_{L,\max}/\langle C_T \rangle \sim St \cdot k$ for the added mass force and $\langle C_T \rangle$ by Garrick [335] for purely plunging airfoils. For all chordwise flexible airfoil cases parametrized by (f, h_s^*), $\langle C_F \rangle / \Pi_1$ shows almost a linear correlation with γ. Recall that the inertial force term arising from the plunging boundary condition is small compared to the added mass term because for the plunging chordwise flexible airfoils $\rho^* h_s^* \ll 1$. Compared to the higher motion frequency cases, the thrust generation at the lowest frequency at the five thicknesses has a larger variance, which is not shown in Figure 4.40. A plausible explanation is that the current analysis breaks down due to the presence of the rigid teardrop at lower motion frequencies. When the plunging motion is very slow at the rigid teardrop, the large leading-edge radius produces time-averaged drag that overwhelms the thrust generation from the thin flat plate with small deflection. The airfoil produces drag at $St = 0.085$ on all five thickness ratios as shown in Figure 4.40, which would result in an under-predicted value of $\langle C_F \rangle / \Pi_1$.

For the spanwise flexible wing, although the Reynolds number and the thrust direction relative to the wing flexibility are different than in the chordwise flexible airfoil, a similar analysis could be made by approximating the three-dimensional wing as a beam with the correction factors $l/c_{\mathrm{m}} = AR$ for Π_0 and Π_1 as discussed in Section 4.5. The force coefficient is scaled with the same parameters as for the chordwise flexible airfoils for the same reasons; that is, $\langle C_F \rangle = \langle C_T \rangle / (St/k)$.

In the case of a flapping Zimmerman wing, the wing hovers in air, and so the density ratio is higher than in water. Hence, the inertial force dominates over the added mass force, as previously found [470] [472]. The horizontal force $\langle C_F \rangle$ is found by normalizing $\langle C_L \rangle$ by h_s^* because the vertical force and the horizontal force are proportional to the thickness ratio, if we assume that the pressure differentials are of the order of $O(1)$. Furthermore, the computed lift from the numerical framework represents only the fluid dynamic force without the inertial force of the wing. The inertial force that acts on the wing is estimated by multiplying the factor $\rho^* h_s^* / (St/k)$, which is the ratio between the inertial force ($\sim \rho^* h_s^* k^2$) and the fluid force ($\sim St\, k$) to $\langle C_L \rangle$. The resulting normalization for the vertical axis is then $\langle C_F \rangle = \langle C_L \rangle \rho^* / (St/k)$.

Even though the current case has different kinematics (plunging vs. flapping; forward flight vs. hovering), different density ratios (low vs. high), and structural flexibilities (unidirectional vs. isotropic), the previous trend reemerges, suggesting the generality of this scaling parameter γ. The trend for the flapping isotropic Zimmerman wing hovering case is slightly offset in the vertical direction, suggesting that the resulting lift is lower. An important factor is the influence of the presence of the rigid triangle (see Figure 4.37) that constrains the tip deformation, such that the resulting tip deformation is less than in the setup where the imposed kinematics is actuated at the root of the wing without the rigid triangle.

For the flapping isotropic Zimmerman wing case in hovering motion, we could correlate the lift generation to γ. This result suggests extrapolation of the current scaling analysis to the lift generation of hovering insect flyers. The lift in hovering motion for several insects is approximated by the experimentally measured weights of hawkmoths [226, 227], bumble bees [512], and fruit flies [513] [514]. To calculate the parameters listed in Table 4.3, a flapping rectangular planform has been assumed with constant thickness and density. To compute the effective stiffnesses in the spanwise and the chordwise directions (i.e. $\Pi_{1,\mathrm{s}}$ and $\Pi_{1,\mathrm{c}}$, respectively), the flexural stiffness data presented by Combes and Daniel [402] were used along with their wing lengths. The result is included in Figure 4.42 with the scaling

$$\frac{W}{\frac{1}{2}\rho U_{\mathrm{ref}}^2 c_{\mathrm{m}}^2} = \Pi_1 \frac{St/k}{\rho^* h_s^*} \Psi(\gamma) \sim \Pi_1 \frac{\rho h_a}{\rho_s h_s} \Psi(\gamma). \tag{4–39}$$

Again, the lift approximated with the weights of the insects scales with γ.

The current analysis shows that the time-averaged force, such as the thrust or lift, can be related to the maximum relative tip displacement by normalizing the

Table 4.3. *Kinematic, geometric, fluid, and structural parameters for the hawkmoth, bumble bee, and fruit fly obtained from the literature*

Insect	Hawkmoth	Bumble bee	Fruit fly
c_m [mm]	18.2	3.22	0.96
R [mm]	47.3	10.9	3.0
f [Hz]	26.1	181	240
ϕ_a [deg.]	57.2	72	75
Re [10^3]	6.2	2.2	0.25
K	0.30	0.18	0.19
St	0.25	0.25	0.25
h_s^* [10^{-3}]	2.0	1.0	0.6
$\rho*$ [10^3]	2.0	2.1	1.1
$\Pi_{1,s}$ [10^2]	0.43	1.4	26
$\Pi_{1,c}$	0.53	2.8	211

Source: Willmott and Ellington [226] [227], Buchwald and Dudley [512], Vogel [513], Shevtsova et al. [514], and Combes and Daniel [402].

force by the effective stiffness, Π_1, resulting in a measure equivalent to the static tip displacement. These features are shown in Figure 4.41. Note that the relation is not linear, indicating that c in Eq. (4–34) is actually not a constant. For the spanwise flexible wing case, although the very flexible wing has larger relative deformation, the effective stiffness is 5.6 times smaller than that of the flexible wing. Hence even if the static tip displacement is larger for the very flexible wing at the highest motion frequency, the force corresponding to this deformation is smaller due to a lower effective stiffness. Furthermore, plotting the $\langle C_T \rangle / \Pi_1$ against the maximum

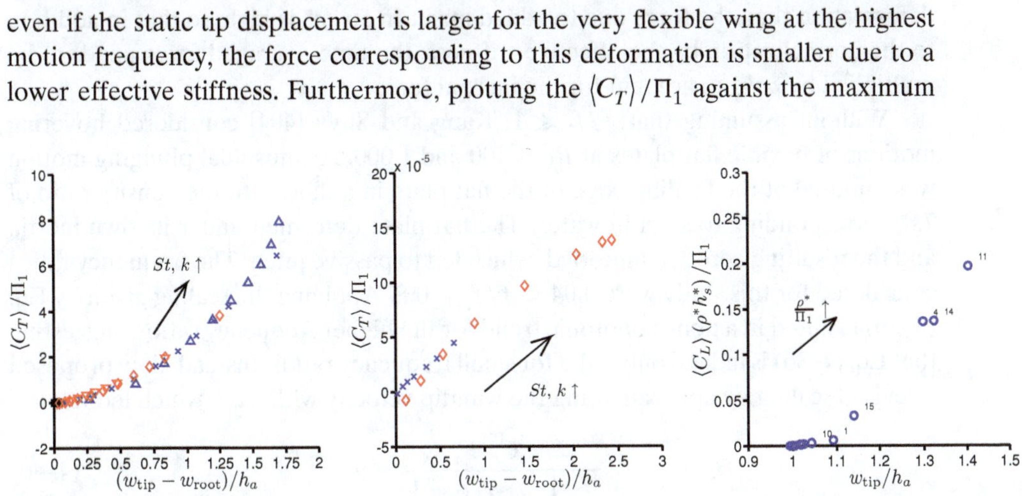

(a) Chordwise flexible:

△ $h_s^* = 0.56 \times 10^{-3}$;

× $h_s^* = 0.85 \times 10^{-3}$;

◊ $h_s^* = 1.13 \times 10^{-3}$;

▽ $h_s^* = 1.41 \times 10^{-3}$;

□ $h_s^* = 4.23 \times 10^{-3}$

(b) Spanwise flexible:

× $\Pi_1 = 121, \rho^* = 7.8$

◊ $\Pi_1 = 38, \rho^* = 2.7$

(c) Isotropic Zimmerman

Figure 4.41. Time-averaged force (thrust or lift) coefficient normalized by the effective stiffness plotted against the maximum relative tip deformations for the current computations. The numbers shown in (c) next to the markers indicate their case numbers. From Kang et al. [351].

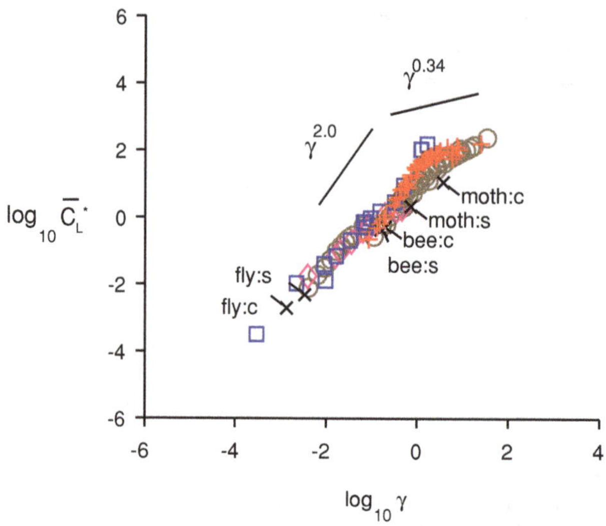

Figure 4.42. Normalized time-averaged force coefficients as function of γ, following Eq. (4–40). The data points for the forward flight cases, the Zimmerman wing cases, and the insects are the same as in Figure 4.40. From Kang and Shyy [359].

relative tip deformation collapses both curves on top of each other. It is shown in Eq. (4–35) that for $\rho^* h_s^* \gg 1$ the inertial force will have greater influence on the wing deformation than the force due to added mass. In air, ρ^* is high, so that in addition to the normalization by the effective stiffness, the time-averaged force needs to be multiplied by $\rho^* h_s^*$ to account for the inertial force.

Without assuming that $f/f_1 \ll 1$, Kang and Shyy [404] considered hovering motions of flexible flat plates at $Re = 100$ and 1,000. A sinusoidal plunging motion was imposed at the leading edge of the flat plate in a fluid with the density ratio of 7.8, corresponding to steel in water. The flat plate deformed under its own inertia and the resulting aerodynamic load, which led to passive pitch. The frequency ratios considered for this study were $0.04 < f/f_1 < 0.8$. Applying the scaling given by Eq. (4–36) resulted in a non-monotonic trend for the higher frequency ratios, suggesting that Eq. (4–36) is indeed only valid for small frequency ratios. Instead, they proposed a revised scaling by approximating the wingtip velocity with $2\pi\gamma$, which led to

$$\frac{< C_F >}{\Pi_1 \left\{1 - \left(\frac{f}{f_1}\right)^2\right\}} \sim \gamma. \tag{4–40}$$

Eq. (4–40) was able to scale the lift generated by a hovering flexible flat plate including passive pitch, where $\beta_1 = \Pi_1 \left\{1 - \left(\frac{f}{f_1}\right)^2\right\}$. An interesting property of the scaling given by Eq. (4–40) is that when $f/f_1 \ll 1$, the denominator in the LHS becomes Π_1, resulting in Eq. (4–36).

In summary, the thrust generation consists of contributions from both leading-edge suction and the pressure projection of the chordwise deformed rear foil. When the rear foil's flexibility increases, the thrust of the teardrop element decreases, as does the effective angle of attack. Within a certain range, as chordwise flexibility increases, even though the effective angle of attack and the net aerodynamic force are reduced due to chordwise shape deformation, both mean and instantaneous thrust are enhanced due to the increase in the projected area normal to the flight trajectory. Furthermore, for the spanwise flexible case, correlations of the motion from the root to the tip play a role. Within a suitably selected range of spanwise flexibility, the effective angle of attack and thrust forces of a plunging wing are enhanced due to the wing deformations.

4.5.2 Scaling and Lift Generation of Hovering Flexible Wing of Insect Size

To illustrate the implications of flexible wing further, Kang and Shyy [404] argued that the effects of wing flexibility can be explored from two angles: (i), one can study the equilibrium shape compliance in accordance with the stress distributions around the body. The resulting modification of the angle-of-attack and the wing shape can potentially better adapt, reducing tendencies toward, e.g., massive flow separation and stall. (2), structural dynamics during transient process can offer further time-dependent behavior other than static shape compliance. The interplay of the fluid flow with the shape compliance and transient response is nonlinear and intriguing [557] even at low Reynolds numbers relevant to insect flyers, i.e. $Re = O(10^2)$.

To isolate these sources of aerodynamic force modification, Kang and Shyy [404] considered a flexible plunging wing operated in translational motion only by prescribing a sinusoidal displacement on the leading-edge (LE) of the wing (Fig. 4.43a). Consequently, the resulting wing camber deformations and rotation α, the angle between the trailing-edge (TE) and LE, are purely passive due to dynamic balance between the aerodynamic loading, elastic restoring force, and inertia of the wing. In order to elicit the effects of flexibility with minimum complexity and ambiguity, α was approximated with a first-order harmonic (FH) approximation using α_m and α_e, the angles at the end and mid-of-the-strokes, respectively, as $a_{FH} = 90 - a_a \cos(2\pi t^* + \varphi)$ by calculating the angular amplitude α_a and phase lag ϕ. Specifically they imposed α and α_{FH} on a rigid flat plate without camber. To elucidate the interplay between the fluid physics and shape compliance they first considered the flexible wing and the rigid wing with α. As shown in Section 4.5.2, the cycle averaged lift coefficients $\overline{C}_L$ versus γ covers a wide variety of wing flexibility and flapping characteristics relevant to insects as well as artificially-devised flexible wings (Fig. 4.43b). Parameter γ can be interpreted as the non-dimensional wing tip displacement relative to the LE. The chordwise wing deformations act as passive pitch with an angle-of-attack, which is a key measure of the aerodynamics and directly affects lift (Fig. 4.43c).

Additional forces arise at the stroke reversals for rigid wings. When actively rotating wings exert force on the surrounding fluid, the fluid adds to or subtracts from the lift due to delayed stall, or translational mechanisms as described in Section 3.3. Their benefits depend on the phase difference between translation and rotation, advanced rotation being the optimal. Flexible wings also depict these three phase modes via passive rotation (Fig. 4.43d). The phase lag ϕ strongly correlates to f/f_1, yielding the advanced, symmetric, and delayed rotation modes with increasing f/f_1 (Fig. 4.43e). However, the large pressure differentials that exist on actively rotating rigid wings near the TE due to lift enhancing rotational effects are relaxed

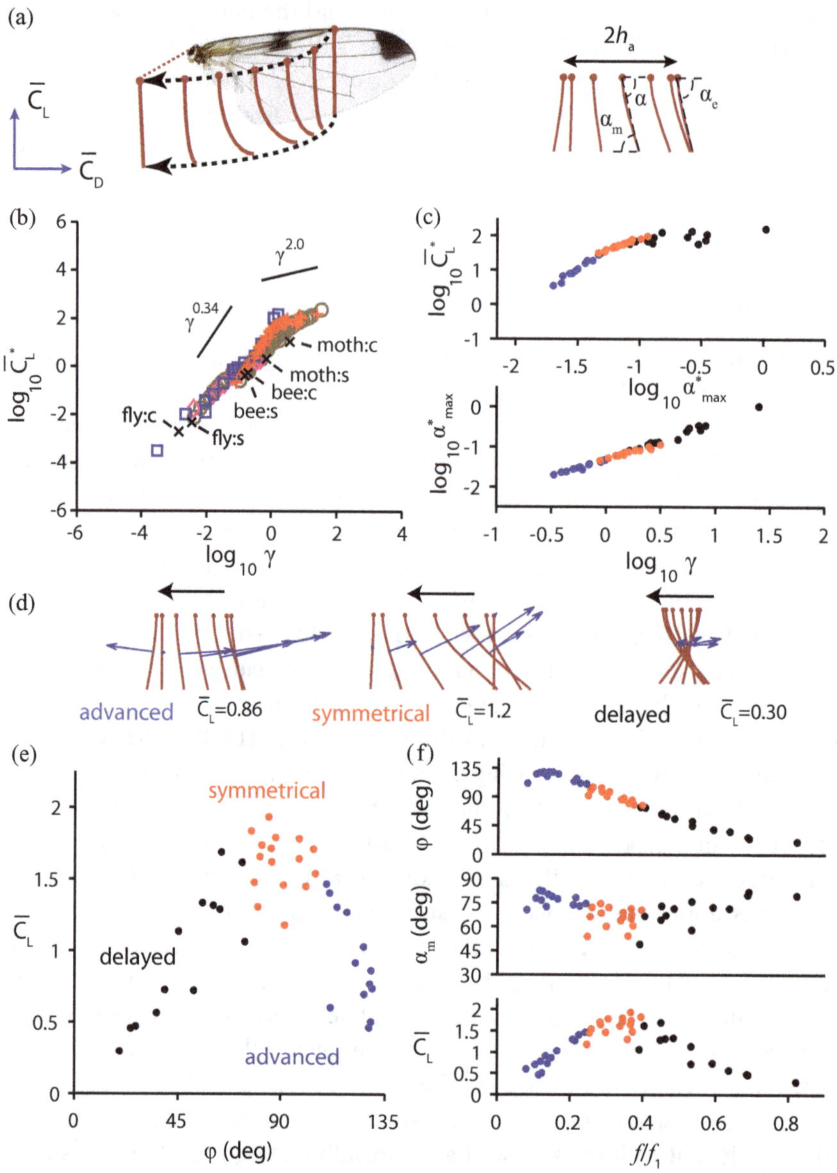

Figure 4.43. Aerodynamic forces generated by a flexible wing. (a) Schematic of wing motion for an advanced rotation case. (b) Scaling of the normalized lift $\overline{C}_L^*$ by γ. (c) Correlations between $\overline{C}_L^*$, maximum angle α_{max}* normalized in the same manner as for $\overline{C}_L^*$, and γ. (d) Schematics illustrating the advanced, symmetric, and delayed phase modes. (e) φ, α_m, and $\overline{C}_L^*$ as a function of f/f_1. (f) $\overline{C}_L^*$ as a function of φ. From Kang and Shyy [404].

by the compliant nature of flexible wing. Instead of generating rotational forces, the wing streamlines its shape, such that the wing shape and motion are in equilibrium with the fluid forces, similar to the drag-reducing reconfiguration of flexible bodies. Passive rotational angle α is purely due to deformation where the amplitude and the phase scale with γ and f/f_1. Moreover, α_m is measured at the instant of maximum translational velocity, hence indicative of the forces related to the translational mechanisms. For advanced rotations ($f/f_1 < 0.25$), α_m is greater than 70 deg. The

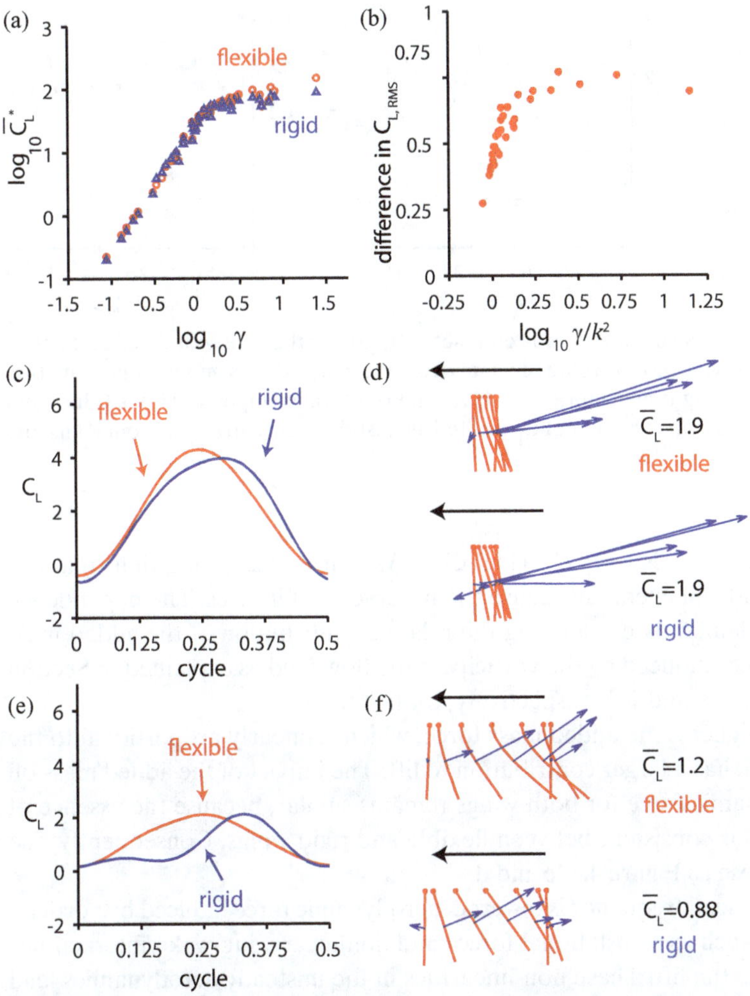

Figure 4.44. Difference in lift between flexible and rigid wings. (a) $\overline{C}_L^*$ on the flexible and rigid wings as a function of γ (b) RMS difference in the lift as a function of $\log_{10}\gamma/k^2$. Time histories of lift for the cases and schematics of wing shapes with the largest and smallest differences in lift are shown in (c,d) and (e,f), respectively. Both cases exhibit symmetric rotations with (c) a high reduced frequency and with (e) a low reduced frequency for the flexible (red), rigid (blue) wings and their instantaneous wing shapes (d, f). From Kang and Shyy [404].

wing orientation is almost vertical, producing small lift. On the other hand, deformations can become substantial for delayed rotations modes with $f/f_1 > 0.4$, but since the translation is out-of-phase with the passive rotation, the resulting α_m and hence the lift remain smaller than for symmetric rotations. The translational forces peak when α_m is around 45 to 50 deg, which corresponds to $0.25 < f/f_1 < 0.4$ (Fig. 4.43e). The resulting lift is of similar magnitude to that of rigid wings with active rotations. However, the lift is now optimal for symmetric rotation modes (Fig. 4.43f). This is also observed in the wing kinematics of fruit flies and honeybees.

To further assess the effects of flexibility, Kang and Shyy [404] compared the lift generation on the flexible wing and the rigid wing rotating with α_{FH}. Both wings generate lift that scales with γ (Fig. 4.44a). Most flexible wings yield higher lift with

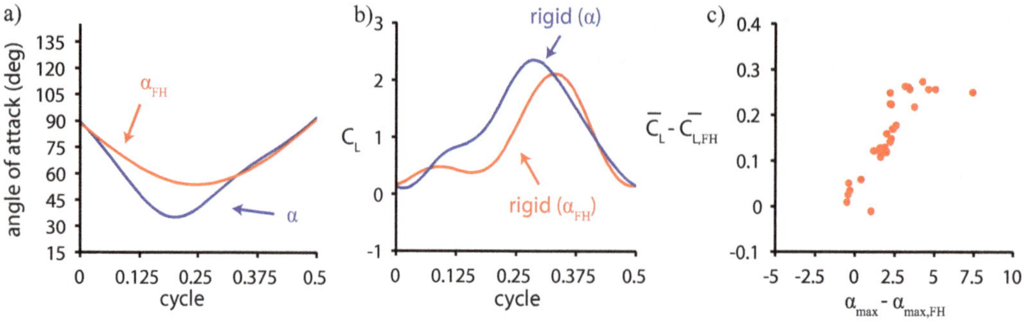

Figure 4.45. Non-linear structural motion enhances lift. (a) For the case with significant deformation, the resulting wing motion (blue) deviates from the first-order harmonic approximation (red). (b) Corresponding time histories of lift for the rigid wings imposed with α (blue) and α_{FH} (red) motions. (c) Enhancement of $\overline{C}_L^*$ plotted against difference in α_{max}. From Kang and Shyy [404].

a difference that depends on γ/k^2 (Fig. 2B). As γ increases, wing deformations become larger and will eventually cause the observed differences. The dependence on k^2 can be explained by considering the relative contribution of the added mass force and the forces induced by the vorticity in the flow field as explained in Section 3.6.4, which scale as k and $1/k$, respectively, for hover.

As such, at higher k the added mass force, which is linearly proportional to the wing acceleration, has a larger contribution to lift. The impact of the added mass on the total aerodynamic force for both wings remains similar, because the essence of the wing motions is consistent between flexible and rigid wings. Consequently, the resulting lift is close as Figure 4.44c and d.

When the reduced frequency is lowered, aerodynamic force induced by vorticity in the flow field, such as translational forces and nonlinear wing-wake interactions, starts to dominate the lift. These non-linearities in the unsteady aerodynamics lead to intriguing consequences caused by small differences in the resulting non-linear wing motion and the camber deformation. For example, Figure 4.44e illustrates the case corresponding to the largest difference in the lift. Lift on the flexible wing is considerably higher during the midstroke compared to that of the rigid wing (Fig. 4.44e, f), such that the lift generated by the flexible wing is superior: 1.2 versus 0.88.

The simple representation α_{FH} in the study of flapping wings is a popular approximation for rigid wings as shown in Section 3.4.1, despite the observations of hovering fruit flies and honeybees that are clearly far from sinusoidal [228]. Similar to these insects, the resulting flexible wing motion significantly deviates from being a sinusoid, depending on the wing properties (Fig. 4.45a). In general, the lift is higher for rigid wings with α. The dynamic balance between the fluid and the wing structure results in a rotation with larger deformation indicated by $\alpha_{\text{max}} = \max(\alpha)$ near the midstroke (Fig. 4.45b). The rigid wing following α generates higher lift that correlates to the difference between α_{max} and $\alpha_{\text{max,FH}}$, the angular amplitude in the simple harmonic motion, as illustrated in Figure 4.45c for the most cases. Inversely, this means that better performance can be obtained by departing from a simple sinusoidal rotational motion.

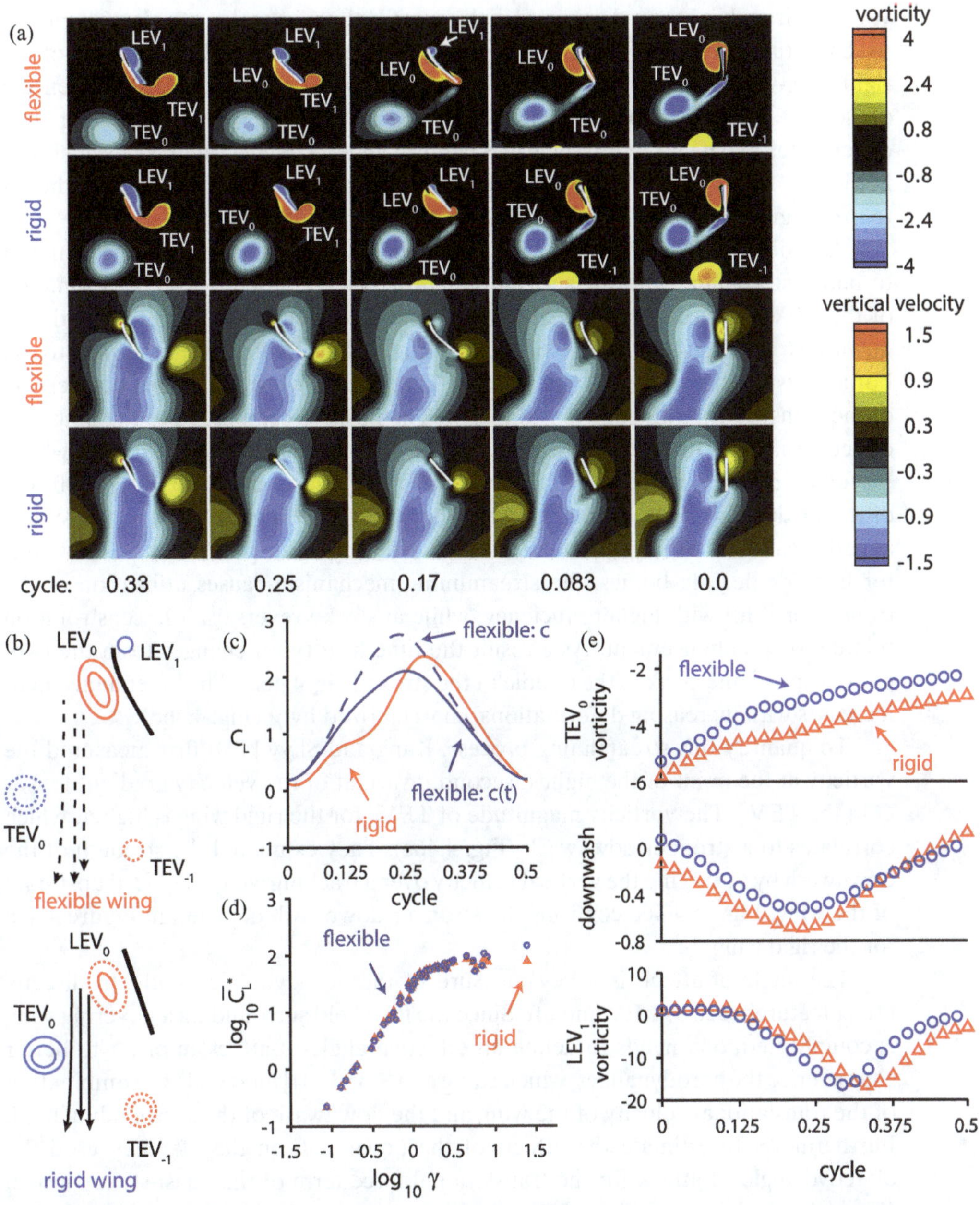

Figure 4.46. Effects of streamlining for a symmetric rotation mode by comparing to the rigid wing rotating with α. (a) Vorticity and vertical velocity fields. The wing thickness is exaggerated for clarity. (b) Schematic illustration of the wing-wake interaction. (c) Time history of lift normalized by c or $c(t^*)$ (8). (d) $\overline{C}_L^*$ as a function of γ. (e) Time histories of the vorticity of TEV_0 and LEV_1 and downwash. From Kang and Shyy [404].

An important mechanism responsible for lift enhancement of flexible wing over its rigid counterpart is due to wing-wake interactions, as highlighted in Figure 4.46. Kang and Shyy [404] considered a flexible and its rigid counterpart, rotating with α to focus solely on the effects of camber deformations. The LEV and the TEV shed in the previous motion stroke, denoted as LEV_0 and TEV_0, respectively, form a vortex pair and induce a downward wake around the midstroke, see also Chapter 3, which in turn interacts with the wing during its return stroke. The outcome of the rigid wing-wake interactions varies. Under favorable conditions, added momentum causes the lift to increase during the first portion of the stroke. The wing then experiences two peaks in lift: the wake-capture and the delayed stall peaks as shown in Figure 4.46b. However, the wake can also form a downward flow, causing the lift to drop significantly during midstroke.

A flexible wing, on the other hand, can reconfigure its shape, adjusting its camber to better streamline with the surrounding flow [404]. Consequently, the formation of the TEVs reduces for the flexible wing and the negative impact of the induced downwash is mitigated (Fig. 4.46a, b), leading to higher lift during the midstroke. For a locust, positive camber was measured in the hind wing due to the compression of the veins via trailing-edge tension [224]. The positive camber led to the umbrella effect, enhancing lift in the downstroke [224, 521]. Compared to the locust, whose Reynolds number is $Re \approx 4 \times 10^3$ [224], Figure 4.46 focuses on $Re = 100$ and different detailed wing characteristics. These lead to different observations: The negative camber has in general less favorable aerodynamic performance. However, for flapping flexible bodies, the streamlining mechanism causes utilization of the translational lift with higher efficiency, while at stroke reversals, it lessens rotation related force enhancement. As a result the time history lift changes from the two-peak shape to one-peak in the middle of the stroke (Fig. 4.46c). The lift enhancement increases with increasing deformations, characterized by γ (Fig. 4.46d).

To quantify this streamlining process, Kang and Shyy [404] first measured the vorticity at the point of the highest second invariant of the velocity gradient tensor Q in the TEV_0. The vorticity magnitude of TEV_0 for the rigid wing is higher, which correlates to a stronger downwash (Fig. 4.46c). They estimated the strength of the downwash by averaging the vertical velocity over a tracking window placed upstream of the wing. Figure 4.46e confirms the stronger downwash depicted in Figure 4.46a for the rigid wing.

The angle-of-attack is a key measure of the aerodynamics, and can directly affect features such as LEV and lift. Since the flow field surrounding a hovering wing is complicated, one needs to define an effective angle-of-attack in order to better characterize the aerodynamics, which Kang and Shyy [404] modeled as a combination of the translational velocity of the wing and the downwash of the surrounding fluid. Furthermore, to estimate the effects of the downwash on the lift, they used the effective angle-of-attack for the translational force term of the quasi-steady model (Fig. 4.47a). The correlation of the resulting quasi-steady lift to the Navier-Stokes solutions substantially improves compared to the estimation without the correction for the downwash (Fig. 4.47b, c). The enhanced agreement suggests that the drop in lift during the midstroke is indeed caused by the wing-wake interaction for the rigid wing (Fig. 4.47c). The flexible wing alleviates the downwash by streamlining its wing shape and outperforms its rigid counterpart with a difference in lift coefficient up to 0.3. The shape compliance and the dynamic responses of the flexible structure together contribute to the lift enhancement. However, without knowing the shape of a deformed flexible wing, the quasi-steady model cannot satisfactorily predict the aerodynamic force. Moreover, its performance can improve noticeably if we know how to correct the effective, instantaneous AoA a priori. Without comprehensive knowledge of the instantaneous flow field, such corrections are difficult to make.

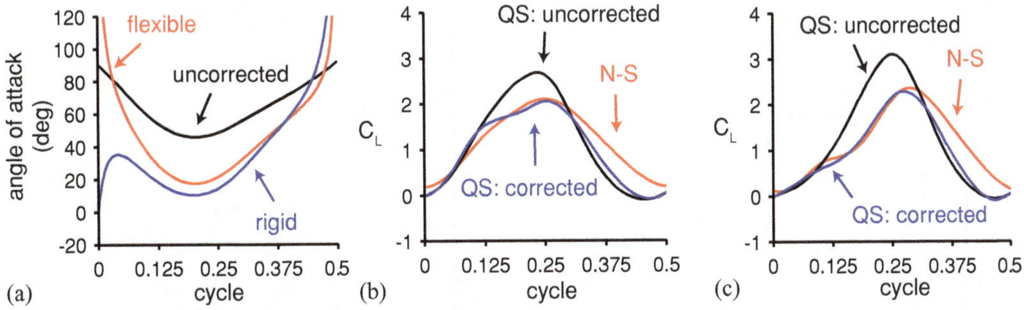

Figure 4.47. Effects of downwash illustrated with quasi-steady model predictions. (a) Angles of attack and (b,c) quasi-steady model predictions for the uncorrected angle-of-attack based on the wing's speed and orientation (black) and for the angles corrected with the downwash shown in Fig. 4.46 for the flexible (b) and the rigid (c) wings (blue). From Kang and Shyy [404].

4.5.3 Power Input and Propulsive Efficiency

The propulsive efficiency defined as

$$\eta = \frac{\langle C_T \rangle}{\langle C_P \rangle}, \tag{4–41}$$

where $\langle C_p \rangle$ is the time-averaged power input for a purely plunging wing computed as

$$\langle C_p \rangle = \langle C_L \dot{h} \rangle. \tag{4–42}$$

Note that the time-averaged power due to inertia vanishes for sinusoidal motions [336] since

$$\langle C_{P,\text{inertia}} \rangle \sim \langle \ddot{h}\,\dot{h} \rangle = \int_0^1 \ddot{h}\,\dot{h}\,dt \sim \int_0^1 \cos(2\pi t)\,\sin(2\pi t)\,dt = 0. \tag{4–43}$$

The propulsive efficiency for the chordwise flexible airfoils described in Section 4.4.2.1 and for the spanwise flexible wings described in Section 4.4.2.2 are plotted against the St in Figure 4.48. For comparison purposes the experimental measurements [336] [502] for both cases are included. For the chordwise flexible airfoils, η increases with decreasing h_s^*. Furthermore, η first increases with increasing motion frequency (i.e., the Strouhal number) since the plunge amplitude is kept constant, but it then plateaus, reaching some optimal efficiency. The thinnest airfoil generates the highest η. The experimental measurements illustrated in Figure 4.45b show a similar trend; however, there is an offset compared to the computed values. Again, uncertainties involved in the computational modeling or experimental setup may play a role. Moreover, as is shown later, the magnitude of $\langle C_P \rangle$ is an order of magnitude smaller than $\langle C_T \rangle$; hence even a small uncertainty in the power input measurement will lead to a large difference in the resulting η. For the spanwise flexible wings (Figure 4.48c, d), similar trends are found.

Before discussing the scaling of the power input and the propulsive efficiency, note that the scaling for the thrust generation for the chordwise flexible airfoils and the spanwise flexible wings is $\langle C_T \rangle / \beta_1 \sim \gamma^{1.17}$ with $\beta_1 = \Pi_1/(k/St)$. The power of γ has changed slightly compared to the previously determined value of 1.19 (Fig. 4.40)

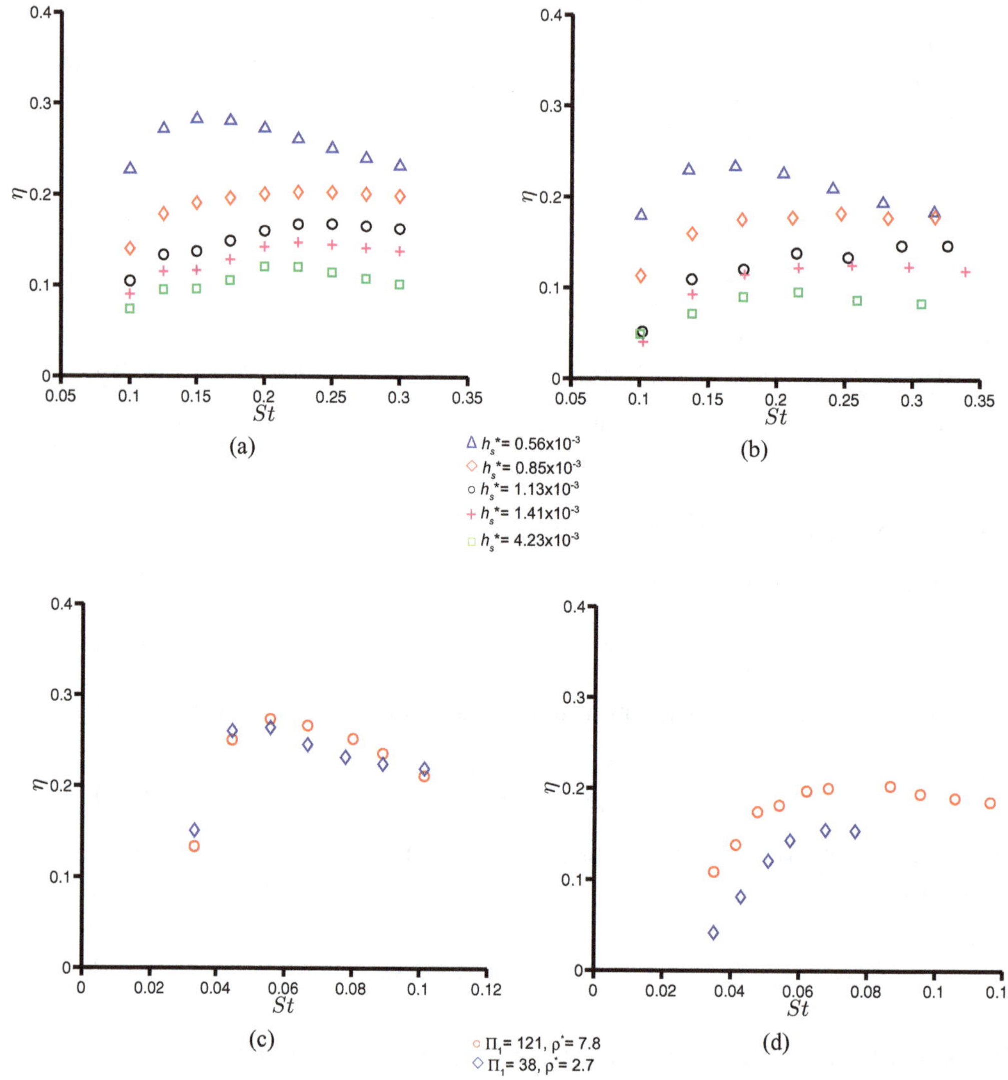

Figure 4.48. Propulsive efficiency plotted against the Strouhal number for the chordwise flexible airfoil cases (a)–(b) and the spanwise flexible wing cases (c)–(d).

because the data points from the isotropic Zimmerman wing cases and the insects are excluded. Although the value is different, the qualitative trend of the propulsive efficiency to be discussed later remains the same.

An interesting point concerning the power input is that the fluid dynamic force has been modeled as an added mass term, which is proportional to the acceleration of the wing motion. If the wing were rigid, just as for the power input due to the inertial force, we would have

$$\langle C_{P,\text{added mass, rigid}} \rangle \sim \langle \ddot{h}\,\dot{h} \rangle = 0, \qquad (4\text{–}44)$$

which is clearly not the case; see Figure 4.49. For small *St* the power input scales as St^2, but as the *St* increases, either the thickness ratio for the chordwise flexible airfoils or the different structural properties for the spanwise flexible wings affect the

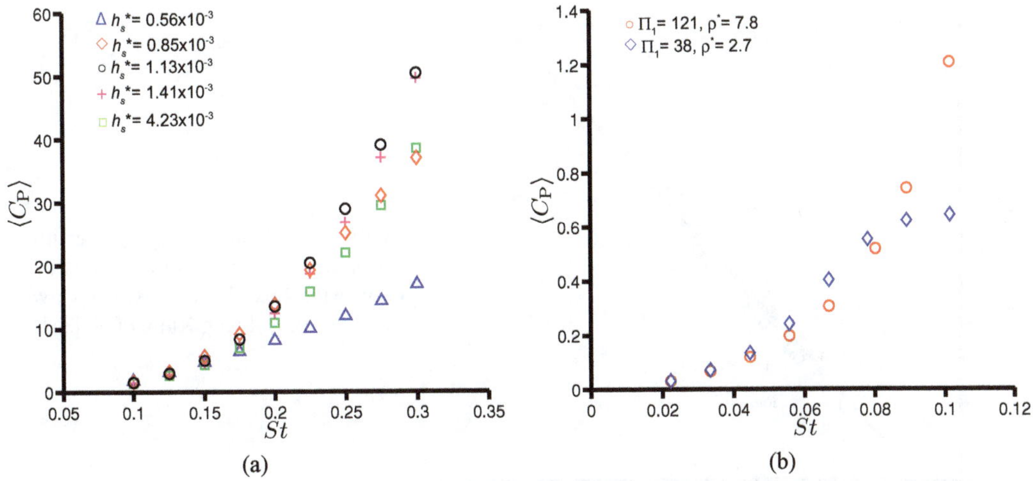

Figure 4.49. Time-average power input plotted against the Strouhal number (St). From Kang et al. [351].

resulting power. The fact that the power required is non-zero means that the resulting instantaneous lift on the wing should have a phase lag relative to the imposed motion. A major source for the phase lag is the wing deformation. By acknowledging the wing deformation given in Eq. (4–34), the time-averaged power input coefficient due to added mass can be approximated as in the first mode:

$$\begin{aligned}\langle C_{P,\text{added mass}}\rangle &= \int_0^1 \Pi_0(\ddot{T}+\ddot{h})\dot{h}\,dt \\ &= \pi^2 \frac{Q_1\gamma^2 k_1^8 \Pi_1^2}{k^2\left(1+\dfrac{4}{n}\rho^* h_s^*\right)} \sim \frac{\Pi_1^2}{k^2\left(1+\dfrac{4}{\pi}\rho^* h_s^*\right)}\gamma^2 \\ &= \beta_2\gamma^2,\end{aligned} \tag{4–45}$$

where the integral is approximated as

$$\begin{aligned}\int_0^1 \cos(\omega_1 t)\,\sin(2\pi t)\,dt &= \frac{2\pi(1+\cos\omega_1)}{\omega_1^2-4\pi^2} \approx -\frac{\pi\omega_1^2}{\omega_1^2-4\pi^2} \\ &= -\frac{\Pi_1}{4\pi}\frac{\gamma}{St\,k\left(1+\dfrac{4}{\pi}\rho^* h_s^*\right)}.\end{aligned} \tag{4–46}$$

For $\Pi_1 \gg \Pi_0$ the scaling for $\langle C_P\rangle$ reduces to $St^2(1+4\rho^* h_s^*/\pi)$; hence $\langle C_P\rangle \sim St^2$ in water, such as in the experimental setup considered in this case [336] [502], or for fixed density ratios and thickness ratios of the wing, consistent with the previous literature [456] [502] and Figure 4.49. Because the scales of $\langle C_P\rangle$ vary enormously, $\langle C_P\rangle/\beta_2$ is plotted against γ in log-scale in Figure 4.50. A linear fit with $R^2 = 0.98$

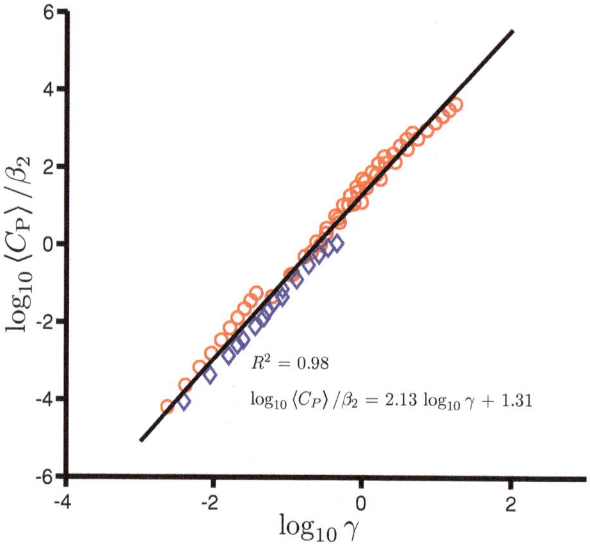

Figure 4.50. Time-average power input normalized by β_2 plotted againstγ. o: chordwise flexible airfoils;◊: spanwise flexible wings. From Kang et al. [351].

indicates that the power input scales with $\gamma^{2.13}$. The scaling for the propulsive efficiency now follows from the scaling for the thrust; that is, $\langle C_T\rangle/\beta_1 \sim \gamma^{1.17}$ with $\beta_1 = \Pi_1/(k/St)$ and $\langle C_P\rangle/\beta_2 \sim \gamma^2$ with $\beta_2 = \dfrac{\Pi_1^2}{k^2(1+\frac{4}{\pi}\rho^* h_s^*)}$ as

$$\eta = \frac{\langle C_T\rangle}{\langle C_P\rangle} \sim \frac{\beta_1\gamma^{1.17}}{\beta_2\gamma^2} \rightarrow \frac{\eta}{\beta_3} \sim \gamma^{1.17}. \tag{4–47}$$

where $\beta_3 = \dfrac{St\,k}{\Pi_1}\dfrac{\left(1+\frac{4}{\pi}\rho^* h_s^*\right)}{\gamma^2}$. The resulting scaling is shown in Figure 4.51.

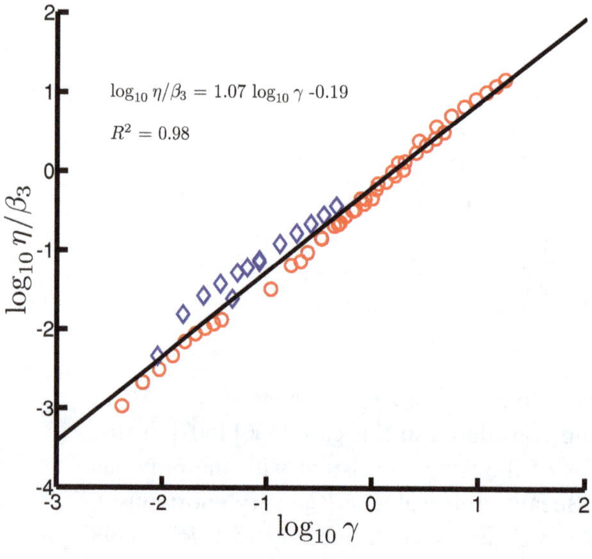

Figure 4.51. Propulsive efficiency normalized by β_3 plotted againstγ. o : chordwise flexible airfoils;◊: spanwise flexible wings. From Kang et al. [351].

4.5.4 Implications of the Scaling Parameters on the Aerodynamic Performance of Flapping Flexible Wings

The time-averaged force $\langle C_F \rangle$ and the propulsive efficiency η could be related to the resultant force on the wing depending on the situation, such as fluid/inertial force, with/without free-stream, or thrust/lift/weight. The current result enables us to estimate the order of magnitude of the time-averaged force generation and and its efficiency for a flexible flapping wing using a priori known parameters.

Furthermore, the scaling can guide the design of flapping wing MAVs. For example, to support a given weight W_{body} of a vehicle body in air, the scaling Eq. (4–36) reduces to

$$W_{\text{body}} \sim \frac{\rho_s^1 R^{3.19} f^{2.38} \phi_a^{2.19} c_{\text{m}}^{1.57}}{h_s^{0.38} E^{0.19}} \rightarrow f \sim \frac{m^{0.42} g^{0.42} h_s^{0.16} E^{0.080}}{\rho_s^{0.42} R^{1.34} \phi_a^{0.92} c_m^{0.66}}, \tag{4–48}$$

assuming $\Pi_1 \gg \Pi_0$. The condition $\Pi_1 \gg \Pi_0$ is satisfied when $f_1 \gg f$ and simplifies the algebra; however, it poses constraints on the range of the structural properties, such that the natural frequency of the wing is higher than the motion frequency. Equation (4–48) shows that increasing either the wing area, motion frequency, or the flapping amplitude helps generate sufficient lift to sustain hovering flight. In contrast, by making the wing softer (i.e., reducing the Young's modulus or wing thickness), wing deformations will increase, leading to higher lift generation. However, softening the wing further will violate the frequency ratio assumption, $f_1/f \gg 1$. The relation given for the flapping frequency is similar to that identified by Pennycuick [59]. Note that the wing weight is assumed to be negligible compared to the body weight in Eq. (4–48).

Another implication of the scaling is the interesting behavior of the role of h_s^* for the chordwise flexible airfoil cases shown in Section 4.3.2. It was observed that the $\langle C_T \rangle$ increased first and then decreased with decreasing h_s^*; see Figure 4.31. Using the current scaling, which is repeated as

$$\langle C_T \rangle \sim \Pi_1 \frac{St}{k} \gamma^{1.17} = \Pi_1 \frac{h_a}{c_m} \left\{ \frac{St\, k}{\Pi_0 \left(\frac{k_1^4}{4\pi^2} \frac{\Pi_1}{\Pi_0} - 1 \right)} \right\}^{1.17}, \tag{4–49}$$

consider first the situation that $f_1 \gg f$, i.e., h_s^* is not small. Then the denominator in γ can be approximated as $\Pi_0 \left\{ \frac{k_1^4}{4\pi^2} \frac{\Pi_1}{\Pi_0} - 1 \right\} \approx \Pi_1$, yielding

$$\langle C_T \rangle \sim St^2 \left(\frac{St\, k}{\Pi_1} \right)^{0.17}, \tag{4–50}$$

hence by reducing the thickness ratio, $\Pi_1 \sim h_s^{*3}$ will decrease, leading to the observed enhanced $\langle C_T \rangle$. However, by decreasing h_s^* further, the frequency ratio f_1/f will be eventually of the same order of magnitude, resulting in different physical behavior. If, say $f_1/f = O(1)$, but not in the resonance region, then the denominator in γ will scale as $\Pi_0 \left\{ \frac{k_1^4}{4\pi^2} \frac{\Pi_1}{\Pi_0} - 1 \right\} \approx \Pi_0$. Then, the resulting scaling will be

$$\langle C_T \rangle \sim St^2 \left(\frac{St\, k}{\Pi_0} \right)^{0.17} \frac{\Pi_1}{\Pi_0}. \tag{4–51}$$

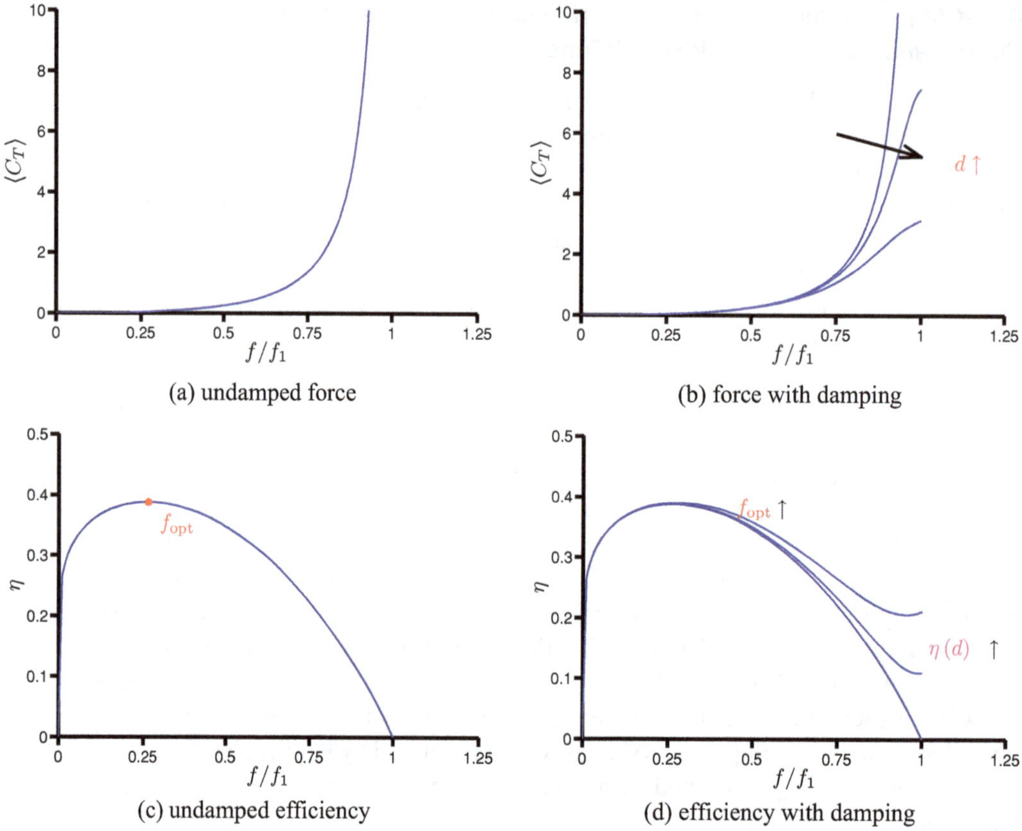

Figure 4.52. Force and propulsive efficiency plotted against the frequency ratio f/f_1.

Since $\Pi_1 \sim h_s^{*3}$ and $\Pi_0 \sim h_s^*$, we have $\langle C_T \rangle \sim h_s^{*1.83}$. This is consistent with the trend shown in Figure 4.31 that reducing the thickness ratio further will reduce the thrust.

Furthermore, for the thrust scaling for flapping flexible wings in water in forward motion, Eq. (4–37) can be rewritten as

$$\langle C_T \rangle \sim St^2 \left(\frac{\rho h_a}{\rho_s h_s}\right)^{0.17} \left(\frac{f_1}{f}\right)^2 \left\{\frac{1}{\left(\frac{f_1}{f}\right)^2 - 1}\right\}^{1.17}, \tag{4–52}$$

and since $St \sim \omega$, $k \sim \omega$, and $\Pi_0 \sim k^2 \sim \omega^2$, the thrust has a maximum at the resonance frequency. In reality, there is damping in the system, either structural or aerodynamic [456]. Although the effects of damping are not considered here, applying those effects for linear oscillators (e.g., [515]) yields the resonance frequency slightly below the natural frequency of the wing, with a finite value for the maximum relative wingtip deformation and hence the thrust: see Figure 4.52. In Figure 4.52b the effects of damping have been incorporated by approximating the denominator term as

$$\sqrt{\left\{1 - \left(\frac{f}{f_1}\right)^2\right\}^2 + \left(d\frac{f}{f_1}\right)^2}, \tag{4–53}$$

Table 4.4. *Optimal frequency ratios reported in the literature*

Literature	f/f_1	Description
Vanella et al. [492]	0.33	Hovering, airfoil, Navier-Stokes equation coupled with a torsion spring model
Yin and Luo [516]	0.4–0.5	Hovering, airfoil, Navier-Stokes equation coupled with a membrane model
Ramananarivo et al. [454]	0.5–0.6	Self-propelled flapper, experiment
Kang et al. [351]	0.41	Scaling analysis

where d is some small damping coefficient. This finding is consistent with the previous findings that the optimal propulsive performance is found near but slightly below the natural frequency of the wing [222] [453] [454] [508].

The propulsive efficiency scaling Eq. (4–47) can be rewritten in terms of the frequency ratio f/f_1 as

$$\eta \sim \left\{1-\left(\frac{f}{f_1}\right)^2\right\}^{0.83}\left(\frac{f}{f_1}\right)^{0.34}, \tag{4–54}$$

which has a local maximum at $f = 0.41f_1$. Depending on the scaling of the force, the optimal frequency can be found as

$$\frac{\partial\eta}{\partial f} = 0 \rightarrow \frac{f}{f_1}|_{\text{opt}} = \sqrt{\beta - 1}, \tag{4–55}$$

where β is the exponent of γ in the force scaling (i.e., 1.17 in Eq. (4–37) or 1.19 in Eq. (4–47), which yields the optimal frequency of 0.41 of the natural frequency. This indicates that the optimal efficiency is not achieved at the resonance frequency, but that the optimal frequency is some fraction of the natural frequency of the wing, which is also consistent with previous findings in the literature [453] [454] [492] [510] [516]; see Table 4.4. A cautionary note should be made, however, that Eq. (4–52) is derived from Eq. (4–36) and assumes $f/f_1 \ll 1$. Hence, strictly speaking, the conclusion that the maximum force is generated at the resonance frequency is not valid. The revised scaling proposed by Kang and Shyy [404] by considering hovering flexible flat plates cases with frequency ratios up to 0.8 would yield the frequency ratio of 0.6 without including the effects of damping.

Figure 4.52c depicts Eq. (4–55) as a function of the frequency ratio and shows that the propulsive efficiency increases with the increasing frequency ratio until reaching the optimal efficiency, and then it drops to zero at the resonance frequency. Since the undamped linear oscillator is unable to represent the resonance behavior correctly, an arbitrary damping has to be included in the system as before. Figure 4.52b shows the effects of including damping, where d is taken to be 0.0, 0.2, and 0.4: the optimal frequency ratio increases with increasing d, as does the efficiency at the resonance frequency. Note also the similarity between Figure 4.52b and the computed propulsive efficiency curves shown in Figure 4.48. A qualitative comparison with the measurements reported in Ramananarivo, Godoy-Diana, and Thiria [454] shows that, although the precise details differ, the overall qualitative trend is similar.

For example, for a 2 percent thickness wing with a rectangular planform made of aluminum hovering in air, the optimal frequency of the flapping motion is 5.4 Hz

Table 4.5. *Summary of the proposed scaling*

	Forward locomotion, water		Hovering, air
Force	$\beta_1\gamma^{1.19}$	$\Pi_1 h_a^* \left(\frac{\rho h_a}{\rho_s h_s}\right)^{1.19} \left\{\frac{1}{\left(\frac{f_1}{f}\right)^2 - 1}\right\}^{1.19}$	$\Pi_1 h_a^{*1.19} \left(\frac{\rho h_a}{\rho_s c_m}\right) \left\{\frac{1}{\left(\frac{f_1}{f}\right)^2 - 1}\right\}^{1.19}$
Efficiency	$\beta_3\gamma^{1.17}$	$\beta_3\gamma^{1.17}$	

Source: [351].

when the wing has a chord length of 20 cm and span of 50 cm with a flapping amplitude of 30°. Scaling down the geometry of the wing 10 times to the chord length of 2 cm and span of 5 cm, and keeping the aspect ratio the same, the optimal flapping frequency increases to 54 Hz. The resulting propulsive force coefficient and the propulsive efficiency remain the same for both cases. However, the dimensional propulsive force and the power required are 100 times smaller for the smaller sized wing, proportional to the square of the chord. Yet, for the same aspect ratio and thickness ratio, the volume of the wing is proportional to the cubic power of the chord. Consequently, for the same material, the mass of the smaller wing is 1,000 times smaller. The current scaling shows, consistent with Shyy et al. [245], that smaller flyers need to flap faster from the efficiency point of view, but their relative payload capacity increases because their weight reduces at a much faster rate compared to larger flyers.

Finally, the scaling parameters for diverse flow and kinematics conditions are summarized in Table 4.5. For forward flight in water, the effective stiffness Π_1, the normalized plunge amplitude h_a^*, the mass ratio $\mu_s = \rho h_a/(\rho_s h_s)$, and the frequency ratio f/f_1 dictate the propulsive force and the efficiency. For hovering in air, for the studied kinematics, the role of the mass ratio is taken over by the factor $\rho^* h_a^* = \mu_s/h_s^*$. Determining the efficiency for the hovering motion in air is left to future work in this study, and the resulting scaling is only predicted by following the same argument as for the propulsive efficiency in forward flight in air. Notice that the factor $\rho^* h_a^* = \rho h_a/(\rho_s c_m)$ is much smaller than $\mu_s = \rho h_a/(\rho_s h_s)$, resulting in a much lower propulsion for the hovering symmetric flap/plunge motion in air, since the thickness ratio is usually only of the order of 0.01. This order estimation matches well with the values shown in Ramananarivo, Godoy-Diana, and Thiria [454]. For the hovering Zimmerman wing the ratio between the propulsive force C_L and the force required for input power, C_T, was of the order of h_s^*. This scaling suggests that insects may require different mechanisms, such as active or passive pitching motion with a reinforced leading edge, which is commonly observed in many insect wings (see Fig. 4.4). Furthermore, non-symmetric motions, such as figure-of-eight motion (see Fig. 1.11) where the wing experiences forward motion, may yield higher efficiency [517].

4.6 Biological Flyers and Flexible Wings

Recently, efforts have been made to directly investigate the aerodynamic performance of biological flyers while accounting for the effect of flexible wing structures.

For example, Agrawal and Agrawal [518] investigated the benefits of insect wing flexibility on flapping wing aerodynamics based on experiments and numerical simulations at *Re* of 7.0×10^3. They compared the performance of two synthetic wings: (i) a flexible wing based on a bio-inspired design of the hawkmoth wing and (ii) a rigid wing of similar geometry. The results demonstrated that the bio-inspired flexible wing generated more thrust than the rigid wing in all wing kinematic patterns considered. They emphasized that the results provided motivation for exploring the advantages of passive deformation through wing flexibility.

Singh and Chopra [519] experimentally measured the thrust generated for a number of wing designs undergoing flapping motion at *Re* of 1.5×10^4. The key conclusions that stemmed from this study were that the inertial loads constitute the major portion of the total loads acting on the flapping wings tested on the mechanism and that, for all the wings tested, the thrust drops at higher frequencies. Further, it was observed that at such frequencies, the lightweight and highly flexible wings used in the study exhibit significant aeroelastic effects.

Hamamoto et al. [520] studied FSI analysis on a deformable dragonfly-like wing in hover and examined the advantages and disadvantages of flexibility at an *Re* of 1.0×10^3. They tested three types of flapping flight: a flexible wing driven by dragonfly flapping motion, a rigid wing (stiffened version of the original flexible dragonfly wing) driven by dragonfly flapping motion, and a rigid wing driven by modified flapping based on tip motion of the flexible wing. They found that the flexible wing with nearly the same average energy consumption generates almost the same amount of lift force as the rigid wing with modified flapping motion. In this case, the motion of the tip of the flexible wing provides equivalent lift to that provided by the motion of the root of the rigid wing. However, the rigid wing requires 19 percent more peak torque and 34 percent more peak power, indicating the usefulness of wing flexibility.

Young et al. [521] conducted numerical investigations on a tethered desert locust, *Schistocerca gregaria*. Their results showed that time-varying wing twist and camber are essential for the maintenance of the attached flow at the Reynolds number, 4×10^3. The authors emphasized that, although high-lift aerodynamics are typically associated with massive flow separation and large LEVs, when high lift is not required, attached flow aerodynamics can offer greater efficiency. Their results further showed that, in designing robust lightweight wings that can support efficient attached flow, it is important to build wings that undergo appropriate aeroelastic wing deformation through the course of a wing beat.

Tanaka et al. [522] investigated the effect of wing flexibility in hoverflies using an at-scale mechanical model. They suggested that at-scale models operating in air have the potential to simulate the aerodynamic phenomena of compliant flapping wings because their structure, inertia, operating frequency, and trajectories are similar to those of insects in free flight. For this purpose, an at-scale polymer wing mimicking a hoverfly was fabricated using a custom micro-molding process. It had venation and corrugation profiles that mimic those of a hoverfly wing, and its measured flexural stiffness was comparable to that of the natural wing. To emulate the torsional flexibility at the wing-body joint, a discrete flexure hinge was created. A range of flexure stiffness was chosen to match the torsional stiffness of pronation and supination in a hoverfly wing. The polymer wing was compared with a rigid, flat,

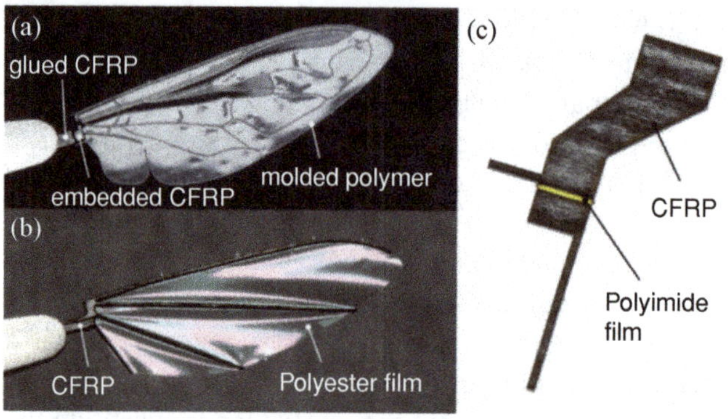

Figure 4.53. Photos of an at-scaled model wing: (a) hoverfly mimic; (b) a rigid carbon fiber model; and (c) example of hinge. The wing length is 11.7×10^3 [m]. From Tanaka et al. [522].

carbon-fiber wing using a flapping mechanism driven by a piezoelectric actuator as shown in Figure 4.53. Both wings exhibited passive rotation around the wing hinge; however, these rotations were reduced in the case of the compliant polymer wing due to chordwise deformations during flapping that caused a reduced effective AoA. Maximum lift was achieved when the stiffness of the hinges was similar to that of a hoverfly in both wing cases and the magnitude of measured lift was sufficient for hovering. These results suggest that hoverflies could exploit intrinsic compliances to generate desired motions of the wing and that, for the same flapping motions, a rigid wing could be more suitable for producing large lift.

In the next several subsections, we summarize recently reported studies based on biological models.

4.6.1 Implications of Anisotropic Wing Structure on Hovering Aerodynamic: Hawkmoths

As discussed earlier, there are few studies of the aerodynamics of flapping wings associated with anisotropic wing structure; in addition, the 3D wing shape and the timing of deformation during flapping flight have not yet been studied well. Recently, Nakata and Liu [523] conducted a computational analysis of hawkmoth hovering flight based on a realistic wing-body model, which takes into account the anisotropy of a hawkmoth wing. They investigated how the wing deformation and modified kinematics due to the inherent structural flexibility affect unsteady fluid physics and aerodynamic performance. In their study, a reasonably representative wing-body morphological model was built as shown in Figure 4.54. Note that although hawkmoths are four-winged, for simplicity, they modeled the fore- and hindwings as a single pair of wings because of the highly synchronized motion observed in flapping flight. Hawkmoths' wing structure is mainly supported by wing veins and membrane. The wing veins are clustered and thickened around the wing base and leading edge, as illustrated in Figure 4.54a, and are tapered toward the wingtip and trailing edge [403] [524]. A thin flexible membrane is placed between the veins; the directional arrangement of the wing veins and the difference of bending stiffness between the veins and membrane result in a high anisotropy of the flexural stiffness of hawkmoth wings [402]. Wing and body kinematic models were constructed based

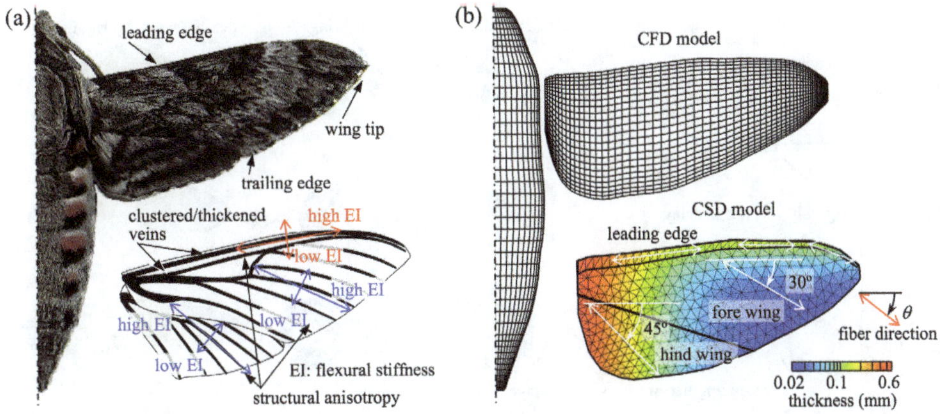

Figure 4.54. (a): A hawkmoth *Agrius Convolvuli* with a generalized wing venation including fore- and hind-wings. (b) A computational model for CFD and CSD analysis. From Nakata and Liu [523].

on the experimental data of the hovering hawkmoth, *Manduca*, and on kinematic parameters [247] [523] described in Section 3.7. Note that the body was assumed to be stationary because the body motion in hovering flight is negligibly small [226].

4.6.1.1 In-Flight Deformation of a Hawkmoth's Wing

Figure 4.55 shows the comparison of the instantaneous and time-averaged results associated with flexible and rigid wings. Figure 4.56 displays the time-varying wing shape and deformation in terms of the spanwise bending, the twist, and the camber. From Figure 4.55a the flexible wing shows an advanced phase in the feathering angle, but a delayed phase in the positional angle at the wingtip, with respect to prescribed motion of the rigid wing at the wingtip. Also from Figure 4.55b, the translational and rotational velocities in the cross-section of $0.8R$ increase remarkably, in particular before stroke reversal. Those results occur because of large spanwise bending and twisting when the wing translates and because of small peaks before the subsequent stroke reversal as illustrated in Figure 4.56. Furthermore, although the spanwise bending and the twist angle vary smoothly, there exists a rapid increase in distal area. The flexible wings show pronounced deformation in spanwise bending and twist immediately after stroke reversal. The maximum of nose-down twist in the distal area (forewing) is approximately 12° when predicted by computation and $15^\circ \sim 20^\circ$ when measured by experimentally [226]. One can see some positive camber, less than 2 percent, that is relatively small compared with the measured cambers of large insects such as locusts [525]. Overall, such deformation leads to significant changes in wingtip kinematics.

4.6.1.2 Aerodynamic Performance of Flexible Wings

How do these deformations affect the force generation? Time courses of vertical and horizontal force coefficients and aerodynamic force vectors generated by flexible and rigid wings are plotted in Figure 4.55c and d. Although both flexible and rigid wings show a plateau in the vertical aerodynamic force production at early downstroke, the flexible wing obviously creates more force than the rigid wing, and the force vectors contribute more to the vertical force components (Fig. 4.55c, d). Instantaneous

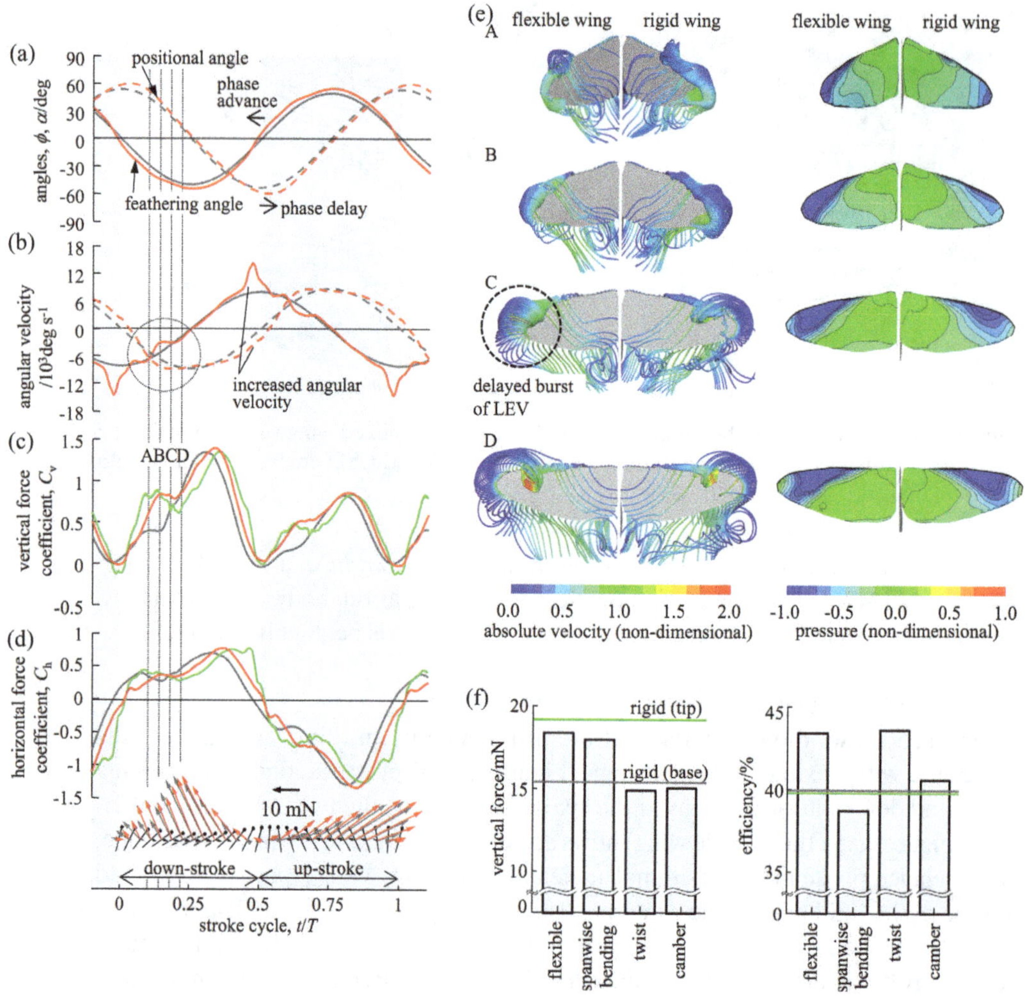

Figure 4.55. Time courses of (a) positional and feathering angles at 0.8 R. Red line, flexible wing; gray line, rigid wing. (b) angular velocities, (c) vertical and (d) horizontal force coefficients of flexible and rigid wings. Red line, flexible wing; gray line, rigid wing (base); green line, rigid wing (tip). (e) Instantaneous streamlines and pressure contours on upper surfaces of flexible and rigid wings at instants A, B, C, and D as shown in (c). (f) Time-averaged vertical aerodynamic force and efficiency generated by a flexible wing, a rigid wing, and three rigid wing modes with prescribed bending, twisting, and camber. From Nakata and Liu [523].

streamlines and pressure contours on the wing surfaces are illustrated in Figure 4.55e during an interval where four instants are marked as A, B, C, and D (Fig. 4.55e). At early downstroke when the wing proceeds to late pronation and undergoes a pitch-down rotation, the LEV and the TEV grow in size and in strength, stretching from the wing base toward the wingtip; the TEV subsequently detaches from the wing, forming a starting vortex but connecting to the LEV and the TiV at the wingtip at A (Fig. 4.55e). Subsequently, the TiV of the rigid wing becomes unstable at C, gradually separating and shedding from the wing surface, which correspondingly results in a shrink pattern of low-pressure contours at the wingtip. In contrast, the LEV and the TiV on the flexible wing seem to be more stable than those of the rigid wing, with an enlarged low-pressure region at B and C (Fig. 4.55e). When the TiV

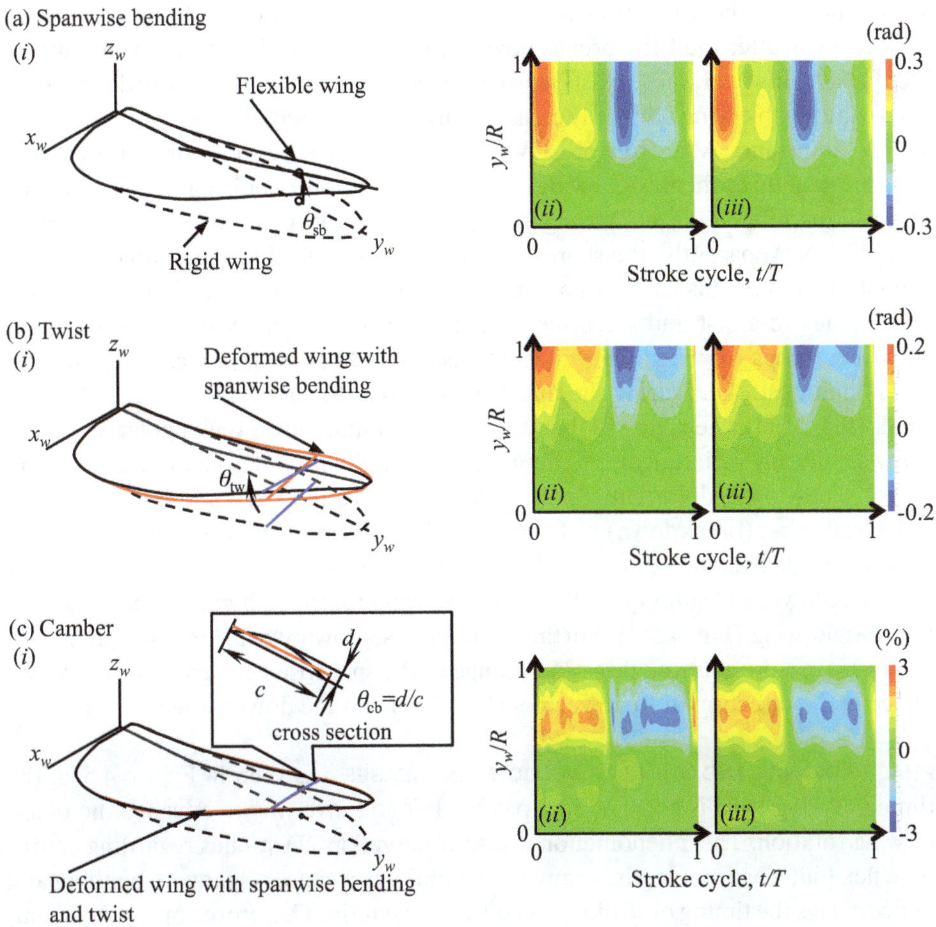

Figure 4.56. (i): Definition of the simulated (a) spanwise bending angle, (b) twist angle, and (c) camber in the flexible wing. (ii) Time courses and spanwise distribution of the three kinds of deformation of a flexible wing and (iii) their interpolated deformation. From Nakata and Liu [523].

breaks down and separates from the wing, the LEV still keeps growing with a strong low-pressure region at D (Fig. 4.55e) until the vertical aerodynamic forces generated by both flexible and rigid wings become almost even immediately after the wing turns to decelerate. However, it is interesting to find that, at late downstroke, the flexible wing eventually reaches a higher force peak other than the rigid wing (Fig. 4.55c). When the wing approaches early supination, the aerodynamic force decreases owing to the breakdown or shedding of the LEV and the translational deceleration. Here, the flexible wing can also create larger forces than the rigid wing.

4.6.1.3 Lift Enhancement by Wing Flexibility

DELAYED BURST OF THE LEADING-EDGE VORTEX. Aerodynamic and inertial forces applied on a flapping wing can result in passive wing deformations, which are likely responsible for stabilizing and hence delaying the breakdown of the LEV/TiV during wing translation. As illustrated in Figure 4.55c, both flexible and rigid wings show a similar high peak of vertical force immediately after the wing turns to decelerate.

This is because the LEV keeps growing and attaching coherently onto the wing surface, even after the LEV breaks down with the TV shedding off the wing surface [225] [388]. However, there exists a pronounced discrepancy at the early downstroke where the flexible wing obviously creates more vertical forces than the rigid wing (Fig. 4.55c). At instance A, a stronger LEV as a portion of a horseshoe vortex is observed near the wingtip of the flexible wing, which grows rapidly over instances of B, C, and D, resulting in a larger and stronger negative pressure region on the wing surface (Fig. 4.55e). Apparently, the spanwise bending of the flexible wing induced during pronation creates this LEV at an earlier timing than in the rigid wing (Fig. 4.56a), which leads to a fast and steep increase in vertical force at A-B (Fig. 4.55c). The LEV then keeps growing for a while up to instance D before approaching the middle down- and upstroke. During the interval, although the inertial force becomes very small (Fig. 4.56i), the spanwise bending and twist and hence the angular velocities show significant variations near the wingtip (B and C; the marked circle in Figure 4.55b). These wing deformations very likely stabilize the LEV and hence result in a delayed burst (breakdown) at D compared with that of the rigid wing at C (Fig. 4.55e). This delayed burst even further influences the development of the LEV after the breakdown, and subsequently the flexible wing reaches a higher force peak than in the rigid wing (Fig. 4.55c). Furthermore, a nose-down twist (Figs. 4.55 and 4.56) can result in a pronounced direction change of the spanwise wing cross-sections, and hence in the direction of force vectors (Fig. 4.55) and the downwashes.

PHASE ADVANCE AND ANGULAR VELOCITY INCREASES. As seen in Figure 4.55a, the timing of wing twist is adjusted in a passive but adaptive way to advance the phase of wing rotation. This phenomenon is also observed in 2D studies regarding chordwise flexibility. Moreover, the spanwise flexibility can cause a spanwise bending and hence delays the timing of stroke reversal at the wingtip. Therefore appropriate combination of the chord- and spanwise deformation leads to a relative phase advance of rotation in flexible wings, which can strengthen the vortex ring and the downwash, as well as the rotational circulation, while modifying the wing attitude to benefit from the wake capture [201].

In addition, the wing deformations seem to lead to increasing angular velocities of positional and feathering angles, mostly in the distal area of flapping wings (Fig. 4.55), which can augment the circulation around insect wings [291]. Obviously, the relative phase advance and the angular velocity increase correspond with a larger vertical force (Fig. 4.55), rather than that of the rigid wing during wing rotation. Hence the spanwise bending is responsible for most vertical forces produced immediately before stroke reversal (Fig. 4.55). Furthermore, the rigid wing model with the deformed wing kinematics prescribed (Fig. 4.55) does provide concrete evidence that not only the three-dimensional wing configuration but also the variation in wing kinematics can enhance the aerodynamic force production. In relation to the downwash and the force production, the stronger downward flow (Fig. 4.57) is created in a timely fashion when the wing experiences a rapid stroke reversal where larger forces are created (Fig. 4.55), which was also confirmed experimentally by Mountcastle and Daniel [473].

As noted by Nakata and Liu [523], a rigid wing model with the prescribed wingtip kinematics can create a large vertical aerodynamic force comparable with that of the

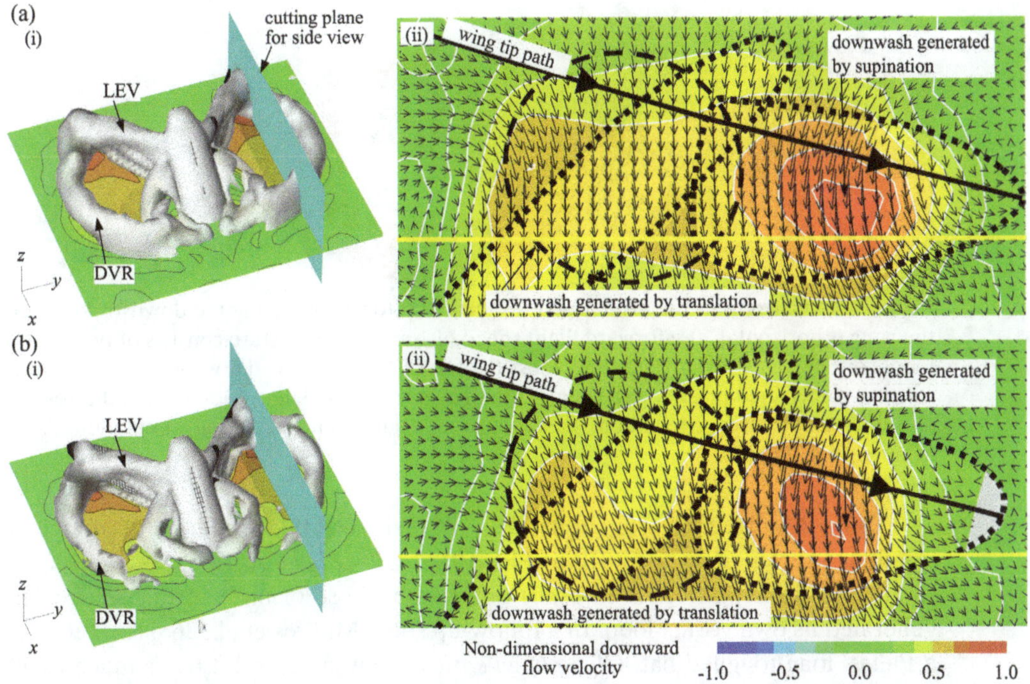

Figure 4.57. Wake structures generated at downstroke about (a) flexible and (b) rigid wings. (i) Iso-vorticity surface is plotted with a magnitude of 1.5; downwash at a horizontal plane and (ii) at a cutting plane of 0.6 R from the wing root is visualized in terms of velocity vectors and downward velocity contours. Note that DVR stands for the downstroke vortex ring. From Nakata and Liu [523].

flexible wing; however, this force results in a significant drop in aerodynamic efficiency of 39.8 percent. Furthermore, the efficiency enhancement attained by flexible wings requires more input power. This is because the wing kinematics at the wingtip is modified passively – but favorably – due to the elastic wing deformation, albeit with an input of relatively inefficient wing kinematics at the wing base. Note that the wing base in hovering has low velocity and is ineffective for aerodynamic force production. This suggests that there may be an optimal distribution of wing kinematics between the wing base and wingtip, which should be more efficient in creating vertical aerodynamic forces; see also the discussion in Section 4.5. This points to the importance of the dynamic spanwise distribution of wing kinematics in enhancing aerodynamic efficiency and thus confirms that the elastic wing deformation is an effective way to achieve higher aerodynamic performance in insect flapping flight.

4.7 Aerodynamics of Bat Flight

As mentioned in the introduction to this chapter, bat wings possess more than two dozen joints, highly deforming bones, and anisotropic membrane skin with variable stiffness. How these wing structures and aerodynamic loading interact and affect the aerodynamics is of primary interest. In this section we highlight the aerodynamics of several species of free-flying bats at several flight speeds as observed from the experimental studies.

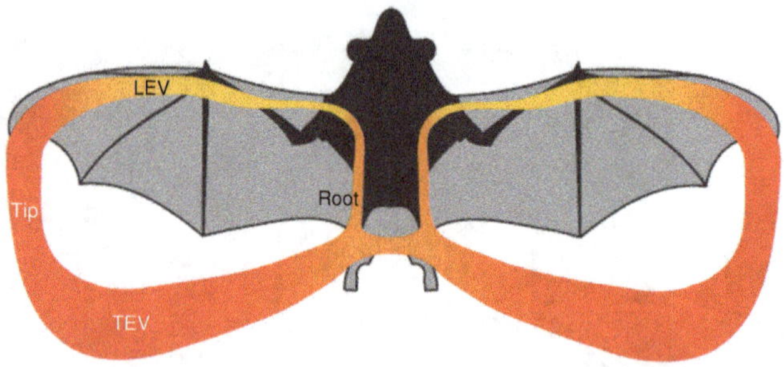

Figure 4.58. Cartoon of the primary vortex structure for a bat during the downstroke when the wing is horizontal, at a forward flight speed of 1 m/s. The structure consists of two closed loops, one for each wing. The loop consists of a LEV on top of the wing, connected to a starting vortex shed in the wake via a TiV and a root vortex. The color coding indicates the absolute value of local circulation; yellow is low circulation and red is high circulation. From Muijres et al. [268].

Using PIV, Hendenström et al. [267] [269] showed that the wake structures of small bat species ($Re = 2.6 \times 10^3$ to 1.8×10^4 in *Glossphaga soricina*, $Re = 3.2 \times 10^3$ to 2.3×10^4 in *Leponycteris curasoa*) differ from those of birds in that each wing generated its own vortex loop. In a follow-up study Muijres et al. [268] showed that a Pallas' long-tongued bat, *Glossphaga soricina* can increase lift by as much as 40 percent using attached LEVs in slow forward flight (1 m/s), resulting in a maximum lift coefficient of 4.8 ($Re = 5 \times 10^3$, $St = 1.36$). The flow passing over the LEV reattaches behind the LEV smoothly to the wing, despite the exceptionally large local AoA and wing camber. Figure 4.58 shows a proposed cartoon model of the vortex system around the bat wing during the downstroke; these vortex structures are similar to those discussed in relation to the hovering flight of insects (see Section 3.7.1.1) [225].

For a wide range of flight speed from 1.3–5.5 m/s of a medium-sized bat, *Cynopterus brachyotis*, Hubel et al. [243] [262] performed synchronized time-resolved measurements of the wing kinematics (body and wing joint) and wake structures in the spanwise plane with a sampling rate of 200 Hz in a closed-return wind tunnel ($Re = 1.7 \times 10^4$ to 2.7×10^4, $St = 0.25$ to 0.57). They consistently identified four vortex structures that are typical for both lower and higher speed flight in *Cynopterus brachyotis*. The wake structure is dominated by a strong TiV that develops during the downstroke and remains until almost the end of the upstroke. However, although the portion of the wing-beat cycle without the TiV is very short, the wake structure can still be considered a vortex ring rather than a continuous vortex structure. The proportion of the cycle without the TiV increases with speed; this contrasts with the continuous vortex or ladder structure at higher speeds that has been proposed for bird flight based on both observations and theory [230] [526]–[528]. As shown in Figure 4.59, simultaneously with the onset of the TiV at the beginning of the downstroke, a counter-rotating near-body vortex develops, indicating that the left and right wings operate independently, with no or little lift generation over the body. The overall wake structure model illustrated in Figure 4.59 globally resembles those proposed by Hendenström et al. [269], but not in details because different bat species are used and the experimental conditions (e.g., free-stream turbulence intensity, section size, and so forth) were not identical.

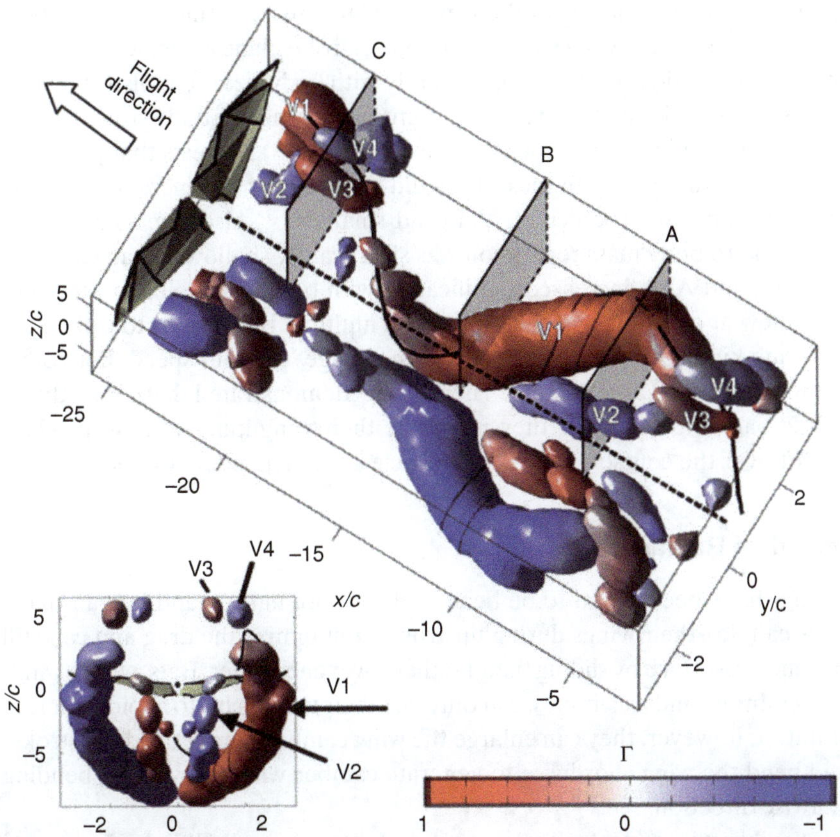

Figure 4.59. Reconstruction of the wake structure (upper) for approximately one and a half wing-beat cycles. Iso-surfaces of the transverse swirl are shown based on the 2D PIV images. The wake is displayed in the body reference coordinate in the *y-z* plane, and time was converted into space to allow the spatial display of the wake structure behind the bat. The wake structure was mirrored along the body trajectory lie and colored based on the circulation of the found dominant vortex structures (V1-V4: blue: clockwise rotation, red: counterclockwise rotation). From Hubel et al. [243].

As already discussed in Section 1.1 and earlier in this section, important issues in natural animal flight that have been recently recognized are the wing beat to wing beat, flight to flight, and individual variability. Recently Riskin et al. [529] compared the wing kinematics of 27 bats representing six pteropodid species, varying more than 40 times in body mass (0.0278–1.152 kg), to determine whether wing posture and overall wing kinematics scaled as predicted according to the existing theories. The smallest species flew in a wind tunnel and the other five species in a flight corridor. Seventeen kinematic markers on the midline and left side of the body were tracked in three dimensions. The investigators used phylogenetically informed reduced major axis regression to test for allometry. Their data showed that maximum wingspan and maximum wing area scale with more positive allometry, and wing loading scales with more negative allometry than has been reported in previous studies [4] that were based on measurements from specimens stretched out flat on a horizontal surface (see Table 1.3). They observed that larger bats open their wings more fully than small bats do in flight and, that for bats, body measurements alone cannot be used to predict the conformation of the wings in flight. Several kinematic variables, including downstroke ratio, wing stroke amplitude, stroke plane angle, wing camber, and Strouhal number, do not change significantly with body size, demonstrating that many aspects of wing kinematics are similar across this range of body sizes.

Whereas the aerodynamic theory (see also Section 1.3) suggests that preferred flight speed should increase with mass, they did not observe this increase. Instead, large bats had higher lift coefficients than did small bats. Also, the slope of the wing-beat period to body mass regression was significantly shallower than expected under isometry, and AoA increased significantly with body mass. None of the bats in their data flew at constant speed, so they used multiple regressions to isolate the changes in wing kinematics that correlated with changes in flight speed, horizontal acceleration, and vertical acceleration. Their results demonstrated that for medium- to large-sized bats, the ways that they modulate their wing kinematics to produce thrust and lift over the course of a wing-beat cycle are independent of body size.

4.8 Concluding Remarks

Flexible wings have been found to be beneficial for both natural and human-made flyers. Birds can flex their wings during upstroke to minimize the drag and can still maintain a smooth surface by sliding their feathers over each other. Bats, whose wings consist of membrane and arm bones, can only flex their wings a bit to avoid structure failure or flutter; however, they can enlarge the wing camber during the downstroke. Insects can bend the wing chordwise to generate camber while preventing bending in the spanwise direction.

Fixed, flexible wings can facilitate steadier, better controlled flight. In a gusty environment, a flexible wing can provide a more consistent lift-to-drag ratio than can a rigid wing by adaptively adjusting the camber in accordance with the instantaneous flow field. By responding to the aerodynamic loading variations, a membrane wing can also adaptively conduct passive camber control to delay stall. Moreover, a flexible wing can adjust the wing shape to its motion to mitigate the loss of lift due to the non-linear wing-wake interaction, which can be significant for a rigid wing in hover.

A membrane wing is found to exhibit flutter, whose frequency is about an order of magnitude higher than that of the vortex shedding frequency. The flutter exists even under a steady-state free-stream condition. Such intrinsic vibrations result from coupled aerodynamics and structural dynamics.

We also highlighted the impacts of the anisotropic nature of flexibility on resulting flight performance by considering the chordwise, spanwise, isotropic, and anisotropic wing flexibilities individually. Chordwise and spanwise flexibilities interact with surrounding air, and depending on the imposed kinematics and wing flexibility, they can enhance propulsive force. Studies on anisotropic wing structures are more recent and can provide insightful observations that can be applied to MAV development. The impact of structural flexibility on aerodynamics undergoing plunging motion in forward flight can be highlighted as follows:

1. The thrust generation consists of contributions due to both leading-edge suction and the pressure projection of the chordwise deformed rear foil. When the rear

foil's flexibility increases, the thrust of the teardrop element decreases, as does the effective angle of attack.
2. Within a certain range, as chordwise flexibility increases, even though the effective angle of attack and the net aerodynamic force are reduced due to chordwise shape deformation, both mean and instantaneous thrust are enhanced due to the increase in the projected area normal to the flight trajectory.
3. For the spanwise flexible case, correlations of the motion from the root to the tip play a role. Within a suitably selected range of spanwise flexibility, the effective angle of attack and thrust forces of a plunging wing are enhanced due to the wing deformations.

Furthermore, we have demonstrated that as the Reynolds number drops to $O(10^2)$–$O(10^3)$, which corresponds to flyers such as fruit flies and honey bees, the viscous effects are significant, and the flexibility of the wing structure can noticeably change the instantaneous, effective angle of attack via the large-scale vertical flows. The structural flexibility can also mitigate the induced downward jet via shape deformation. Both effects can result in enhanced aerodynamic performance. Furthermore, lift generated on the flexible wing scales with the relative tip deformation parameter, whereas the optimal lift is obtained when the wing deformation synchronizes with the imposed translation, which we also observe in fruit flies and honey bees. Hence, under such modest Reynolds number, the observation that synchronized rotation is aerodynamically preferable for flexible wings is different from what we observe for the rigid wing. It is recalled that in Chapter 3, we have concluded that appropriate combinations of advanced rotation and dynamic stall associated with large AoAs can produce more favorable lift. These findings clearly highlight the effect of wing flexibility.

Since the various scaling parameters vary with the length and time scales in different proportionality, the scaling invariance of both fluid dynamics and structural dynamics as the size changes is fundamentally difficult and challenging. It also seems that there is a desirable level of structural flexibility to support desirable aerodynamics. Significant work needs to be done to better understand the interaction between structural flexibility and aerodynamic performance under unpredictable wind gust conditions. Dimensional analysis and non-dimensionalization of the governing equations for the fluid and the wing structure lead to a system of non-dimensional parameters, such as Reynolds number (Re), reduced frequency (k), Strouhal number (St), aspect ratio (AR), effective stiffness (Π_1), effective AoA (α_e), thickness ratio (h_s^*), the density ratio (ρ^*), and finally the force coefficients. Compared to the set of parameters in rigid flapping wing aerodynamics, three additional non-dimensional parameters were introduced: Π_1, ρ^*, and h_s^*. From the scaling arguments, the time-averaged force coefficient could be related to non-dimensional relative tip deformation. The tip deformation is an outcome of the interplay between the imposed kinematics and the response of the wing structure dictated by the wingtip amplitude and the phase lag. The amplitude of the maximum relative wingtip deformation, γ, was obtained from the non-dimensional beam analysis and is only a function of the a priori known non-dimensional paramters. By considering the energy balance of the wing, the time-averaged force normalized by the effective stiffness was related to γ. The time-averaged force can be related to the resultant force on the wing depending on the situation, such as fluid/inertial force, with/without free-stream, or thrust/lift/weight. These results enable us to estimate the order of magnitude of the time-averaged force generation for a flexible flapping wing using a priori known parameters. Furthermore, for propulsive efficiency it was seen that the optimal efficiency was obtained for the motion frequency that is lower than the natural frequency. The current scaling shows that smaller flyers need to flap faster from the efficiency point of view, but the relative payload capacity increases because their weight reduces at a much faster rate compared to larger flyers.

The study of hawkmoth hovering suggests that flexibilty plays a role in both the resulting wing kinematics and aerodynamic force generation. However, because of the limited number of studies regarding fluid-structure interaction in biological flyer-like models, further investigations are needed to derive general conclusions regarding the role of wing flexibility in their flights.

5 Future Perspectives

Scaling laws have clearly established that MAVs and small natural flyers cannot operate with the same lift and thrust generation mechanisms as larger aircraft and that moving wings are needed to conduct the flight mission. Furthermore, these flyers can substantially benefit from active and passive morphing for flight performance enhancement and control, just like the flyers that we see in nature. For example, Lee et al. [530] investigated longitudinal flight dynamics of a bio-inspired ornithopter with a reduced-order aerodynamics model, including wing flexibility effects. They showed that this ornithopter is robust to external disturbances due to its trimmed flight dynamic characteristics that limit cycle oscillation. Furthermore the mean forward flight speed increases almost linearly with the flapping frequency. Inspiration from nature will give us insight on how to manage the complex flight environment. Still we need to keep in mind there is no single perfect mechanism that all flyers should adopt. It is ultimately up to the scientists and engineers to figure out the best combination of the available techniques for a given flyer to suit its size, weight, shape, and flight environment, including wind gust, and mission characteristics.

A variety of wing kinematics and body/leg maneuvers are observed in flyers of various sizes and even in different flight missions of the same flyer; these include takeoff, forward flight of varying speeds, wind gust response, hover, perch, threat avoidance, station tracking, and payload variations. Flying insects are known to execute aerial maneuvers through subtle manipulations of their wing motions. For example, as reported by Bergou et al. [531], fruit flies asymmetrically change the spring rest angles to generate rowing motion of their wings during sharp-turning flight. Also, Combes and Dudley [532] have observed that bees flying in outdoor turbulent air become increasingly unstable about their roll axis as air speed and flow variability increase. The bees are reported to extend their hind legs ventrally at higher speeds, improving roll stability, but also increasing body drag and associated power requirements by 30 percent. Our knowledge of these complex dynamics and our understanding of the influence of the wing-body interaction in the flapping wing aerodynamics are largely incomplete.

A topic that needs to be addressed more comprehensively is the interaction between fore- and hindwings, and the resulting force and flight control implications. For example, it is well known that dragonflies have the ability to control aerodynamic performance by modulating the phase lag between forewings and hindwings. Wang

Figure 5.1. The relative sizes of the insect's fore- and hindwings are different between species, and consequently, the phase relationship of their wing movement also varies.

and Russell [533] filmed the wing motion of a tethered dragonfly and computed the aerodynamic force and power as a function of the phase. They found that the out-of-phase motion as seen in steady hovering uses nearly minimal power to generate the required force to balance the weight, and the in-phase motion seen in takeoffs provides an additional force for acceleration. Thomas et al. [258] and Wang and Sun [534] studied the fore- and hindwing interactions via flow visualization and computational simulations, respectively. Brackenbury [535] used high-speed flash photography to analyze wing movements in more than 30 species of butterflies. He observed that early in the upstroke the wings show pronounced ventral flexure that, combined with inertial lag in the posterior parts of both wing pairs and delayed supination in the hindwing, leads to the formation of a funnel-like space between the wings. As shown in Figure 5.1, the relative sizes of fore- and hindwings differ between species, and consequently, the phase relationship of their wing movement also varies. The wing-to-wing interactions are complicated due to a large number of parameters involved. Depending on the flight mission (takeoff, climbing cruising, landing), the fluid physics, such as the leading-edge vortex, wake, and tip vortices; the wing geometry including size, shape, and aspect ratio; and the structural flexibility of the wings all play roles in coupled manners. Wings and body interactions and relative movement are so far largely open research topics.

Much work remains to be done in modeling robust aerodynamics models that can be used to develop control strategies and designs of flapping wing MAVs. As discussed in Chapter 3 most investigations attempting to reduce the complex aerodynamics of flapping wings are fragmented. There is a need for a coherent systematic effort involving simplified analytic theories, high-fidelity numerical simulations, and experimentalists to discern the "quasi-steadiness" of the forces on a flapping wing and to document the validity of these simplified models for various configurations and dimensionless parameters, such as k, St, phase lag, and so on. As already reviewed, Gogulapati and Friedmann [367] offer an updated approach to treat the flapping wing aerodynamics in a simplified framework. Progress is being made in this direction; however, more comprehensive guidelines are needed to establish viscous flow features such as leading-edge vortices and wakes, which depend on these dimensionless parameters. Furthermore, if a wing is flexible, then the viscous and flexible structure interaction becomes significantly more complicated. Gogulapati et al. [536] have made an effort in this direction while awaiting the development of a more rigorous approach.

As discussed in Chapter 4, the translational forces resulted from the quasi-steady model for the uncorrected (based on nominal AoA), corrected with the downwash for the flexible, and the rigid wings are highlighted in the figure as well. Clearly, the modifications in (i) effective AoA and (ii) shape deformation affect the aerodynamic performance significantly. The quasi-steady model's performance can improve noticeably if we know how to correct the effective, instantaneous AoA as well as shape deformation. Without knowing the instantaneous flow field comprehensively, such corrections are difficult to make. More efforts are needed in establishing better guidelines to use such simplified aerodynamic models more effectively.

Regarding the dynamics and stability of a flight vehicle in association with flapping wing aerodynamics, Orlowski and Girard [537] presented a recent overview of various analyses of flight dynamics, stability, and control. Although efforts are being made to use the multi-body flight dynamics model to predict the behavior of flapping wing MAVs, the majority of the flight dynamics research still involves the standard aircraft (6DOF) equations of motion [537]. Furthermore, the investigations of the stability of flapping wing MAVs have so far been essentially limited to hover and steady forward flight, with most studies focusing on linear, time-invariant aspects on the basis of reference flight conditions. Based on such approaches, flapping wing MAVs have been found to be intrinsically unstable in an open loop setting [538]–[544]. Moreover, control of flapping wing MAVs has been largely investigated by neglecting the mass of the wings on the position and orientation of the central body [545]–[549]. As summarized in Chapter 1, the ratio between the wing mass and the mass of the vehicle can be of order $O(1)$, which means that the motion of the wing can affect vehicle dynamics and stability. Natural flyers are often sensor-rich, and these sensors can offer necessary information needed for real-time flight maneuvers in uncertain and unpredictable flight environments. For example, vision-based sensing techniques can be very helpful for flight control as well as for estimating the aeroelastic states of the vehicle.

It is established in Chapter 4 that local flexibility can significantly affect aerodynamics in both fixed and flapping wings. Preliminary research has been reported the vehicle stabilization via passive shape deformation due to flexible structures [550]. Furthermore, as already discussed, insects' wing properties are anisotropic, with the spanwise bending stiffness about 1 to 2 orders of magnitude larger than the chordwise bending one. As the vehicle size changes, the scaling parameters cannot all maintain invariance due to the different scaling trends associated with them. This means that, for measurement precision, instrumentation preference, and the like, one cannot do laboratory tests of a flapping wing design using different sizes or flapping frequencies [450] [551]. A closely coordinated computational and experimental framework is needed to facilitate the exploration of the vast design space (which can have $O(10^2)$ or more design variables including geometry, material properties, kinematics, flight conditions, and environmental parameters) while searching for optimal and robust designs.

Natural flyers and swimmers share many similarities in terms of the physical mechanisms for locomotion. Although lift is more important for flight than for swimming, pitch, plunge, and flexible structures are all clearly observable in force generation. The difference in density between air and water directly influences the structural properties and some other parameters. However, from a scaling viewpoint, these differences can be linked via the same dimensionless parameters. Much of the same scaling laws are equally applicable to swimming as to flight [37] [38] [41] [186] [552] [553].

Further, cross-fertilization has been regularly achieved in areas related to locomotion, energetics, morphology, and hydrodynamics. Of course, more interactions are to be expected. For example, Weihs [554] used the slender body theory from aerodynamics to study the turning mechanism in fishes. He showed that the turning process includes three stages, distinguished by different movements of the center of mass. In the first and third stages the center of mass moves in straight lines in the initial and final directions of swimming, whereas in the middle period it moves along an approximately circular connecting arc. The forces and moments acting on the fish can be satisfactorily predicted by treating the vortex wake by the circulation shed from the fins. Weihs [555] also showed that fish can swim more efficiently by alternating periods of accelerated motion and powerless gliding. His analysis of the mechanics of swimming showed that large savings of more than 50 percent in the energy required to traverse a given distance can be obtained by such means. In calculations based on measured data for salmon and haddock, the possibility of range increases of up to three times the range at constant speed was demonstrated. Wu [553] reviewed the forward swimming motion, bird flight approximated by oscillating wings, and insect flight with the high-lift generation via LEVs. From these studies he derived mechanical and biological principles for unified studies on the energetics in deriving metabolic power for animal locomotion. Studies such as these can offer significant insight into analyzing and designing flapping wing flyers.

Biological and physical scientists and engineers can benefit from close collaboration to better understand features that enable natural flight. For instance, although wings produce lift required for flight in natural systems, not all wing features are flight related. Honeybees employ short-amplitude high-frequency strokes when they hover [556]; these strokes generate high lift as these flyers consume floral nectar containing high energy, which enables them to carry loads or perform high-power-required maneuvers or missions. However, these strokes are shown to be aerodynamically inefficient [216] [351] [404]. Furthermore, bird wings, bat wing membranes, insect wings, and the like have some interesting but widely varied material properties, which can be used for flapping MAV development. These bio-inspired mechanisms include, for example, joints and distributed actuation to enable flapping and morphing. Another topic of much interest but that has so far not been adequately covered is the flight envelope from takeoff to landing. For flapping flyers, key advantages are the variable speed and flapping kinematics along with shape deformation, resulting in highly maneuverable flight characteristics. Furthermore, depending on each species'

Figure 5.2. The Formosan (Taiwan) Barbet (Megalaima nuchalis), nests in tree cavities, and, as shown in the upper row, forms a noticeably inflated body contour immediately prior to takeoff from the cavity. As shown in the lower row, it also balances delicately between speed and position control to land.

biological development, its key features in takeoff and landing/perching can vary more than the flapping patterns in comparison to other flyers of comparable sizes. An interesting example is the Formosan (Taiwan) Barbet (*Megalaima nuchalis*), which nests in tree cavities (see Fig. 5.2). Consequently, they take off with distinctive body movement to generate the initial thrust, including a noticeably inflated body contour immediately prior to take-off. The landing requires a balancing act between speed and position control (see Fig. 5.2.).

Birds enjoy maneuvering flexibility by combining wings and tails, especially during take-off and landing. However, depending on the individual cases, the relative size of the tail varies significantly. While the parrot and magpie exhibit relatively longer tails than most birds, some birds feature noticeably short tails. These characteristics have major implications on aerodynamics, active and flexible structures, and above all, flight control. Figure 5.3 highlights these features. Another interesting features related to bird flight are that the shape of a wing can be apparently less than streamlined. As shown in Figure 5.4, during takeoff, feather can intrude into the surrounding flows, whose impact is yet to be quantitatively characterized.

Another example is related to hovering. In Chapter 1, it is reported that many birds and bats hover with the stroke plane inclined about 30 degrees to the horizontal, and the resultant downward stroke force with a lift-to-drag ratio of 1.7 is vertical [70]. However, flight environment, especially wind, and vehicle size, can cause strong impact on the hovering pattern. Two small birds, the Ruby-throated Hummingbird (7–9 cm total length) and the Light-vented Bulbul (*Pycnonotus sinensis*), also known as the Chinese Bulbul (15–20 cm long), can be used as examples here. The flapping of a hummingbird is on a horizontally-inclined stroke plane with a symmetric

Figure 5.3. Birds enjoy maneuvering flexibility by combining wings and tails, especially during take-off and landing. However, depending on the individual cases, the relative size of the tail varies significantly. While parrot and magpie exhibit relatively longer tails, some birds feature noticeably shorter tails, resulting in apparently different flight control patterns.

Figure 5.4. During takeoff, feather in certain portion of the bird wing can intrude into the surrounding flow field.

Figure 5.5. Hummingbirds utilize wing-wail coordination and varied inclination toaccommodate wind.

figure-eight pattern. Furthermore, as shown in Figure 5.5, due to its small size and light weight, a hummingbird frequently utilizes wing-tail combination and flexible wing structures to accommodate wind gust. Unlike the hummingbird, the bulbul flaps asymmetrically while hovering. As shown in Figure 5.6, in the downward stroke, bulbuls' horizontally inclined wings move while being rotated. During the upward stroke, the wings are initially flexed and then spread out horizontally. Consequently, lift is generated during the downward stroke; the forward and backward thrusts created during different strokes cancel each other. In short, hummingbirds and bulbuls, both small in size, use distinctively different flapping kinematics, wing morphology and structural flexibility while hovering. Many outstanding issues related to flapping

(a) Wing flexed in the early phase of upstroke

(b) Wings horizontally Inclined in the early phase of downstroke

(c) Wings fully spread out and vertically inclined toward the end of downstroke

Figure 5.6. Flapping pattern of a Chinese Bulbul while hovering.

wing aerodynamics are to be investigated even for commonly observed phenomena. Such complex and varied flight patterns vividly show how much room we have to further learn from nature. By working across the scientific disciplines, scientists and engineers will be able better equipped to uncover the magic of natural flyers' amazing performance as well as contribute to the advancement of the human-engineered MAVs.

References

[1] K. P. Dial, "Inside look at how birds fly: Experimental studies of the inertial and external process controlling flight," in *Thirty-Eighths Symposium Proceedings Experimental Test Pilots*, Lancaster, California, pp. 301–314, 1994.

[2] L. Da Vinci, *Codex on the flight of birds*, 1505.

[3] J. D. Anderson Jr., *Inventing flight: The Wright brothers and their predecessors*: Johns Hopkins University Press, 2004.

[4] U. M. Norberg, *Vertebrate flight: Mechanics, physiology, morphology, ecology and evolution*: Springer-Verlag, 1990.

[5] J. H. McMasters and M. J. Henderson, "Low speed single element airfoil synthesis," *Technical Soaring*, vol. 6, pp. 1–21, 1980.

[6] G. E. Goslow, K. P. Dial, and F. A. Jenkins, "Bird flight: Insights and complications," *BioScience*, vol. 40, pp. 108–115, 1990.

[7] P. Shipman, *Taking wing: Archaeopteryx and the evolution of bird flight*: Simon and Schuster, 1999.

[8] B. W. Tobalske and K. P. Dial, "Flight kinematics of black-billed magpies and pigeons over a wide range of speeds," *Journal of Experimental Biology*, vol. 199, pp. 263–280, 1996.

[9] G. C. Aymar, *Bird flight*: Garden City Publishing Co., 1938.

[10] J. H. Storer, *The flight of birds*: Cranbrook Institute of Science, 1948.

[11] A. Magnan, *La locomotion chez les animaux*: Hermann and cie, 1934.

[12] J. H. McMasters, "The flight of the bumblebee and related myths of entomological engineering," *American Scientist*, pp. 164–169, 1989.

[13] E. Franco, D. N. Pekerek, J. Peng, and J. O. Dabiri, "Geometry of unsteady fluid transport during fluid-structure interactions," *Journal of Fluid Mechanics*, vol. 589, pp. 125–145, 2007.

[14] D. R. Warrick, B. W. Tobalske, and D. R. Powers, "Aerodynamics of the hovering hummingbirds," *Nature*, vol. 435, pp. 1094–1097, 2005.

[15] J. M. McMicheal and M. S. Francis, "Micro air vehicles – toward a new dimension in flight," available at www.fas.org/irp/program/collect/docs/mav_auvsi.htm.

[16] D. Pines and F. Bohorquez, "Challenges facing future micro-air-vehicle development," *Journal of Aircraft*, vol. 43, pp. 290–305, 2006.

[17] P. G. Ifju, D. A. Jenkins, S. Ettinger, Y. Lian, W. Shyy, and M. R. Waszak, "Flexible-wing-based micro air vehicles," in *AIAA*, p. 0705, 2002,

[18] K. D. Jones and F. M. Platzer, "Bio-inspired design of flapping-wing micro air vehicles – An engineers's perspective," *in 44th AIAA Aerospace Science Meeting and Exhibit, 9–12 January*, Reno, Nevada, AIAA-2006-37, 2006.

[19] Y. Kawamura, S. Souda, and C. P. Ellington, "Clapping-wing micro air vehicles of insect size," in *Bio-mechanisms of Swimming and Flying*, eds. N. Kato and S. Kamimura: Springer, 2007, ch. 26, pp. 319–322.

[20] G. Warwick, "UAVForge contest seeks better, cheaper approach for design," *Aviation Week*, pp. 1–3, Oct. 2011.

[21] S. P. B. Lissaman, "Low Reynolds number airfoils," *Annual Review of Fluid Mechanics*, vol. 15, pp. 222–239, 1983.

[22] D. E. Alexander, *Nature's flyers: Birds, insects, and the biomechanics of flight*: Johns Hopkins University Press, 2004.

[23] D. E. Alexander, *Why don't jumbo jet flap their wings?: Flying animals, flying machines, and how they are different*: Rutgers University Press, 2009.

[24] A. Azuma, *The biokinetics of flying and swimming* (2nd ed.): American Institute of Aeronautics and Astronautics, 2006.

[25] A. A. Biewener, *Animal locomotion*: Oxford University Press, 2003.

[26] A. K. Brodsky, *The evolution of insect flight*: Oxford University Press, 1994.

[27] R. Dudley, *The biomechanics of insect flight: Form, function, evolution*: Princeton University Press, 1999.

[28] D. L. Grodnitsky, *Form and function of insect wings: The evolution of biological structures*: Johns Hopkins University Press, 1999.

[29] H. Tennekes, *The simple science of flight*: MIT Press, 1996.

[30] J. J. Videler, E. J. Stamhuis, and G. D. E. Povel, "Leading-edge vortex lift swifts," *Science*, vol. 306, pp. 1960–1962, 2004.

[31] S. Vogel, *Life in moving fluids: The physical biology of flow*: PWS Kent Publishers, 1981.

[32] A. J. Ward-Smith, *Biophysical aerodynamics and the natural environment*: John Wiley and Sons Ltd, 1984.

[33] S. Dalton, *The miracle of flight*: Merrell, 2006.

[34] J. W. S. Pringle, *Insect flight* (Reprint ed.): Cambridge University Press, 2010.

[35] C. L. Henderson, *Birds in flight: The art and science of how birds fly*: Voyageur Press, 2008.

[36] D. Floreano, J.-C. Zufferey, M. V. Srinivasan, and C. P. Ellington, *Flying insects and robots*: Springer-Verlag, 2009.

[37] T. Y.-T. Wu, C. J. Brokaw, and C. Brennen, eds., *Swimming and flying and nature*: Plenum Press, 1975.

[38] T. J. Pedley, *Scaling effects in animal locomotion*: Academic Press, 1977.

[39] L. Maddock, Q. Bone, and J. M. V. Rayner, eds., *Mechanics and physiology of animal swimming*: Cambridge University Press, 1994.

[40] M. J. Lighthill, "Hydrodynamics of aquatic animal propulsion," *Annual Review of Fluid Mechanics*, vol. 1, pp. 413–446, 1969.

[41] T. Y.-T. Wu, *Hydromechanics of swimming of fishes and cetaceans*: Academic, 1971.

[42] S. Childress, *Mechanics of swimming and flying*: Cambridge University Press, 1981.

[43] T. Maxworthy, "Experiments on the Weis-Fogh mechanism of lift generation by insects in hovering flight. Part 1. Dynamics of the 'fling,'" *Journal of Fluid Mechanics*, vol. 93, pp. 47–63, 1979.

[44] J. Anderson, *Introduction to flight* (7th ed.): McGraw-Hill Science/Engineering/Math, 2011.

[45] J. Katz and A. Plotkin, *Low-speed aerodynamics* (2nd ed.): Cambridge University Press, 2001.

[46] R. S. Shevell, *Fundamentals of flight*: Prentice-Hill, 1983.

[47] M. Thollesson and U. M. Norberg, "Moments of inertia of bat wings and body," *Journal of Experimental Biology*, vol. 158, pp. 19–35, 1991.

[48] A. Magnan, *Le vol des insectes*: Hermann, 1934.

[49] A. Magnan, "Les characteristiques des oiseaux," *Annales Des Sciences Naturelle*, vol. 5, pp. 125–334, 1922.

[50] S. Dhawan, "Bird flight, Sadhara," *Academy Proceedings in Engineering Sciences*, vol. 16, pp. 275–352, 1991.
[51] J. B. Anders, "Biomimetic flow control," in *AIAA 2000–2543*, Denver, 2000, pp. 1–28.
[52] M. J. Lighthill, "*Introduction to the scaling of aerial locomotion*," in *Scale effects in animal locomotion*, ed. T. J. Pedley: Academic Press, 1977, pp. 365-404.
[53] C. J. Pennycuick, *Newton rules biology: A physical approach to biological problems*: Oxford University Press, 1992.
[54] K. Schmidt-Nielsen, *Scaling: Why is animal size so important*: Cambridge University Press, 1984.
[55] C. H. Greenewalt, "The flight of birds: The significant dimensions, their departure from the requirements for dimensional similarity, and the effect on flight aerodynamics of that departure," *Transactions of the American Philosophical Society*, vol. 65, pp. 1–67, 1975.
[56] J. M. V. Rayner, "Form and function in avian flight," in *Current ornithology*, ed. R. F. Johnston: Plenum, 1988, pp. 1–66.
[57] T. Liu, "Comparative scaling of flapping- and fixed-wing flyers," *AIAA Journal*, vol. 44, pp. 24–33, 2006.
[58] P. Jackson, *Jane's all the world's aircraft*: Jane's Information Group, 2001.
[59] C. J. Pennycuick, "Wingbeat frequency of birds in steady cruising flight: New data and improved predictions," *Journal of Experimental Biology*, vol. 199, pp. 1613–1618, 1996.
[60] C. J. Pennycuick, *Bird flight performance: A practical calculation manual*: Oxford University Press, 1989.
[61] S. J. Kirkpatrick, "Scale effects on the stresses and safety factors in the wing bones of birds and bats," *Journal of Experimental Biology*, vol. 19, pp. 195–515, 1994.
[62] C. J. Pennycuick, "Mechanics of flight," in *Avian biology*, eds. D. S. Farner and J. R. King: Academic Press, 1975, pp. 1–75.
[63] P. M. Alexander, *Principles of animal locomotion*: Princeton University Press, 2002.
[64] R. T. Jones, *Wing theory*: Princeton University Press, 1990.
[65] C. P. Ellington, "The aerodynamics of hovering insect. I. The quasi-steady analysis," *Philosophical Transactions of the Royal Society of London. Series B*, vol. 305, pp. 1–15, 1984.
[66] J. McGahan, "Flapping flight of the andean condor in nature," *Journal of Experimental Biology*, vol. 58, pp. 239–253, 1973.
[67] R. H. J. Brown, "The flight of birds," *Journal of Experimental Biology*, vol. 30, pp. 90–103, 1948.
[68] T. Weis-Fogh, "Quick estimates of flight fitness in hovering animals, including novel mechanisms for lift production," *Journal of Experimental Biology*, vol. 59, pp. 169–230, 1973.
[69] M. Stolpe and K. Zimmer, "Der schwirrflug des kolibri im zeitlupenfilm," *Journal of Ornithology*, vol. 87, pp. 136–155, 1939.
[70] C. P. Ellington, "The aerodynamics of hovering insect. III. Kinematics," *Philosophical Transactions of the Royal Society of London, Series B*, vol. 305, pp. 41–78, 1984.
[71] C. P. Ellington, "Unsteady aerodynamics of insect flight," *Symposia of the Society for Experimental Biology*, 1995, pp. 109–129.
[72] A. V. Hill, "The dimensions of animals and their muscular dynamics," *Science Progress*, vol. 38, pp. 209–230, 1950.
[73] C. J. Pennycuick, "Predicting wingbeat frequency and wavelength of birds," *Journal of Experimental Biology*, vol. 150, pp. 171–185, 1990.
[74] R. M. Alexander, "Mechanics of bipedal locomotion," in *Perspectives in experimental biology 1*, ed. P. S. Davis: Pergamon Press, 1976, pp. 493–504.
[75] C. J. Pennycuick, M. Klaassen, A. Kvist, and A. Lindstrom, "Wingbeat frequency and the body drag anomaly: Wind-tunnel observations on a thrush nightingale and a teal," *Journal of Experimental Biology*, vol. 199, pp. 2757–2765, 1996.
[76] P. Q. Collins and J. M. Graham, "Human flapping – Wing flight under reduced gravity," *Aeronautical Journal*, vol. 98, pp. 177–184, 1994.

[77] C. J. Pennycuick, "The mechanics of bird migration," *Ibis*, vol. 111, pp. 525–556, 1969.

[78] C. J. Pennycuick, "Mechanical constraints on the evolution of flight," in *The origin of birds and the evolution of flight*, ed. K. Padian: California Academy of Science, 1986, pp. 83–98.

[79] J. M. V. Rayner, "A vortex theory of animal flight. Part. 2. The forward flight of birds," *Journal of Fluid Mechanics*, vol. 91, pp. 731–763, 1979.

[80] U. M. Norberg, "Aerodynamics, kinematics, and energetics of horizontal flapping flight in the long-eared bat Plecotus Auritus," *Journal of Experimental Biology*, vol. 65, pp. 179–212, 1976.

[81] G. Goldspink, "Energy cost of locomotion," in *Mechanics and energetics of animal locomotion*, eds. R. M. Alexander and G. C. Chapman: Chapman and Hall, 1977.

[82] R. M. Alexander, "The U J and L of bird flight," *Nature*, vol. 390, p. 13, 1977.

[83] B. W. Tobalske, T. L. Hedrick, K. P. Dial, and A. A. Biewener, "Comparative power curves in bird flight," *Nature*, vol. 421, pp. 363–366, 2003.

[84] F. W. Schmitz, *Aerodynamik des Flugmodells*: Verlag, 1983.

[85] W. C. Brown, "Boston low-speed wind tunnel, and wind tunnel: Characteristics of indoor airfoils," *Journal of International Aeromodeling*, pp. 3–7, 1939.

[86] H. Weiss, "Wind tunnel: Effect of wing spar size," *Journal of International Aeromodeling*, pp. 5–7, 1939.

[87] R. H. Liebeck, "Laminar separation bubbles and airfoil design at low Reynolds numbers," in *AIAA Applied Aerodynamics Conference*, Palo Alto, 1992, pp. 441–456.

[88] M. S. Selig, J. J. Guglielmo, A. P. Broeren, and P. Giguere, *Summary of low-speed airfoil data*, Vol. 1: SoarTech Publications, 1995.

[89] M. S. Selig, C. A. Lyon, P. Giguere, C. N. Ninham, and J. J. Guglielmo, *Summary of low-speed airfoil data*, Vol. 2: SoarTech Publications, 1996.

[90] M. S. Selig, J. J. Guglielmo, A. P. Broeren, and P. Giguere, "Experiments on airfoils at low Reynolds numbers," in *34th AIAA Aerospace Sciences Meeting and Exhibit*, 1996.

[91] F.-B. Hisao, C.-F. Liu, and Z. Tang, "Aerodynamic performance and flow structure studies of a low Reynolds number airfoil," *AIAA Journal*, vol. 27, pp. 129–137, 1989.

[92] B. M. Jones, "Stalling," *Journal of the Royal Aeronautical Society*, vol. 38, pp. 747–770, 1934.

[93] W. B. Roberts, "Calculation of laminar separation bubbles and their effect on airfoil performance," *AIAA Journal*, vol. 18, pp. 25–31, 1980.

[94] B. H. Carmichael, "Low Reynolds number airfoil survey," National Aeronautics and Space Administration, NASA Contractor Report, 1981.

[95] I. Tani, "Low-speed flows involving bubble separations," *Progress in Aerospace Sciences*, no. 5, pp. 70–103, 1964.

[96] M. Drela, "XFOIL: An analysis and design system for low Reynolds number airfoils," in *Low Reynolds number aerodynamics*, ed. T.J. Mueller, Vol. 54, Lecture Notes in Engineering, Springer-Verlag, 1989, pp. 1–12.

[97] A. D. Young and H. P. Horton, "Some results of investigation of separation bubbles," in *AGARD*, 1966, pp. 779–811.

[98] L. F. Crabtee, "Effect of leading edge separation on thin wings in two-dimensional incompressible flow," *Journal of Aeronautical Sciences*, vol. 24, pp. 597–604, 1957.

[99] R. F. Huang, W. Shyy, S. W. Lin, and F.-B. Hsiao, "Influence of surface flow on aerodynamic loads of a cantilever wing," *AIAA Journal*, vol. 34, pp. 527–532, 1996.

[100] R. Hillier and N. J. Cherry, "The effects of stream turbulence on separation bubbles," *Journal of Wind Engineering and Industrial Aerodynamics*, vol. 8, pp. 49–58, 1981.

[101] M. Kiya and K. Sasaki, "Free-stream turbulence effects on a separation bubble," *Journal of Wind Engineering and Industrial Aerodynamics*, vol. 14, pp. 375–386, 1983.

[102] C. D. Wilcox, *Turbulence modeling for CFD*: DCW Industries, 2000.

[103] T. Herbert, "Parabolized stability equations," *Annual Review of Fluid Mechanics*, vol. 29, pp. 245–283, 1997.

[104] M. Lesieur and O. Metais, "New trends in large-eddy simulations of turbulence," *Annual Review of Fluid Mechanics*, vol. 28, pp. 45–82, 1996.

[105] P. Moin and K. Mahesh, "Direct numerical simulation: A tool in turbulence research," *Annual Review of Fluid Mechanics*, vol. 30, pp. 539–578, 1998.

[106] R. J. Volino and D. G. Bohl, "Separated flow transition mechanism and prediction with high and low freestream turbulence under low pressure turbine conditions," in *ASME Turbo Expo 2004: Power for Land, Sea, and Air*, 2004.

[107] R. E. Mayle, "The role of laminar-turbulent transition in gas turbine engines," *Journal of Turbomachinery*, vol. 113, pp. 509–537, 1991.

[108] T. J. Praisner and J. P. Clark, "Predicting transition in turbomachinery. Part-I-A: Review and new model development," in *ASME Turbo Expo 2004: Power for Land, Sea, and Air*, 2004.

[109] S. K. Roberts and M. I. Yaras, "Effects of surface roughness geometry on separation bubble transition," *Journal of Turbomachinery*, vol. 128, pp. 349–357, 2006.

[110] A. R. Wazzan, J. C. Gazley, and A.M.O Smith, "Tollmien-Schlichting waves and transition: Heated and adiabetic wedge flows with application to bodies of revolution," *Progress in Aerospace Sciences*, vol. 18, pp. 351–392, 1979.

[111] A. M. O. Smith, "Transition, pressure gradient, and stability theory," in *IX International Congress of Applied Mechanics*, Brussels, 1956, pp. 234–283.

[112] J. L. van Ingen, "A suggested semi-empirical method for the calculation of the boundary layer transition region," Delft University of Technology, Internal Report 1956.

[113] R. Radespiel, K. Graage, and O. Brodersen, "Transition predictions using Reynolds-averaged Navier-Stokes and linear stability analysis methods," in *AIAA Fluid Dynamics, Plasma Dynamics and Lasers Conference*, Honolulu, 1991.

[114] H. W. Stock and W. Haase, "A feasibility study of eN transition prediction in Navier-Stokes methods for airfoils," *AIAA Journal*, vol. 37, pp. 1187–1196, 1999.

[115] X. He, I. Senocak, W. Shyy, S. S. Thakur, and S. Gangadharan, "Evaluation of laminar-turbulent transition and equilibrium near wall turbulence models," *Numerical Heat Transfer, Part A*, vol. 37, pp. 101–112, 2000.

[116] W. Yuan, M. Khalid, J. Windte, U. Scholz, and R. Radespiel, "An investigation of low-Reynolds-number flows past airfoils," in *23rd AIAA Applied Aerodynamics Conference*, Toronto, 2005, pp. 1–19.

[117] Y. Lian and W. Shyy, "Laminar-turbulent transition of a low Reynolds number rigid or flexible airfoil," *AIAA Journal*, vol. 45, pp. 1501–1513, 2007.

[118] X. Zheng, C. Liu, F. Liu, and C. Yang, "Turbulence transition simulation using the k-omega model," *International Journal for Numerical Methods in Engineering*, vol. 42, pp. 907–926, 1998.

[119] C. D. Wilcox, "Simulation of transition with a two-equation turbulence model," *AIAA Journal*, vol. 32, pp. 247–255, 1994.

[120] D. S. Holloway, D. K. Walters, and J. K. Leylek, "Prediction of unsteady, separated boundary layer over a blunt body for laminar, turbulent, and transitional flow," *International Journal for Numerical Methods in Fluids*, vol. 45, pp. 1291–1315, 2004.

[121] E. Dick and J. Steelant, "Coupled solution of the steady compressible Navier-Stokes equations and the k-omega turbulence equations with a multigrid method," *Applied Numerical Mathematics*, vol. 23, pp. 49–61, 1997.

[122] E. Dick and J. Steelant, "Modeling of bypass transition with conditioned Navier-Stokes equations coupled to an intermittency transport equation," *International Journal for Numerical Methods in Fluids*, vol. 23, pp. 193–220, 1996.

[123] Y. B. Suzen and P. G. Huang, "Modeling of flow transition using an intermittency transport equation," *Journal of Fluid Engineering*, vol. 122, pp. 273–284, 2000.

[124] I. Mary and P. Sagaut, "Large eddy simulation of flow around an airfoil near stall," *AIAA Journal*, vol. 40, pp. 1139–1145, 2002.

[125] A. R. Wazzan, T. Okamura, and A. M. O. Smith, "Spatial and temporal stability charts for the Falkner-Skan boundary layer profiles," Douglas Aircraft Co, 1968.

[126] M. R. Malik, "COSAL – a black-box compressible stability analysis code for transition prediction in three-dimensional boundary layers," National Aeronautics and Space Administration, Technical Report 1982.

[127] G. Schrauf, "COAST3 – a compressible stability code, user's guide and tutorial," Daimler Benz Aerospace Airbus GmbH, Technical Report 1998.

[128] C. Gleyzes, J. Cousteix, and J. L. Bonnet, "Theoretical and experimental study of low Reynolds number transitional separation bubbles," in *Proceedings of the Conference on Low Reynolds Number Airfoil Aerodynamics*, ed. T. J. Mueller: University of Notre Dame Press, 1985, pp. 137–152.

[129] F. M. White, *Viscous fluid flow*: McGraw Hill, 1991.

[130] L. M. Mack, "Transition prediction and linear stability theory," in *AGARD Laminar-Turbulent Transition*, 1977, pp. 1–22.

[131] J.L. van Ingen, "Some introductory remarks on transition prediction methods based on linear stability theory," in *Transitional boundary layers in aeronautics*, eds. R.A.W.M. Henkes and J.L. van Ingen: Elsevier, pp. 209–224, 1996.

[132] T. Cebeci, "Essential ingredients of a method for low Reynolds number-airfoils," *AIAA Journal*, vol. 27, pp. 1680–1688, 1988.

[133] K. K. Chen and N. A. Thyson, "Extension of Emmon's spot theory to flows on blunt bodies," *AIAA Journal*, vol. 9, pp. 821–825, 1971.

[134] M. Ol, B. R. McAuliffe, E. S. Hanff, U. Scholz, and C. Kaehler, "Comparison of laminar separation bubble measurements on a low Reynolds number airfoil in three facilities," in *35th Fluid Dynamics Conference and Exhibit*, Toronto, 2005.

[135] R. Radespiel, J. Windte, and U. Scholz, "Numerical and experimental flow analysis of moving airfoils with laminar separation bubbles," *AIAA Journal*, vol. 45, pp. 1346–1356, 2007.

[136] M.C. Galbraith and M. R. Visbal, "Implicit large-eddy simulation of low Reynolds number flow past the SD7003 airfoil," in *46th AIAA Aerospace Sciences Meeting and Exhibit*, Reno, 2008.

[137] A. Uranga, P.-O. Drela, M. Persson, and J. Peraire, "Implicit large eddy simulation of transition to turbulence at low Reynolds numbers using a discontinuous Galerkin method," *International Journal for Numerical Methods in Engineering*, vol. 87, pp. 232–261, 2011.

[138] R. Hain, C. J. Kahler, and R. Radespiel, "Dynamics of laminar separation bubbles at low-Reynolds-number aerofoils," *Journal of Fluid Mechanics*, vol. 639, pp. 129–153, 2009.

[139] M. Okamoto, K. Yasuda, and A. Azuma, "Aerodynamic characteristics of the wings and body of a dragonfly," *Journal of Experimental Biology*, vol. 199, pp. 281–294, 1996.

[140] S. Sunada, T. Yasuda, and K. Yasuda, K., "Comparison of wing characteristics at an ultralow Reynolds number," *Journal of Aircraft*, vol. 39, pp. 331–338, 2002.

[141] B. G. Newman and H. T. Low, "Two-dimensional impervious sails: Experimental results compared with theory," *Journal of Fluid Mechanics*, vol. 144, pp. 455–462, 1984.

[142] R. H. Buckholz, "The functional role of wing corrugation in living systems," *Journal of Fluid Engineering*, vol. 108, pp. 93–97, 1986.

[143] A. B. Kesel, "Biologisches vorbild insektenflugel mehrkriterienoptimierung ultraleichter trangflachen," in *BIONA-report 12 A*, eds. W. Nachtigall and A. Wisser A., 1998.

[144] A. B. Kesel, "Aerodynamic characteristics of dragonfly wing sections compared with technical airfoils," *Journal of Experimental Biology*, vol. 203, pp. 3125–3135, 2000.

[145] A. Vargas, R. Mittal, and H. Dong, "A computational study of the aerodynamic performance of a dragonfly wing section in gliding flight," *Bioinspiration and Biomimetics*, vol. 3, p. 026004, 2008.

[146] W.-K. Kim, J. H. Ko, H. C. Park, and D. Byun, "Effects of corrugation of the dragonfly wing on gliding performance," *Journal of Theoretical Biology*, vol. 260, pp. 523–530, 2009.

[147] W. Shyy, F. Kleverbring, M. Nilsson, J. Sloan, and B. Carrol, "Rigid and flexible low Reynolds number airfoils," *Journal of Aircraft*, vol. 36, pp. 523–523, 1999.

[148] M. S. Selig and M. D. Maughmer, "Multipoint inverse airfoil design method based on conformal mapping," *AIAA Journal*, vol. 30, pp. 1162–1170, 1992.

[149] J. T. Murphy and H. Hu, "An experimental study of a bio-inspired corrugated airfoil for micro air vehicle applications," *Experiments in Fluids*, vol. 49, pp. 531–546, 2010.

[150] M. M. O'Meara and T. J. Mueller, "Laminar separation bubble characteristics on an airfoil at low Reynolds numbers," *AIAA Journal*, vol. 25, p. 1033, 1987.

[151] T. J. Mueller, L. J. Pohlen, P. E. Conigliaro, and B. J. J. Jansen, "The influence of free-stream disturbances on low Reynolds number airfoil experiments," *Experiments in Fluids*, vol. 1, pp. 3–14, 1983.

[152] H. J. Obremski and A. A. Fejer, "Transition in oscillating boundary layer flow," *Journal of Fluid Mechanics*, vol. 29, pp. 93–111, 1967.

[153] H. J. Obremski and M. V. Morkovin, "Application of a quasi-steady stability model to periodic boundary layer flows," *AIAA Journal*, vol. 7, pp. 1298–1301, 1969.

[154] Y. Lian and W. Shyy, "Numerical simulations of membrane wing aerodynamics for micro air vehicle applications," *Journal of Aircraft*, vol. 42, pp. 865–873, 2005.

[155] D. Viieru, R. Albertani, W. Shyy, and P. Ifju, "Effect of tip vortex on wing aerodynamics of micro air vehicles," *Journal of Aircraft*, vol. 42, pp. 1530–1536, 2005.

[156] R. Albertani, P. Huber, P. Ifju, R. Lind, and J. Jackowski, "Wind tunnel testing of micro air vehicles at low Reynolds numbers," SAE Technical Paper 2004.

[157] R. M. Cummings, S. A. Morton, S. G. Siegel, and S. Bosscher, "Numerical prediction and wind tunnel experiment for pitching unmanned combat air vehicles," in *41st Aerospace Sciences Meeting and Exhibit*, Reno, 2003.

[158] Y. Lian and W. Shyy, "Three-dimensional fluid-structure interactions of a membrane wing for micro air vehicle applications," *Journal of Aircraft*, vol. 208, pp. 3075–3092, 2003.

[159] T. J. Mueller and J. D. Delaurier, "Aerodynamics of small vehicles," *Annual Review of Fluid Mechanics*, vol. 35, pp. 89–111, 2003.

[160] Y. Lian, W. Shyy, D. Viieru, and B. H. Zhang, "Membrane wing aerodynamics for micro air vehicles," *Progress in Aerospace Sciences*, vol. 39, pp. 425–465, 2003.

[161] Y. Lian, "Membrane and adaptive-shaped fluid-structure interactions of membrane wing for micro air vehicle applications," University of Florida, PhD thesis 2003.

[162] G. E. Torres and T. J. Mueller, "Aerodynamics characteristics of low aspect ratio wings at low Reynolds numbers," in *Fixed and flapping wing aerodynamics for micro air vehicle applications*, ed. T. J. Mueller: American Institute of Aeronautics and Astronautics, 2001.

[163] Y. Lian, W. Shyy, P. Ifju, and E. Verron, "A membrane wing model for micro air vehicles," *AIAA Journal*, vol. 41, pp. 2492–2494, 2003.

[164] S. Leibovich, "The structure of vortex breakdown," *Annual Review of Fluid Mechanics*, vol. 10, pp. 221–246, 1978.

[165] D. Viieru, Y. Lian, W. Shyy, and P. Ifju, "Investigation of tip vortex on aerodynamic performance of low Reynolds number flight vehicles," in *33rd AIAA Fluid Dynamics Conference and Exhibit*, Orlando, 2003.

[166] L. Prandtl and O. G. Tietjens, *Fundamentals of hydro- and aeromechanics*: Dover, 1957.

[167] J. Tang and K.-Q. Zhu, "Numerical and experimental study of flow structure of low-aspect-ratio wing," *Journal of Aircraft*, vol. 41, pp. 1196–1201, 2004.

[168] R. Katzmayr, "Effect of periodic change of angle of attack on behavior of airfoils," National Aeronautics and Space Administration, Technical Report 1922.

[169] A. Betz, "Ein beitrag zur erklarung des segelfuges," *Zeitschrift fur Flugtechnik and Motorluftschiffahrt*, vol. 3, pp. 269–272, 1912.

[170] R. Knoller, "Die gesetze des luftwiderstandes," *Flung-und Motorechnik*, vol. 3, pp. 1–7, 1909.

[171] Y. E. Polonskiy, "Some questions on the flapping wing," *Inzschenerniy Sbornik*, vol. 8, pp. 49–60, 1950.

[172] J. B. Bratt, "Flow patterns in the wake of an oscillating airfoil," Aeronautical Research Council Technical Report R and M, 1953.

[173] K. D. Jones, C. M. Dohring, and F. M. Platzer, "Experimental and computational investigation of the Knoller-Betz effect," *AIAA Journal*, vol. 36, pp. 1240–1246, 1998.

[174] J. M. Anderson, K. Streitlien, D. S. Barret, and M. S. Triantafyllou, "Oscillating foils of high propulsive efficiency," *Journal of Fluid Mechanics*, vol. 360, pp. 41–72, 1998.

[175] P. Freymuth, "Thrust generation by an airfoil in hover modes," *Experiments in Fluids*, vol. 9, pp. 17–24, 1990.

[176] M. M. Koochesfahani, "Vortical patterns in the wake of an oscillating airfoil," *AIAA Journal*, vol. 27, pp. 1200–1205, 1989.

[177] B. Satyanarayana and S. Davis, "Experimental studies of unsteady trailing-edge conditions," *AIAA Journal*, vol. 16, pp. 125–129, 1978.

[178] R. L. Bass, J. E. Johnson, and J. F. Unruh, "Correlation of lift and boundary-layer activity on an oscillating lifting surface," *AIAA Journal*, vol. 20, pp. 1051–1056, 1982.

[179] J. C. S. Lai and F. M. Platzer, "Jet characteristics of a plunging airfoil," *AIAA Journal*, vol. 37, pp. 1529–1537, 1999.

[180] J. C. S. Lai and F. M. Platzer, "Characteristics of a plunging airfoil at zero freestream velocity," *AIAA Journal*, vol. 39, pp. 531–534, 2001.

[181] K. D. Jones, T. C. Lund, and F. M. Platzer, "Experimental and computational investigation of flapping-wing propulsion for micro air vehicles," in *Fixed and flapping wings aerodynamics for micro air vehicles applications*, ed. T. J. Mueller: American Institute of Aeronautics and Astronautics, 2001, pp. 307–336.

[182] S. Taneda, "Visual study of unsteady separated flows around bodies," *Progress in Aerospace Sciences*, vol. 17, pp. 287–348, 1976.

[183] M. W. Rosen, *Water flow about a swimming fish*: University of California in Los Angeles, 1959.

[184] R. Gopalkrishnan, M. S. Triantafyllou, G. S. Triantafyllou, and D. S. Barret, "Active vorticity control in a shear flow using a flapping foil," *Journal of Fluid Mechanics*, vol. 274, pp. 1–21, 1994.

[185] K. Streitlien and G. S. Triantafyllou, "On thrust estimates for flapping foils," *Journal of Fluids and Structures*, vol. 12, pp. 47–55, 1998.

[186] M. S. Triantafyllou, G. S. Triantafyllou, and D. K. P. Yue, "Hydrodynamics of fishlike swimming," *Annual Review of Fluid Mechanics*, vol. 32, pp. 33–53, 2000.

[187] S. I. Devin, V. M. Zavyalov, and B. K. Korovich, "On the question of unsteady aerodynamic forces acting upon a wing of finite aspect ratio at large amplitudes of oscillation and large Strouhal numbers," *Voprosy Sudostroeniya Ser: Proekirovanie Sudov, Vyp*, vol. 1, pp. 34–41, 1972.

[188] H. D. Ham, "Aerodynamic loading on a two-dimensional airfoil during dynamic stall," *AIAA Journal*, vol. 6, pp. 1927–1934, 1968.

[189] P. W. Harper and R. E. Flanigan, "The effect of rate of change of angle of attack on the maximum lift of a small model," National Aeronautics and Space Administration, Technical Report 1950.

[190] F. D. Harris and R. R. Pruyn, "Blade stall: Half fact, half fiction," *Journal of the American Helicopter Society*, vol. 13, pp. 27–48, 1968.

[191] W. J. McCroskey and R. K. Fisher, "Detailed aerodynamic measurements on model rotor in the blade stall regime," *Journal of the American Helicopter Society*, vol. 17, pp. 20–30, 1972.

[192] K. D. Jones and F. M. Platzer, "Design and development considerations for biologically inspired flapping-wing micro air vehicles," *Experiments in Fluids*, vol. 13, pp. 26–32, 2009.

[193] D. J. Willis, E. R. Israeli, P. O. Persson, M. Drela, and J. Peraire, "A computational framework for fluid structure interaction in biologically inspired flapping flight," in *AIAA Applied Aerodynamics Meeting*, Miami, 2007, pp. 2007–3803.

[194] P.-O. Persson, D. J. Willis, and J. Peraire, "Numerical simulation of flapping wings using a panel method and a high-order Navier-Stokes solver," *International Journal for Numerical Methods in Engineering*, Feb. 2012, DOI: 10.1002/nme.3288.

[195] J. Katz and A. Plotkin, *Low-speed aerodynamics: From wing theory to panel methods*: McGraw-Hill College, 1981.

[196] J. D. Delaurier, "An aerodynamic model for flapping wing flight," *Aeronautical Journal*, vol. 97, pp. 125–130, 1993.

[197] M. J. C. Smith, "Simulating moth wing aerodynamics: towards the development of flapping-wing technology," *AIAA Journal*, vol. 34, pp. 1348–1355, 1996.

[198] M. S. Vest and J. Katz, "Unsteady aerodynamics model of flapping wings," *AIAA Journal*, vol. 34, pp. 1435–1440, 1996.

[199] C. P. Ellington, C. van den Berg, A. P. Willmott, and A. L. R. Thomas, "Leading-edge vortices in insect flight," *Nature*, vol. 384, pp. 626–630, 1996.

[200] H. Liu and K. Kawachi, "A numerical study of insect flight," *Journal of Computational Physics*, vol. 146, pp. 124–156, 1998.

[201] M. H. Dickinson, F.-O. Lehmann, and S. P. Sane, "Wing rotation and the aerodynamic basis of insect flight," *Science*, vol. 284, pp. 1954–1960, 1999.

[202] K. D. Jones and F. M. Platzer, "An experimental and numerical investigation of flapping-wing propulsion," in *37th Aerospace Sciences Meeting*, Reno, 1999.

[203] K. D. Jones and F. M. Platzer, "Experimental investigation of the aerodynamic characteristics of flapping-wing micro air vehicles," in *41st Aerospace Sciences Meeting and Exhibit*, Reno, 2003.

[204] Z. J. Wang, "Two dimensional mechanism for insect hovering," *Physical Review Letters*, vol. 85, pp. 2216–2219, 2000.

[205] W. Shyy, M. Berg, and D. Ljungqvist, "Flapping and flexible wings for biological and micro air vehicles," *Progress in Aerospace Sciences*, vol. 35, no. 5, pp. 455–505, 1999.

[206] C. J. Pennycuick, *Modelling the flying bird*: Academic Press, 2008.

[207] T. J. Mueller, *Fixed and flapping wing aerodynamics for micro air vehicle applications*: AIAA, 2001.

[208] A. Azuma, "Locomotion momentum and local circulation methods for fixed, rotary and beating wings," University of Tokyo, 1983.

[209] M. J. Lighthill, "On the Weis-Fogh mechanism of lift generation," *Journal of Fluid Mechanics*, vol. 60, pp. 1–17, 1973.

[210] G. R. Spedding, "The aerodynamics of flight," *Advances in Comparative Physiology*, vol. 11, pp. 51–111, 1992.

[211] P. Chai and R. Dudley, "Limits to flight energetics of hummingbirds hovering in hypodense and hypoxic gas mixtures," *Journal of Experimental Biology*, vol. 199, pp. 2285–2295, 1996.

[212] R. Dudley and C. P. Ellington, "Mechanics of forward flight in bumblebees II. Quasi-steady lift and power requirements," *Journal of Experimental Biology*, vol. 148, pp. 53–88, 1990.

[213] C. P. Ellington, "Unsteady aerodynamics of insect flight," *Symposium of the Society for Experimental Biology*, vol. 40, pp. 109–129, 1995.

[214] M. Cloupeau, "Direct measurements of instantaneous lift in desert locust; comparison with Jensen's experiments on detached wings," *Journal of Experimental Biology*, vol. 80, pp. 1–15, 1979.

[215] P. J. Wilkin and H. M. William, "Comparison of the aerodynamic forces on a flying sphingid moth with those predicted by quasi-steady theory," *Physiological Zoology*, vol. 66, pp. 1015–1044, 1993.

[216] S. P. Sane and M. H. Dickinson, "The control of flight force by a flapping wing: Lift and drag production," *Journal of Experimental Biology*, vol. 204, pp. 2607–2626, 2001.

[217] Z. J. Wang, J. M. Birch, and M. H. Dickinson, "Unsteady forces and flows in low Reynolds number hovering flight: Two-dimensional computations vs. robotic wing experiment," *Journal of Experimental Biology*, vol. 207, pp. 449–460, 2004.

[218] N. Vandenberghe, J. Zhang, and S. Childress, "Symmetry breaking leads to forward flapping flight," *Journal of Fluid Mechanics*, vol. 506, pp. 147–155, 2004.

[219] N. Vandenberghe, S. Childress, and J. Zhang, "On unidirectional flight of a free flapping wing," *Physics of Fluids*, vol. 18, p. 014102, 2006.

[220] S. Alben and M. Shelley, "Coherent locomotion as an attracting state for a free flapping body," *Proceedings of the National Academy of Sciences*, vol. 102, pp. 11163–11166, 2005.

[221] X. Zhang, S. Ni, and S. Wang, "Effects of geometric shape on the hydrodynamics of a self-propelled flapping foil," *Physics of Fluids*, vol. 21, p. 103302, 2009.

[222] J. Zhang, N.-S. Liu, and X.-Y. Lu, "Locomotion of a passively flapping flat plate," *Journal of Fluid Mechanics*, vol. 659, pp. 43–68, 2010.

[223] S. E. Spagnolie, L. Moret, M. J. Shelley, and J. Zhang, "Surprising behaviors in flapping locomotion with passive pitching," *Physics of Fluids*, vol. 22, p. 041903, 2010.

[224] S. M. Walker, A. L. R. Thomas, and G. K. Taylor, "Deformable wing kinematics in the desert locust: How and why do camber, twist and topography vary through the stroke?," *Journal of the Royal Society Interface*, vol. 6, pp. 735–747, 2008.

[225] H. Liu and H. Aono, "Size effects on insect hovering aerodynamics: An integrated computational study," *Bioinspiration and Biomimetics*, vol. 4, pp. 1–13, 2009.

[226] A. P. Willmott and C. P. Ellington, "The mechanics of flight in the hawkmoth Manduca sexta. I. Kinematics of hovering and forward flight," *Journal of Experimental Biology*, vol. 200, pp. 2705–2722, 1997.

[227] A. P. Willmott and C. P. Ellington, "The mechanics of flight in the hawkmoth Manduca sexta. II. Aerodynamic consequences of kinematic and morphological variation," *Journal of Experimental Biology*, vol. 200, pp. 2723–2745, 1997.

[228] S. N. Fry, R. Sayaman, and M. H. Dickinson, "The aerodynamics of free-flight maneuvers in Drosophila," *Science*, vol. 300, pp. 495–498, 2003.

[229] J. M. Wakeling and C. P. Ellington, "Dragonfly flight. II. Velocities, accelerations and kinematics of flapping flight," *Journal of Experimental Biology*, vol. 200, pp. 557–582, 1997.

[230] T. L. Hedrick, "Software techniques for two- and three-dimensional kinematic measurements of biological and biomimetic systems," *Bioinspiration and Biomimetics*, vol. 3, pp. 1–6, 2008.

[231] X. Tian, J. Iriarte, K. Middleton, R. Galvao, E. Israeli, A. Roemer, A. Sullivan, A. Song, S. Swartz, and K. Breuer, "Direct measurements of the kinematics and dynamics of bat flight," *Bioinspiration and Biomimetics*, vol. 1, no. 2006–2865, pp. S10–S18, 2006.

[232] D. K. Riskin, D. J. Willis, J. Iriarte-Diaz, T. L. Hedrick, M. Kostandov, J. Chen, D. H. Laidlaw, K. Breuer, and S. M. Swartz, "Quantifying the complexity of bat wing kinematics," *Journal of Theoretical Biology*, vol. 254, pp. 604–615, 2008.

[233] I. D. Wallace, N. J. Lawson, A. R. Harvey, J. D. C., and A. J. Moore, "High-speed photogrammetry system for measuring the kinematics of insect wings," *Applied Optics*, vol. 45, pp. 4165–4173, 2006.

[234] L. Ristroph, G. J. Berman, A. J. Bergou, Z. J. Wang, and I. Cohen, "Automated hull reconstruction motion tracking (HRWT) applied to sideways maneuvers of free-flying insects," *Journal of Experimental Biology*, vol. 212, pp. 1324–1335, 2009.

[235] L. Zeng, Q. Hao, and K. Kawachi, "A scanning projected line method for measuring a beating bumblebee wing," *Optical Communication*, vol. 183, pp. 37–43, 2000.

[236] L. Zeng, Q. Hao, and K. Kawachi, "Measuring the body vector of a free flight bumblebee by the reflection beam method," *Measurement Science and Technology*, vol. 12, pp. 1886–1890, 2001.

[237] L. Zeng, H. Matsumoto, and K. Kawachi, "A fringe shadow method for measuring flapping angle and torsional angle of a dragonfly wing," *Measurement Science and Technology*, vol. 7, pp. 776–781, 1996.

[238] D. Song, H. Wang, L. Zeng, and C. Yin, "Measuring the camber deformation of a dragonfly wing using projected comb fringe," *Review of Scientific Instruments*, vol. 72, pp. 2450–2454, 2001.

[239] H. Wang, L. Zeng, H. Liu, and C. Yin, "Measuring wing kinematics, flight trajectory and body attitude during forward flight and turning maneuvers in dragonflies," *Journal of Experimental Biology*, vol. 206, pp. 745–757, 2003.

[240] G. Zhang, J. Sun, D. Chen, and Y. Wang, "Flapping motion measurement of honeybee bilateral wings using four virtual structured-light sensors," *Sensors and Actuator A: Physical*, vol. 148, pp. 19–27, 2008.

[241] S. M. Walker, A. L. R. Thomas, and G. K. Taylor, "Deformable wing kinematics in free-flying hoverflies," *Journal of the Royal Society Interface*, vol. 7, pp. 131–142, 2010.

[242] G. Wu and L. Zeng, "Measuring the kinematics of a free-flying hawkmoth by a comb-fringe projected method," *Acta Mechanica Sinica*, vol. 26, pp. 67–71, 2010.

[243] T. Y. Hubel, D. K. Riskin, S. M. Swartz, and K. S. Breuer, "Wake structure and wing kinematics: The flight of the lesser dog-faced fruit bat, Cynopterus brachyotis," *Journal of Experimental Biology*, vol. 213, pp. 3427–3440, 2010.

[244] J. Tang, D. Viieru, and W. Shyy, "Effects of Reynolds number, reduced frequency and flapping kinematics on hovering aerodynamics," *AIAA Journal*, vol. 46, pp. 967–976, 2008.

[245] W. Shyy, Y. Lian, J. Tang, D. Viieru, and H. Liu, *Aerodynamics of low Reynolds number flyers*: Cambridge University Press, 2008.

[246] K. V. Rozhdestvensky and V. A. Ryzhov, "Aerohydrodynamics of flapping-wing propulsors," *Progress in Aerospace Sciences*, vol. 39, pp. 585–633, 2003.

[247] H. Liu, C. P. Ellington, K. Kawachi, C. van den Berg, and A. P. Willmott, "A computational fluid dynamics study of hawkmoth hovering," *Journal of Experimental Biology*, vol. 146, pp. 461–477, 1998.

[248] C. van den Berg and C. P. Ellington, "The three-dimensional leading-edge vortex of a hovering model hawkmoth," *Philosophical Transactions of the Royal Society of London, Series B*, vol. 352, pp. 329–340, 1997.

[249] G. K. Taylor, R. L. Nudds, and A. L. R., "Flying and swimming animals cruise at a Stouhal number tuned for high power efficiency," *Nature*, vol. 425, pp. 707–711, 2003.

[250] S. N. Fry and M. H. Dickinson, "The aerodynamics of hovering flight in Drosophila," *Journal of Experimental Biology*, vol. 208, pp. 2303–2318, 2005.

[251] R. Dudley and C. P. Ellington, "Mechanics of forward flight in bumblebees I. Kinematics and morphology," *Journal of Experimental Biology*, vol. 148, pp. 19–52, 1990.

[252] P. Chai and D. Millard, "Flight and size constraints: Hovering performance of large hummingbirds under maximal loading," *Journal of Experimental Biology*, vol. 200, pp. 2757–2763, 1997.

[253] C. H. Greenewalt, *Dimensional relationships for flying animals*: Smithsonian Institution Publication, 1962.

[254] J. M. Birch and M. H. Dickinson, "Spanwise flow and the attachment of the leading-edge vortex on insect wings," *Nature*, vol. 412, pp. 729–733, 2001.

[255] A. K. Brodsky, "Vortex formation in the tethered flight of the peacock butterfly Inachis 10 L and some aspects of insect flight evolution," *Journal of Experimental Biology*, vol. 161, pp. 77–95, 1991.

[256] R. B. Srygley and A. L. R, Thomas, "Unconventional lift-generating mechanisms in free-flying butterflies," *Nature*, vol. 420, pp. 660–664, 2002.

[257] A. P. Willmott, C. P. Ellington, and A. L. R. Thomas, "Flow visualization and unsteady aerodynamics in the flight of the hawkmoth, Manduca secta," *Philosophical Transactions of the Royal Society of London, Series B*, vol. 352, pp. 303–316, 1997.

[258] A. L. R. Thomas, G. K. Taylor, R. B. Srygley, R. L. Nudds, and R. J. Bomphrey, "Dragonfly flight: Free-flight and tethered flow visualizations reveal a diverse array of unsteady lift-generating mechanisms, controlled primarily via angle of attack," *Journal of Experimental Biology*, vol. 207, pp. 4299–4323, 2004.

[259] R. J. Bomphrey, G. K. Taylor, and A. L. R. Thomas, "Smoke visualization of free-flying bumblebees indicates independent leading-edge vortices on each wing pair," *Experiments in Fluids*, vol. 46, pp. 811–821, 2009.

[260] R. J. Bomphey, "Advances in animal flight aerodynamics through flow measurement," *Evolutionary Biology*, vol. 39, pp. 1–11, 2012.

[261] D. L. Altshuler, M. Princevac, H. Pan, and J. Lozano, "Wake patterns of the wings and tail of hovering hummingbirds," *Experiments in Fluids*, vol. 46, pp. 835–846, 2009.

[262] T. Y. Hubel, N. I. Hrristov, S. W. Swartz, and K. Breuer, "Time-resolved wake structure and kinematics of bat flight," *Journal of Experimental Biology*, vol. 46, pp. 933–943, 2009.

[263] B. W. Tobalske, J. W. D. Hearn, S. M. Swartz, and K. S. Breuer, "Aerodynamics of intermittent bounds in flying birds," *Experiments in Fluids*, vol. 46, pp. 963–973, 2009.

[264] R. J. Bomphrey, N. J. Lawson, N. J. Harding, G. K. Taylor, and A. L. R. Thomas, "The aerodynamic of Manduca sexta: Digital particle image velocimetry analysis of the leading-edge vortex," *Journal of Experimental Biology*, vol. 208, pp. 1079–1094, 2005.

[265] R. J. Bomphrey, "Insects in flight: Direct visualization and flow measurements," *Bioinspiration and Biomimetics*, vol. 1, pp. 1–9, 2006.

[266] D. R. Warrick, B. W. Tobalske, and D. R. Powers, "Lift production in the hovering hummingbird," *Proceedings of the Royal Society B*, vol. 276, pp. 3747–3752, 2009.

[267] A. Hendenström, L. C. Johansson, M. Wolf, R. von Busse, Y. Winter, and G. R. Spedding, "Bat flight generates complex aerodynamic tracks," *Science*, vol. 316, pp. 894–897, 2007.

[268] F. T. Muijres, L. C. Johansson, R. Barfield, M. Wolf, G. R. Spedding, and A Hedenstrom, "Leading-edge vortex improves lift in slow-flying bats," *Science*, vol. 319, pp. 1250–1253, 2008.

[269] A. Hendenström, F. T. Muijres, R. von Busse, L. C. Johansson, Y. Winter, and G. R. Spedding, "High-speed stereo PIV measurement of wakes of two bats species flying freely in a wind tunnel," *Experiments in Fluids*, vol. 46, pp. 923–932, 2009.

[270] C. Poelma, W. B. Dickson, and M. H. Dickinson, "Time-resolved reconstruction of the full velocity field around a dynamically-scaled flapping wing," *Experiments in Fluids*, vol. 41, pp. 213–225, 2006.

[271] Y. Lu and G. X. Shen, "Three-dimensional flow structure and evolution of the leading-edge vortices on a flapping wing," *Journal of Experimental Biology*, vol. 211, pp. 1221–1230, 2008.

[272] D. Kim and M. Gharib, "Characteristics of vortex formation and thrust performance in drag-based paddling propulsion," *Journal of Experimental Biology*, vol. 214, pp. 2283–2291, 2011.

[273] F.-O. Lehmann, S. P. Sane, and M. H. Dickinson, "The aerodynamic effects of wing-wing interaction in flapping insect wings," *Journal of Experimental Biology*, vol. 208, pp. 3075–3092, 2005.

[274] H. Wagner, "Uber die enstehung des dynamischen auftriebes von tragflugeln," *Zeitschrift fur Angewandte Mathematik und Mechanik*, vol. 5, pp. 17–35, 1925.

[275] M. H. Dickinson and K. G. Götz, "Unsteady aerodynamic performance of model wings at low Reynolds numbers," *Journal of Experimental Biology*, vol. 174, pp. 45–64, 1993.

[276] R. M. H. Beckwith and H. Babinsky, "Impulsively started flat plate flow," *Journal of Aircraft*, vol. 46, no. 6, pp. 2186–2188, 2009.

[277] W. J. McCroskey, K. W. McAlister, L. W. Carr, and S. L. Pucci, "An experimental study of dynamic stall on advanced airfoil section," NASA, 1982.

[278] R. G. Bradley, C. W. Smith, and W. O. Wary, "An experimental investigation of leading edge vortex augmentation by blowing," NASA, 1974.

[279] J. F. Campbell, "Augmentation of vortex lift by spanwise blowing," *Journal of Aircraft*, vol. 13, pp. 727–732, 1976.

[280] S. Sunada, K. Kawachi, I. Watanabe, and A. Azuma, "Fundamental analysis of three-dimensional 'near fling,'" *Journal of Experimental Biology*, vol. 183, pp. 217–248, 1993.

[281] E. L. Houghton and P. W. Carpenter, *Aerodynamics for engineering students*: Butterworth-Heinemann, 2003.

[282] L. David, T. Jardin, P. Braud, and A. Farcy, "Time-resolved scanning tomography PIV measurements around a flapping wing," *Experiments in Fluids*, vol. 52, pp. 857–864, 2012.

[283] J. M. Birch, W. B. Dickson, and M. H. Dickinson, "Force production and flow structure of the leading edge vortex on flapping wings at high and low Reynolds numbers," *Journal of Experimental Biology*, vol. 207, pp. 1063–1072, 2004.

[284] D. Kim and M. Gharib, "Experimental study of three-dimensional vortex structures in translating and rotating plates," *Experiments in Fluids*, vol. 49, pp. 329–339, 2010.

[285] M. Escudier, "Vortex breakdown: Observations of plunging and pitching airfoils," *Progress in Aerospace Sciences*, vol. 25, pp. 88–323, 1988.

[286] M. G. Hall, "Vortex breakdown," *Annual Review of Fluid Mechanics*, vol. 4, pp. 195–218, 1972.

[287] O. de Vries, "On the theory of the horizontal-axis wind turbine," *Annual Review of Fluid Mechanics*, vol. 15, pp. 77–96, 1983.

[288] M. Kramer, "Die zunahme des maximalauftriebes von tragflugeln bei plotzlicher anstellwinkelvergrosserung," *Zeitschrift fur Flugtechnik and Motorluftschiffahrt*, vol. 23, pp. 185–189, 1932.

[289] M. Sun and J. Tang, "Lift and power requirements of hovering flight in Drosophila virilis," *Journal of Experimental Biology*, vol. 205, pp. 2413–2427, 2002.

[290] R. Ramamurti and W. Sandberg, "Simulation of flow about flapping airfoils using finite element incompressible flow solver," *AIAA Journal*, vol. 39, pp. 253–260, 2001.

[291] M. Sun and J. Tang, "Unsteady aerodynamic force generation by a model fruit fly wing in flapping motion," *Journal of Experimental Biology*, vol. 205, pp. 55–70, 2002.

[292] S. P. Sane and M. H. Dickinson, "The aerodynamic effects of wing rotation and a revised quasi-steady model of flapping flight," *Journal of Experimental Biology*, vol. 205, pp. 1087–1096, 2002.

[293] S. Sunada and C. P Ellington, "Approximate added-mass method for estimating induced power for flapping flight," *AIAA Journal*, vol. 38, pp. 1313–1321, 2000.

[294] M. M. F. Osborne, "Aerodynamics of flapping flight with application to insects," *Journal of Experimental Biology*, vol. 28, pp. 221–245, 1951.

[295] F.-O. Lehmann, "The mechanisms of lift enhancement in insect flight," *Naturewisenschaften*, vol. 91, pp. 101–122, 2004.

[296] W. Shyy, P. Trizila, C.-K. Kang, and H. Aono, "Can tip vortices enhance lift of a flapping wing?," *AIAA Journal*, vol. 47, pp. 289–293, 2009.

[297] J. D. Anderson, *Fundamentals of aerodynamics*: McGraw Hill, 1991.

[298] M. Ramasamy and J. G. Leishman, "Phase-locked particle image velocimetry measurements of a flapping wing," *Journal of Aircraft*, vol. 43, pp. 1868–1875, 2006.

[299] M. J. Ringuette, M. Milano, and M. Gharib, "Role of the tip vortex in the force generation of low-aspect-ratio normal flat plates," *Journal of Fluid Mechanics*, vol. 581, pp. 453–468, 2007.

[300] K. Taira and T. Colonius, "Three-dimensional flows around low-aspect-ratio wings at low Reynolds numbers," *Journal of Fluid Mechanics*, vol. 623, pp. 187–207, 2009.

[301] P. Trizila, C.-K. Kang, H. Aono, M. Visbal, and W. Shyy, "Low-Reynolds-number aerodynamics of a flapping rigid flat plate," *AIAA Journal*, vol. 49, no. 4, pp. 806–823, 2011.

[302] M. Gharib, E. Rambod, and K. Shariff, "A universal time scale for vortex ring formation," *Journal of Fluid Mechanics*, vol. 360, pp. 121–140, 1998.

[303] A. R. Ennos, "Inertial and aerodynamic torques on the wings of Diptera in flight," *Journal of Experimental Biology*, vol. 142, pp. 87–95, 1989.

[304] R. J. Wootton and D. J. S. Newman, "Whitefly have the highest contraction frequencies yet recorded in non-fibrillar flight muscles," *Nature*, vol. 280, pp. 402–403, 1979.

[305] K. G. Gotz, "Course-control, metabolism and wing interference during ultralong tethered flight in Drosophila melanogaster," *Journal of Experimental Biology*, vol. 128, pp. 35–46, 1987.

[306] J. Brackenbury, "Wing movements in the bush cricket Tettigonia viridissima and the mantis Ameles spallanziana during natural leaping," *Journal of Zoology*, vol. 220, pp. 593–602, 1990.

[307] R. J. Cooter and P. S. Baker, "Weis-Fogh clap and fling mechanism in locusta," *Nature*, vol. 269, pp. 53–54, 1977.

[308] J. Marden, "Maximum lift production during takeoff in flying animals," *Journal of Experimental Biology*, vol. 130, pp. 235–258, 1987.
[309] D. Kolomenskiy, H, K. Moffatt, M. Farge, and K. Schneider, "Two- and three-dimensional numerical simulations of the clap-fling-sweep of hovering insects," *Journal of Fluid and Structures*, vol. 27, pp. 784–791, 2011.
[310] G. C. H. de Croon, K. M. E. de Clercq, R. R. B. Ruijsink, and C. de Wagter, "Designing, aerodynamics, and vision-based control of the Delfly," *International Journal of Micro Air Vehicles*, vol. 1, pp. 71–97, 2009.
[311] T, Nakata, H. Liu, Y. Tanaka, N. Nishihashi, X. Wang, and A. Sato, "Aerodynamics of a bio-inspired flexible flapping-wing micro air vehicle," *Bioinspiration and Biomimetics*, vol. 6, p. 045002, 2011.
[312] S. P. Sane, "The aerodynamics of insect flight," *Journal of Experimental Biology*, vol. 206, pp. 4191–4208, 2003.
[313] K. M. E. de Clercq, R. de Kat, B. Remes, B. W. van Oudheusden, and H. Bijl, "Aerodynamic experiments on Delfly II: Unsteady lift enhancement," *International Journal of Micro Air Vehicles*, vol. 1, no. 4, pp. 255–262, 2009.
[314] N. V, Queipo, R. T. Haftka, W. Shyy, T. Goel, R, Vaidyanathan, and P. K. Tucker, "Surrogate-based analysis and optimization," *Progress in Aerospace Sciences*, vol. 41, pp. 1–28, 2005.
[315] P. Trizila, C.-K. Kang, M. Visbal, and W. Shyy, "Unsteady fluid physics and surrogate modeling of low Reynolds number flpping airfoils," in *12th AIAA/ISSMO Multidisciplinary Analysis and Optimization Conference*, Victoria, 2008.
[316] K. B. Lua, Lim. T. T., and K. S. Yeo, "Aerodynamic forces and flow fields of a two-dimensional hovering wing," *Experiments in Fluids*, vol. 45, pp. 1047–1065, 2008.
[317] K. M. Miettine, *Nonlinear multiobjective optimization*: Kluwer Academic, 1999.
[318] K. Viswanath and D. K. Tafti, "Effect of frontal gusts on forward flapping flight," *AIAA Journal*, vol. 48, pp. 2049–2062, 2010.
[319] D. Lentink and M. Gerritsma, "Influence of airfoil shape of performance in insect flight," in *33rd AIAA Fluid Dynamics Conference and Exhibit*, 2003, pp. 1–17.
[320] J. R. Usherwood and C. P. Ellington, "The aerodynamics of revolving wings. I. Model hawkmoth wings," *Journal of Experimental Biology*, vol. 205, pp. 1547–1564, 2002.
[321] E. Polhamus, "Prediction of vortex-lift characteristics by a leading-edge suction analogy," *Journal of Aircraft*, vol. 8, pp. 193–199, 1971.
[322] M. A. Ashraf, J. Young, and J. C. S. Lai, "Reynolds number, thickness and camber effects on flapping airfoil propulsion," *Journal of Fluids and Structures*, vol. 27, pp. 145–160, 2011.
[323] M. Ol, L. Bernal, C.-K. Kang, and W. Shyy, "Shallow and deep dynamic stall for flapping low Reynolds number airfoils," *Experiments in Fluids*, vol. 46, no. 5, pp. 883–901, 2009.
[324] C.-K. Kang, H. Aono, P. Trizila, Y. S. Baik, J. M. Rausch, L. P. Bernal, M. Ol, and W. Shyy, "Modeling of pitching and plunging airfoils at Reynolds number between 1×10^4 and 6×10^4," in *27th AIAA Applied Aerodynamics Conference*, San Antonio, 2009.
[325] Y. Baik, J. M. Raush, L. P. Bernal, and M. V. Ol, "Experimental investigation of pitching and plunging airfoils at Reynolds number between 1×10^4 and 6×10^4," in *39th AIAA Fluid Dynamic Conference*, San Antonio, 2009.
[326] M. Ol and et al., "Unsteady aerodynamics for micro air vehicles," NATO Research and Technology Organization, Technical Report AVT-149, 2009.
[327] M. Visbal, "Numerical investigation of deep dynamic stall of a plunging airfoil," *AIAA Journal*, vol. 49, pp. 2152–2170, 2011.
[328] M. Visbal, "Three-dimensional flow structure on a hovering low-aspect-ratio wing," in *49th AIAA Aerospace Sciences Meeting including the New Horizons Forum and Aerospace Exposition*, Orlando, 2011.
[329] H. Wagner, "Über die entstehung des dynamischer auftriebes von tragflügeln," *Z. angew. Math. u. Mech.*, vol. 5, pp. 17–35, 1925.
[330] T. Theodorsen, "General theory of aerodynamic instability and the mechanism of flutter," NACA Technical Report, 1935.

[331] H. G. Küssner, "Zusammenfassender bericht über den instationären auftrieb von flügeln," *Luftfahrtforschung*, vol. 13, p. 410, 1936.

[332] T. von Karman and W. R. Sears, "Airfoil theory for non-uniform motion," *Journal of the Aeronautical Sciences*, vol. 5, no. 10, p. 379, 1938.

[333] R. L. Bisplinghoff, H. Ashley, and R. L. Halfman, *Aeroelasticity*: Dover Publications, inc., 1996.

[334] T. von Karman and J. M. Burgers, *General aerodynamic theory – Perfect fluids*: Springer, 1935.

[335] I. E. Garrick, "Propulsion of a flapping and oscillating airfoil," *NACA* 567, 1936.

[336] S. Heathcote and I. Gursul, "Flexible flapping airfoil propulsion at low Reynolds numbers," *AIAA Journal*, vol. 45, no. 5, pp. 1066–1079, 2007.

[337] J. Young and J. C. S. Lai, "Oscillation frequency and amplitude effects on the wake of a plunging airfoil," *AIAA Journal*, vol. 42, no. 10, pp. 2042–2052, 2004.

[338] W. Hoff, "Der flug der insekten," *Naturwissenschaften*, vol. 7, p. 159, 1919.

[339] T. Weis-Fogh, "Energetics of hovering flight in hummingbirds and in drosophila," *Journal of Experimental Biology*, vol. 56, pp. 79–104, 1972.

[340] C. P. Ellington, "The aerodynamics of hovering insect. V. A vortex theory," *Philosophical Transactions of the Royal Society of London, Series B*, vol. 305, pp. 115–144, 1984.

[341] J. M. V. Rayner, "A new approach to animal flight mechanics," *Journal of Experimental Biology*, vol. 80, pp. 17–54, 1979.

[342] G. R. Spedding, J. M. V. Rayner, and C. J. Pennycuik, "Momentum and energy in the wake of a pigeon (Columba livia) in slow flight," *Journal of Experimental Biology*, vol. 111, pp. 81–102, 1984.

[343] G. R. Spedding, "The wake of a jackdaw (Corvus monedula) in slow flight," *Journal of Experimental Biology*, vol. 125, pp. 287–307, 1986.

[344] G. R. Spedding, M. Rosen, and A. Hendenstrom, "A family of vortex wakes generated by a thrush nightingale in free flight in a wind tunnel over its entire natural range of flight," *Journal of Experimental Biology*, vol. 206, pp. 2313–2344, 2003.

[345] F. Noca, D. Shields, and D. Jeon, "A comparison of methods for evaluating time-dependent fluid dynamic forces on bodies using only velocity fields and their derivatives," *Journal of Fluids and Structures*, vol. 13, pp. 551–578, 1999.

[346] T. Weis-Fogh and M. Jensen, "Biology and physics of locust flight: I. Basic principles in insect flight. A critical review," *Philosophical Transactions of the Royal Society of London, Series B*, vol. 239, pp. 415–458, 1956.

[347] C. P. Ellington, "The aerodynamics of hovering insect. II. The aerodynamics of hovering insect flight," *Philosophical Transactions of the Royal Society of London, Series B*, vol. 305, pp. 17–40, 1984.

[348] C. P. Ellington, "The aerodynamics of hovering insect. IV. Aerodynamic mechanisms," *Philosophical Transactions of the Royal Society of London, Series B*, vol. 305, pp. 79–113, 1984.

[349] C. P. Ellington, "The aerodynamics of hovering insect. VI. Lift and power requirements," *Philosophical Transactions of the Royal Society of London, Series B*, vol. 305, pp. 145–181, 1984.

[350] Y. C. Fung, *An introduction to the theory of aeroelasticity*: Dover, 1969.

[351] C.-K. Kang, H., Aono, C. E. S. Cesnik, and W. Shyy, "Effects of flexibility on the aerodynamic performance of flapping wings," *Journal of Fluid Mechanics*, vol. 689, pp. 32–74, 2011.

[352] B. G. Newman, S. B. Savage, and D. Schouella, "Model test on a wing section of a dragonfly," in *Scale effects in animal locomotion*, ed. T. J. Pedley: Academic, 1977.

[353] J. M. Zanker and K. G. Gotz, "The wing beat of Drosophila Melanogaster. II Dynamics," *Philosophical Transactions of the Royal Society of London, Series B*, vol. 327, pp. 19–44, 1990.

[354] S. Vogel, "Flight in Drosophila: III. Aerodynamic characteristics of fly wings and wing models," *Journal of Experimental Biology*, vol. 46, pp. 431–443, 1967.

[355] F.-O. Lehmann and M. H. Dickinson, "The control of wing kinematics and flight forces in fruit flies," *Journal of Experimental Biology*, vol. 201, pp. 385–401, 1998.
[356] U. M. Norberg, "Hovering flight of the dragonfly Aeschna juncea L.," in *Kinematics and aeronautics*, vol. 2., eds. T. Y.-T. Wu, C. J. Brokaw, and C. Brennen: Plenum, 1975.
[357] J. A. Walker and M. W. Westneat, "Mechanical performance of aquatic rowing and flying," *Philosophical Transactions of the Royal Society of London, Series B*, vol. 410, pp. 323–341, 2000.
[358] J. M. Birch and M. H. Dickinson, "The influence of wing-wake interactions on the productions of aerodynamic forces in flapping flight," *Journal of Experimental Biology*, vol. 206, pp. 2257–2272, 2003.
[359] C. Kang and W. Shyy, "Passive wing rotation in flexible flapping wing," in *30th AIAA Applied Aerodynamics Conference*, New Orleans, 2012, pp. AIAA 2012–2763.
[360] C. B. Pendersen and R. Zbikowski, "An indicial-Polhamus aerodynamic model of insect-like flapping wings in hover," in *Flow Phenomena in Nature*, ed. R. Liebe: WIT Press, 2006, pp. 606–665.
[361] J. M. Wakeling and C. P. Ellington, "Dragonfly flight. III. Lift and power requirements," *Journal of Experimental Biology*, vol. 200, pp. 583–600, 1997.
[362] S. A. Ansari, K. Knowles, and R. Zbikowski, "Insectlike flapping wings in the hover. Part 1: Effect of wing kinematics," *Journal of Aircraft*, vol. 45, pp. 1945–1954, 2008.
[363] S. A. Ansari, R. Zbikowski, and K. Knowles, "Nonlinear unsteady aerodynamic model for insect-like flapping wings in the hover. Part 2: Implementation and validation," *Proceedings of the Institute of Mechanical Engineering. Part G: Journal of Aerospace Engineering*, vol. 220, pp. 169–186, 2006.
[364] J. E. McCune, C. G. Lami, and M. T. Scott, "Nonlinear aerodynamics of two-dimensional airfoils in server maneuver," *AIAA Journal*, vol. 28, pp. 385–393, 1990.
[365] T. S. Tavares and J. E. McCune, "Aerodynamics of maneuvering slender wings with leading-edge separation," *AIAA Journal*, vol. 31, pp. 977–986, 1993.
[366] R. Zbikowski, "On aerodynamic modelling of an insect-like flapping wing in hover for micro air vehicles," *Philosophical Transactions of the Royal Society of London, Series A*, vol. 360, pp. 273–290, 2002.
[367] A. Gogulapati and P. Friedmann, "Approximate aerodynamic and aeroelastic modeling of flapping wings in hover and forward flight," in *52nd AIAA/ASME/ASCE/AHS/ASC Structures, Structural Dynamics and Materials Conference 19th AIAA/ASME/AHS Adaptive Structures Conference*, Denver, 2011.
[368] M. Ol and K. Granlund, "Abstraction of aerodynamics of flapping wings: Is it quasi-steady?," in *50th AIAA Aerospace Sciences Meeting including the New Horizons Forum and Aerospace Exposition*, Nashville, 2012.
[369] Z. J. Wang, "Dissecting insect flight," *Annual Review of Fluid Mechanics*, vol. 37, pp. 183–210, 2005.
[370] P. G. Saffman, *Vortex dynamics*: Cambridge University Press, 1995.
[371] F. Noca, "On the evaluation of time-dependent fluid-dynamic forces on bluff bodies," California Institute of Technology, PhD Thesis 1997.
[372] William Graebel, *Advanced fluid mechanics*: Academic Press, 2007.
[373] M. Visbal, R. E. Gordnier, and M. C Galbraith, "High-fidelity simulations of moving and flexible airfoils at low Reynolds numbers," *Experiments in Fluids*, vol. 46, pp. 903–922, 2009.
[374] T. Theodorsen, "General theory of aerodynamic instability and the mechanism of flutter," NACA report, 1935.
[375] W. J. McCroskey, L. W. Carr, and K. W. McAlister, "Dynamic stall experiments on oscillating airfoils," *AIAA Journal*, vol. 14, pp. 57–63, 1976.
[376] K. Knowles, P. C. Wilkins, S. A. Ansari, and R. W. Zbikowski, "Integrated computational and experimental studies of flapping-wing micro air vehicle aerodynamics," in *3rd International Symposium on Integrating CFD and Experiments in Aerodynamics*, 2007, pp. 1–15.
[377] H. Liu and K. Kawachi, "Leading-edge vortices of flapping and rotary wings at low Reynolds number," in *Fixed and flapping wing aerodynamics for micro air vehicle*

applications, ed. T. J. Mueller: American Institute of Aeronautics and Astronautics, 2001, pp. 275–285.

[378] P. G. Drazin and W. H. Reid, *Hydrodynamic stability*: Cambridge University Press, 1981.

[379] D. Lentink and M. H. Dickinson, "Rotational accelerations stabilize leading edge vortices on revolving fly wings," *Journal of Experimental Biology*, vol. 212, no. 16, pp. 2705–2719, 2009.

[380] A. R. Jones and H. Babinsky, "Unsteady lift generation on rotating wings at low Reynolds numbers," *Journal of Aircraft*, vol. 47, pp. 1013–1021, 2010.

[381] F. Togashi, Y. Ito, M. Murayama, K. Nakahashi, and T. Kato, "Flow simulation of flapping wings of an insect using overset unstructured grid," in *15th AIAA Computational Fluid Dynamics Conference*, Anaheim, 2001.

[382] D. Zuo, W. Chen, S. Peng, and W. Zhang, "Modeling and simulation study of an insect-like flapping-wing micro air vehicle," *Advanced Robotics*, vol. 20, pp. 807–824, 2006.

[383] D. Zuo, W. Chen, S. Peng, and W. Zhang, "Numerical simulation of flapping-wing insect hovering flight at unsteady flow," *International Journal for Numerical Methods in Fluids*, vol. 53, pp. 1801–1817, 2007.

[384] H. Aono, W. Shyy, and H. Liu, "Near wake vortex dynamics of a hovering hawkmoth," *Acta Mechanica Sinica*, vol. 25, pp. 23–36, 2009.

[385] Y. P. Liu and M. Sun, "Wing kinematics measurement and aerodynamics of hovering droneflies," *Journal of Experimental Biology*, vol. 211, pp. 2014–2025, 2008.

[386] Y. P. Liu and M. Sun, "Wing kinematics measurement and aerodynamic force and moments computation of hovering hoverfly," *Bioinformatics and Biomedical Engineering*, pp. 452–455, 2007.

[387] H. Aono, F. Liang, and H. Liu, "Near- and far-field aerodynamics in insect hovering flight: An integrated computational study," *Journal of Experimental Biology*, vol. 211, pp. 239–257, 2008.

[388] H. Liu, "Integrated modeling of insect flight: from morphology, kinematic to aerodynamics," *Journal of Computational Physics*, vol. 228, pp. 439–459, 2009.

[389] U. M. L. Norberg, "Structure, form, and function of flight in engineering and the living world," *Journal of Morphology*, vol. 252, pp. 52–81, 2002.

[390] J. Y. Su, S. C. Ting, Y. H. Chang, and J. T. Yang, "Aerodynamic trick for visual stabilization during downstroke in a hovering bird," *Physical Review E*, vol. 84, p. 012901, 2011.

[391] Y-H Chang, S. C. Ting, C-C Liu, J-T Yang, and C-Y Soong, "An unconventional mechanism of lift production during the downstroke in a hovering bird (Zosterops japonicus)," *Experiments in Fluids*, vol. 51, pp. 1231–1243, 2011.

[392] A. Hedenstrom, M. Rosen, and G. R. Spedding, "Vortex wakes generated by robin Erithacus rubecula during free flight in a wind tunnel," *Journal of Royal Society Interface*, vol. 3, pp. 263–276, 2006.

[393] M. Rosen, G.R. Spedding, and A. Hedenstrom, "Wake structure and wingbeat kinematics of a house-martin Delichon urbica," *Journal of Royal Society Interface*, vol. 4, pp. 659–668, 2007.

[394] G. R. Spedding and A. Hedenstrom, "PIV-based investigations of animal flight," *Experiments in Fluids*, vol. 46, pp. 749–763, 2009.

[395] E. J. Stmhuis and S. Nauwlaerts, "Propulsive force calculations in swimming frogs. II. Applications of a vortex ring model to DPIV data," *Journal of Experimental Biology*, vol. 208, pp. 1445–1451, 2005.

[396] S. C. Ting and J. T. Yang, "Pitching stabilization via caudal fin-wave propagation in a forward-sinking parrot cichlid (Cichlasoma citrinellum x Cichlasoma synspilum)," *Journal of Experimental Biology*, vol. 211, pp. 3147–3159, 2008.

[397] W. Shyy and H. Liu, "Flapping wings and aerodynamic lift: The role of leading-edge vortices," *AIAA Journal*, vol. 45, pp. 2817–2819, 2007.

[398] D. Weihs and E. Barta, "Comb wings for flapping flight at extremely low Reynolds numbers," *AIAA Journal*, vol. 46, pp. 285–288, 2008.

[399] E. Barta, "Motion of slender bodies in unsteady Stokes flow," *Journal of Fluid Mechanics*, vol. 688, pp. 66–87, 2011.
[400] S. M. Swartz, "Allometric patterning in the limb skeleton of bats," *Journal of Morphology*, vol. 234, pp. 277–294, 1997.
[401] S. M. Swartz, M. B. Bennertt, and D. R. Carrier, "Wing bone stresses in free flying bats and the evolution of skeletal design for flight," *Nature*, vol. 359, pp. 726–729, 1992.
[402] S. A. Combes and T. L. Daniel, "Flexural stiffness in insect wings. I: Scaling and the influence of wing venation," *Journal of Experimental Biology*, vol. 206, no. 17, pp. 2979–2987, 2003.
[403] S. A. Combes and T. L. Daniel, "Flexural stiffness in insect wings. II: Spatial distribution and dynamic wing bending," *Journal of Experimental Biology*, vol. 206, pp. 2989–2997, 2003.
[404] C.-K. Kang and W. Shyy, Kang, C. and Shyy, "Scaling and lift generation of hovering flexible wing of insect size," *to be published*
[405] W. Shyy and R. Smith, "A study of flexible airfoil aerodynamics with application to micro aerial vehicles," in *28th AIAA Fluid Dynamics Conference*, Snowmass, 1997.
[406] R. M. Waszak, N. L. Jenkins, and P. Ifju, "Stability and control properties of an aeroelastic fixed wing micro aerial vehicle," in *AIAA Atmospheric Flight Mechanics Conference and Exhibit*, Montreal, 2001.
[407] B. Stanford, P. Ifju, R. Albertani, and W. Shyy, "Fixed membrane wings for micro air vehicles: Experimental characterization, numerical modeling, and tailoring," *Progress in Aerospace Sciences*, vol. 44, pp. 258–294, 2008.
[408] W. Shyy, P. G. Ifju, and D. Viieru, "Membrane wing-based micro air vehicles," *Applied Mechanics Reviews*, vol. 58, pp. 283–301, 2005.
[409] D. Lentink, W. B. Dickson, J. L. van Leeuwen, and M. H. Dickinson, "Leading-edge vortices elevate lift of autorotating plant seeds," *Science*, vol. 324, pp. 1438–1440, 2009.
[410] U. Pesavento and Z. J. Wang, "Falling paper: Navier-Stokes solutions, model of fluid forces, and center of mass elevation," *Physical Review Letters*, vol. 93, pp. 144501–144504, 2004.
[411] A Andersen, U. Pesavento, and Z. J. Wang, "Unsteady aerodynamics of fluttering and tumbling plates," *Journal of Fluid Mechanics*, vol. 541, pp. 65–90, 2005.
[412] A. Andersen, U. Pesavento, and Z. J. Wang, "Analysis of transitions between fluttering, tumbling and steady descent of falling cards," *Journal of Fluid Mechanics*, vol. 541, pp. 91–104, 2005.
[413] C. Shao, Y. Chen, and J. Lin, "Wind induced deformation and vibration of a Platanus Acerifolia Leaf," *Acta Mechanica Sinica*, vol. 28, pp. 583–594, 2012.
[414] M. J. Shelley and J. Zhang, "Flapping and bending bodies interacting with fluid flows," *Annual Reviews in Fluid Mechanics*, vol. 43, pp. 449–465, 2011.
[415] H. Garcia, M. Abdulrahim, and R. Lind, "Roll control for a micro air vehicle using active wing morphing," in *AIAA Guidance, Navigation, and Control Conference and Exhibit*, Austin, 2003.
[416] D. W. Bechert, M. Bruse, W. Hage, and R. Meyer, "Biological surfaces and their technological application – laboratory and flight experiments on drag reduction and separation control," in *28th AIAA Fluid Dynamics Conference*, Snowmass, 1997.
[417] W. Shyy, H. S. Udaykumar, M. R. Madhukar, and W. S. Richard, *Computational fluid dynamics with moving boundaries*: Taylor and Francis, 1996.
[418] P. P. Friedmann, "Renaissance of aeroelasticity and its future," *Journal of Aircraft*, vol. 36, pp. 105–121, 1999.
[419] E. Livne, "Future of airplane aeroelasticity," *Journal of Aircraft*, vol. 40, pp. 1066–1092, 2003.
[420] S. K. Chimakurthi, J. Tang, R. Palacios, C. E. S. Cesnik, and W. Shyy, "Computational aeroelasticity framework for analyzing flapping wing micro air vehicles," *AIAA Journal*, vol. 47, pp. 1865–1878, 2009.

[421] K. Voelz, "Profil und luftriebeines segels," *Zeitschrift fur Angewandte Mathematik und Mechanik*, vol. 30, pp. 301–317, 1950.

[422] B. Thwaites, "The aerodynamic theory of sails, Part I. Two-dimensional sails.," *Philosophical Transactions of the Royal Society of London, Series B*, vol. 261, pp. 402–422, 1961.

[423] J. N. Nielsen, "Theory of flexible aerodynamic surfaces," *Journal of Applied Mechanics*, vol. 30, pp. 435–442, 1963.

[424] L. G. Chamber, "A variational formulation of the Thwaites sail equation," *Quarterly Journal of Mechanics and Applied Mathematics*, vol. 19, pp. 221–231, 1966.

[425] S. Greenhalgh, H. S. Curtiss, and B. F. Smith, "Aerodynamic properties of a two-dimensional inextensible flexible airfoil," *AIAA Journal*, vol. 22, pp. 865–870, 1984.

[426] B. G. Newman, "Aerodynamic theory for membranes and sails," *Progress in Aerospace Sciences*, vol. 24, pp. 1–27, 1987.

[427] T. Sugimoto and J. Sato, "Aerodynamic characteristics of two-dimensional membrane airfoils," *Journal of the Japan Society for Aeronautical and Space Sciences*, vol. 36, pp. 36–43, 1988.

[428] J. M. Vanden-Broeck and J. B. Keller, "Shape of a sail in a flow," *Physics of Fluids*, vol. 24, pp. 552–553, 1981.

[429] J. M. Vanden-Broeck, "Nonlinear two-dimensional sail theory," *Physics of Fluids*, vol. 25, pp. 420–423, 1982.

[430] H. Murai and S. Maruyama, "Theoretical investigation of the aerodynamics of double membrane sailwing airfoil sections," *Journal of Aircraft*, vol. 17, pp. 294–299, 1980.

[431] P. S. Jackson, "A simple model for elastic two-dimensional sails," *AIAA Journal*, vol. 25, pp. 672–682, 1983.

[432] A. D. Sneyd, "Aerodynamic coefficients and longitudinal stability of sail airfoils," *Journal of Fluid Mechanics*, vol. 149, pp. 127–146, 1984.

[433] S. Murata and S. Tanaka, "Aerodynamic characteristics of a two-dimensional porous sail," *Journal of Fluid Mechanics*, vol. 144, pp. 445–462, 1984.

[434] G. de Matteis and L. de Socio, "Nonlinear aerodynamics of a two-dimensional membrane airfoil with separation," *Journal of Aircraft*, vol. 23, pp. 831–836, 1986.

[435] A. Song, X. Tian, E. Israeli, R. Galvao, K. Bishop, S. Swartz, and K. Breuer, "Aeromechanics of membrane wings with implications for animal flight," *AIAA Journal*, vol. 46, pp. 2096–2106, 2008.

[436] P. S. Jackson and G. W. Christie, "Numerical analysis of three-dimensional elastic membrane wings," *AIAA Journal*, vol. 25, pp. 676–682, 1987.

[437] J. T. Oden and T. Sato, "Finite strains and displacements of elastic membrane by the finite element method," *International Journal for Solids and Structures*, vol. 3, pp. 471–488, 1967.

[438] C. H. Jenkins and J. W. Leonard, "Nonlinear dynamic response of membranes: State of the art," *Applied Mechanics Reviews*, vol. 44, pp. 319–328, 1991.

[439] C. H. Jenkins, "Nonlinear dynamic response of membranes: State of the art update," *Applied Mechanics Reviews*, vol. 49, pp. S41-S48, 1996.

[440] E. Verron, G. Marckmann, and B. Pesaux, "Dynamic inflation of non-linear elastic and viscoelastic rubber-like membranes," *International Journal for Numerical Methods in Engineering*, vol. 50, pp. 1233–1251, 2001.

[441] H. Ding, B. Yang, M. Lou, and H. Fang, "New numerical method for two-dimensional partially wrinkled membrane," *AIAA Journal*, vol. 41, pp. 125–132, 2003.

[442] B. Stanford, D. Viieru, R. Albertani, W. Shyy, and P. Ifju, "A numerical and experimental investigation of flexible micro air vehicle wing deformation," in *44th AIAA Aerospace Sciences Meeting and Exhibit*, Reno, 2006.

[443] M. Mooney, "A theory of large elastic deformation," *Journal of Applied Physics*, vol. 11, pp. 582–592, 1940.

[444] A. E. Green and J. E. Adkins, *Large elastic deformations*: Clarendon, 1960.

[445] T. Nakata and H. Liu, "A fluid-structure interaction model of insect flight with flexible wings," *Journal of Computational Physics*, vol. 231, pp. 1822–1847, 2012.

[446] S. K. Chimakurthi, C. E. S. Cesnik, and B. K. Stanford, "Flapping-wing structural dynamics formulation based on a corotational shell finite element," *AIAA Journal*, vol. 49, no. 1, pp. 128–142, 2011.
[447] G. I. Barenblatt, *Scaling*: Cambridge University Press, 2003.
[448] G. I. Barenblatt, *Scaling, self-similarity, and intermediate asymptotics*: Cambridge University Press, 1996.
[449] W. E. Baker, P. S. Westine, and F. T. Dodge, *Similarity methods in engineering dynamics*: Elsevier, 1991.
[450] W. Shyy, H. Aono, S. K. Chimakurthi, P. Trizila, C.-K. Kang, C. E. S. Cesnik, and H. Liu, "Recent progress in flapping wing aerodynamics and aeroelasticity," *Progress in Aerospace Sciences*, vol. 46, no. 7, pp. 284–327, 2010.
[451] D. Ishihara, T. Horie, and M. Denda, "A two-dimensional computational study on the fluid-structure interaction cause of wing pitch changes in dipteran flapping flight," *Journal of Experimental Biology*, vol. 212, no. 1, pp. 1–10, 2009.
[452] D. Ishihara, Y. Yamashita, T. Horie, S. Yoshida, and T. Niho, "Passive maintenance of high angle of attack and its generation during flapping translation in crane fly wing," *Journal of Experimental Biology*, vol. 212, pp. 3882–3891, 2009.
[453] B. Thiria and R. Godoy-Diana, "How wing compliance drives the efficiency of self-propelled flapping flyers," *Physical Review E*, vol. 82, no. 1, p. 015303, 2010.
[454] S. Ramananarivo, R. Godoy-Diana, and B. Thiria, "Rather than resonance, flapping wing flyers may play on aerodynamics to improve performance," *Proceedings of the National Academy of Sciences*, vol. 28, p. 10179–10180, 2011.
[455] W. Shyy, Y. Lian, J. Tang, H. Liu, P. Trizila, B. Stanford, L. Bernal, C. E. S. Cesnik P. P. Friedmann, and P. Ifju, "Computational aerodynamics of low Reynolds number plunging, pitching and flexible wings for MAV applications," *Acta Mechanica Sinica*, vol. 24, pp. 351–373, 2008.
[456] W. Shyy, Jenkins D. A., and R. W. Smith, "Study of adaptive shape airfoils at low Reynolds number in oscillatory flow," *AIAA Journal*, vol. 35, pp. 1545–1548, 1997.
[457] R. Albertani, B. Standford, J. P. Hubner, and P. G. Ifju, "Aerodynamic coefficients and deformation measurements on flexible micro air vehicle wings," *Experimental Mechanics*, vol. 47, pp. 625–635, 2007.
[458] Y. Abudaram, B. Stanford, and P. Ifju, "Wing tunnel testing of load-alleviating membrane wings at low Reynolds numbers," in *47th AIAA Aerospace Sciences Meeting including The New Horizons Forum and Aerospace Exposition*, Orlando, 2009, p. 1468.
[459] P. Rojratsirikul, Z. Wang, and I. Gursul, "Effect of pre-strain and excess length on unsteady fluid-structure interactions of membrane airfoils," in *AIAA*, 2009, pp. 2009–578.
[460] H. Hui, M. Tamai, and J. Murphy, "Flexible-membrane airfoils at low Reynolds numbers," *Journal of Aircraft*, vol. 45, pp. 1767–1778, 2008.
[461] B. Stanford, M. Sytsma, R. Albertani, D. Viieru, W. Shyy, and P. Ifju, "Static aeroelastic model validation of membrane micro air vehicle wings," *AIAA Journal*, vol. 45, pp. 2828–2837, 2007.
[462] R. Galvao, E. Israel, A. Song, X. Tian, K. Bishop, S. Swartz, and K. Breuer, "The aerodynamics of compliant membrane wings modeled on mammalian flight mechanics," in *36th AIAA Fluid Dynamics Conference and Exhibit*, San Francisco, 2006.
[463] Y. Lian and W. Shyy, "Laminar-turbulent transition of a low Reynolds number rigid or flexible airfoil," in *36th AIAA Fluid Dynamic Conferences and Exhibit*, San Francisco, 2006.
[464] B. Stanford and P. Ifju, "Aeroelastic tailoring of fixed membrane wings for micro air vehicles," in *49th AIAA/ASME/ASCE/AHS/ASC Structures, Structural Dynamics, and Materials Conference*, Schaumburg, 2008.
[465] M. Hepperle, "MH AeroTools Online Database," available at www.mh-aerotools.de/airfoils/ribs.htm, March 2007.
[466] M. Argentina and L. Mahadevan, "Fluid-flow-induced flutter of a flag," *Proceedings of the National Academy of Sciences*, vol. 102, pp. 1829–1834, 2005.

[467] R. Ormiston, "Theoretical and experimental aerodynamics of the sail wing," *Journal of Aircraft*, vol. 8, pp. 77–84, 1971.

[468] J. W. Johansson, W. Romberg, P. J. Attar, and R. Parthasarathy, "Experimental characterization of limit cycle oscillations in membrane wing micro air vehicles," *Journal of Aircraft*, vol. 47, pp. 1300–1308, 2010.

[469] R. E. Gordnier, "High fidelity computational simulation of a membrane wing airfoil," *Journal of Fluid and Structures*, vol. 25, pp. 897–917, 2009.

[470] T. L. Daniel and S. A. Combes, "Flexible wings and fins: Bending by inertial or fluid-dynamic forces,?" *Integrative and Comparative Biology*, vol. 42, no. 5, pp. 1044–1049, 2002.

[471] S. A. Combes, "Wing flexibility and design for animal flight," University of Washington, PhD Thesis 2002.

[472] S. A. Combes and T. L. Daniel, "Into thin air: Contributions of aerodynamic and inertial-elastic forces to wing bending in the hawkmoth Manduca sexta," *Journal of Experimental Biology*, vol. 206, no. 17, pp. 2999–3006, 2003.

[473] A. M. Mountcastle and T. L. Daniel, "Aerodynamic and functional consequences of wing compliance," *Experiments in Fluids*, vol. 46, pp. 873–882, 2009.

[474] Q. Zhu, "Numerical simulation of a flapping foil with chordwise or spanwise flexibility," *AIAA Journal*, vol. 45, pp. 2448–2457, 2007.

[475] S. Shkarayev, D. Silin, G. Abate, and R. Albertani, "Aerodynamics of cambered membrane flapping wing," in *48th AIAA Aerospace Sciences Meeting including the New Horizons Forum and Aerospace Exposition*, Orlando, 2010.

[476] H. Hui, A. G. Kumar, G. Abate, and R. Albertani, "An experimental investigation on the aerodynamic performances of flexible membrane wings in flapping flight," *Aerospace Science and Technology*, vol. 4, pp. 575–586, 2010.

[477] D.-K. Kim, J.-H. Han, and K.-J. Kwon, "Wind tunnel tests for a flapping wing model with a changeable camber using macro-fiber composite actuators," *Smart Materials and Structures*, vol. 18, pp. 024008-1–024008-8, 2009.

[478] D. Mueller, H. A. Bruck, and S. K. Gupta, "Measurement of thrust and lift forces associated with drag of compliant flapping wing for micro air vehicles using a new test stand design," *Experimental Mechanics*, pp. 1–11, 2009.

[479] D. Watman and T. Furukawa, "A parametric study of flapping wing performance using a robotic flapping wing," in *IEEE International Conference on Robotics and Automation*, 2009, pp. 3638–3643.

[480] M. F. Platzer, K. D. Jones, J. Young, and J. C. S. Lai, "Flapping wing aerodynamics: Progress and challenges," *AIAA Journal*, vol. 46, pp. 2136–2149, 2008.

[481] R. J. Wood, "Design fabrication and analysis of a 3 dof, 3 cm flapping-wing MAV," in *IEEE/RSJ International Conference on Intelligent Robots and Systems*, San Diego, 2007, pp. 1576–1581.

[482] P. Wu, B. Stanford, E. Sallstrom, L. Ukeiley, and P. Ifju, "Structural dynamics and aerodynamics measurements of biologically inspired flexible flapping wings," *Bioinspiration and Biomimetics*, vol. 6, pp. 016009-1–016009-20, 2011.

[483] J. Katz and D. Weihs, "Hydrodynamic propulsion by large amplitude oscillation of an airfoil with chordwise flexibility," *Journal of Fluid Mechanics*, vol. 88, pp. 485–497, 1978.

[484] J. Pederzani and Haj-Hariri, "Numerical analysis of heaving flexible airfoils in a viscous flow," *AIAA Journal*, vol. 44, pp. 2773–2779, 2006.

[485] M. R. Chaithanya and K. Venkatraman, "Hydrodynamic propulsion of a flexible foil," in *12th Asian Congress of Fluid Mechanics*, 2008.

[486] P. Gopalakrishnan, "Unsteady aerodynamic and aeroelastic analysis of flapping flight," Virginia Polytechnic Institute and State University, PhD Thesis 2008.

[487] P. J. Attar, R. E. Gordnier, J. W. Johansson, W. Romberg, and R. N. Parthasarathy, "Aeroelastic analysis of membrane micro air vehicles – Part II. Computational study of a plunging membrane airfoil," *Journal of Vibration and Acoustics*, vol. 133, pp. 021009-1–021009-6, 2011.

[488] U. Gulcat, "Propulsive force of a flexible flapping thin airfoil," *Journal of Aircraft*, vol. 46, pp. 465–473, 2009.

[489] J.-M. Miao and M.-H. Ho, "Effect of flexure on aerodynamic propulsive efficiency of flapping flexible airfoil," *Journal of Fluids and Structures*, vol. 22, pp. 401–409, 2006.

[490] J. Toomey and J. D. Eldredge, "Numerical and experimental study of the fluid dynamics of a flapping wing with low order flexibility," *Physics of Fluids*, vol. 20, pp. 1–10, 2008.

[491] D. Poirel, Y. Harris, and A. Benaissa, "Self-sustained aeroelastic oscillations of a NACA0012 airfoil at low-to-moderate Reynolds numbers," *Journal of Fluids and Structures*, vol. 24, pp. 700–719, 2008.

[492] M. Vanella, T. Fitzgerald, S. Preidikman, E. Balaras, and B. Balachandran, "Influence of flexibility on the aerodynamic performance of a hovering wing," *Journal of Experimental Biology*, vol. 212, pp. 95–105, 2009.

[493] S. Heathcote, D. Martin, and I. Gursul, "Flexible flapping airfoil propulsion at zero freestream velocity," *AIAA Journal*, vol. 42, pp. 2196–2204, 2004.

[494] S. Michelin and S. G. Llewellyn Smith, "Resonance and propulsion performance of a heaving flexible wing," *Physics of Fluids*, vol. 21, p. 071902, 2009.

[495] L. A. Miller and C. S. Peskin, "A computational fluid dynamic of 'clap and fling' in the smallest insects," *Journal of Experimental Biology*, vol. 208, pp. 195–212, 2005.

[496] L. A. Miller and C. S. Peskin, "Flexible clap and fling in tiny insect flight," *Journal of Experimental Biology*, vol. 212, pp. 3076–3090, 2009.

[497] L. Zhao, X. Deng, and S. P. Sane, "Modulation of leading edge vorticity and aerodynamic forces in flexible flapping wings," *Bioinspiration and Biomimetics*, p. 036007, 2011.

[498] L. Zhao, Q. Huang, X. Deng, and S. P. Sane, "Aerodynamic effects of flexibility in flapping wings," *Journal of the Royal Society Interface*, vol. 7, no. 44, pp. 485–497, 2010.

[499] G. Du and M. Sun, "Effects of unsteady deformation of flapping wing on its aerodynamic forces," *Applied Mathematics and Mechanics*, vol. 29, pp. 731–743, 2008.

[500] K.-B. Lee, J.-H. Kim, and C. Kim, "Aerodynamic effects of structural flexibility in two-dimensional insect flapping flight," *Journal of Aircraft*, vol. 48, pp. 894–909, 2011.

[501] A. M. Mountcastle and T. L. Daniel, "Vortexlet models of flapping flexible wings show tuning for force production and control," *Bioinspiration and Biomimetics*, vol. 5, p. 045005, 2010.

[502] S. Heathcote, Z. Wang, and I. Gursul, "Effect of spanwise flexibility on flapping wing propulsion," *Journal of Fluids and Structures*, vol. 24, no. 2, pp. 183–199, 2008.

[503] P. Liu and N. Bose, "Propulsive performance from oscillating propulsors with spanwise flexibility," *Philosophical Transactions of the Royal Society of London, Series A*, vol. 453, pp. 1763–1770, 1963.

[504] R. E. Gordnier, P. J. Attar, S. K. Chimakurthi, and C. E. S. Cesnik, "Implicit LES simulations of a flexible flapping wing," in *18th AIAA/ASME/AHS Adaptive Structures Conference*, Orlando, 2010.

[505] H. Aono, S. K. Chimakurthi, C. E. S. Cesnik, H. Liu, and W. Shyy, "Computational modeling of spanwise effects on flapping wing aerodynamics," 47th AIAA aerospace sciences meeting including the new horizons forum and aerospace exposition, January 5–8, 2009, Orlando, Florida, AIAA-2009-1270, 2009.

[506] H. Aono, S. K. Chimakurthi, P. Wu, E. Sällström, B. K. Stanford, C. E. S., P. Ifju, L. Ukeiley, and W. Shyy, "A computational and experimental study of flexible flapping wing aerodynamics," in *48th AIAA Aerospace Sciences Meeting including the New Horizons Forum and Aerospace Exposition*, Orlando, 2010.

[507] P. Wu, P. Ijfu, and B. Stanford, "Flapping wing structural deformation and thrust correlation study with flexible membrane wings," *AIAA Journal*, vol. 48, no. 9, pp. 2111–2122, 2010.

[508] H. Masoud and A. Alexeev, "Resonance of flexible flapping wings at low Reynolds number," *Physical Review E*, vol. 81, p. 056304, 2010.

[509] S. Sunada, L. Zeng, and K. Kawachi, "The relationship between dragonfly wing structure and torsional deformation," *Journal of Theoretical Biology*, vol. 193, pp. 39–45, 1998.

[510] J.-S. Chen, J.-Y. Chen, and Y.-F. Chou, "On the natural frequencies and mode shapes of dragonfly wings," *Journal of Sound and Vibration*, vol. 313, pp. 643–654, 2008.
[511] R. D. Mindlin and L. E. Goodman, "Beam vibrations with time-dependent boundary conditions," *Journal of Applied Mechanics*, vol. 17, pp. 353–356, 1950.
[512] R. Buchwald and R. Dudley, "Limits to vertical force and power production in bumblebees (Hymenoptera: Bombus impatiens)," *Journal of Experimental Biology*, vol. 213, pp. 426–432, 2010.
[513] S. Vogel, "Flight in Drosophila: I. Flight performance of tethered flies," *Journal of Experimental Biology*, vol. 44, pp. 567–578, 1966.
[514] E. Shevtsova, C. Hansson, D. H. Janzen, and J. Kjaerandsen, "Stable structural color patterns displayed on transparent insect wings," *Proceedings of the National Academy of Sciences*, vol. 108, no. 213, pp. 668–673, 2011.
[515] S. Timoshenko, D. H. Young, and W. Weaver, JR., *Vibration problems in engineering*: John Wiley & Sons, Ltd., 1974.
[516] B. Yin and H. Luo, "Effect of wing inertia on hovering performance of flexible flapping wings," *Physics of Fluids*, vol. 22, p. 111902, 2010.
[517] Z.J. Wang, "Aerodynamic efficiency of flapping flight: Analysis of a two-stroke model," *Journal of Experimental Biology*, vol. 211, pp. 234–238, 2008.
[518] A. Agrawal and S. K. Agrawal, "Design of bio-inspired flexible wings for flapping-wing micro-sized air vehicle applications," *Advanced Robotics*, vol. 23, no. 7–8, pp. 979–1002, 2009.
[519] B. Singh and I. Chopra, "Insect-based hover-capable flapping wings for micro air vehicles: Experiments and analysis," *AIAA Journal*, vol. 46, pp. 2115–2135, 2008.
[520] M. Hamamoto, Y. Ohta, K. Hara, and T. Hisada, "Application of fluid-structure interaction analysis to flapping flight of insects with deformable wings," *Advanced Robotics*, vol. 21, pp. 1–21, 2007.
[521] J. Young, S. M. Walker, R. J. Bomphrey, G. K. Taylor, and A. L. R. Thomas, "Details of insect wing design and deformation enhance aerodynamic function and flight efficiency," *Science*, vol. 325, pp. 1549–1552, 2009.
[522] H. Tanaka, J. P. Whitney, and R. J. Wood, "Effect of flexural and torsional wing flexibility on lift generation in hoverfly flight," *Integrative and Comparative Biology*, pp. 1–9, 2011.
[523] T. Nakata and H. Liu, "Aerodynamic performance of a hovering hawkmoth with flexible wings: A computational approach," *Proceeding of the Royal Society B: Biological Sciences*, vol. 22, pp. 722–731, 2012.
[524] R. J. Wootton, "Leading edge section and asymmetric twisting in the wings of flying butterflies," *Journal of Experimental Biology*, vol. 180, pp. 105–117, 1993.
[525] S. M. Walker, A. L. R. Thomas, and G. K. Taylor, "Deformable wing kinematics in the desert locust: How and why do camber, twist and topography vary through the stroke?," *Journal of the Royal Society Interface*, vol. 6, pp. 735–747, 2008.
[526] J. M. V. Rayner and R. Gordon, "Visualization and modelling of the wakes of flying birds," in *Biona Report 13, Motion Systems*, eds. R. Blickhan, A. Wisser, and W. Nachtiga, 1998.
[527] B. W. Tobalske, D. R. Warrick, J. P. Clark, D. R. Powers, T. L. Hedrick, G. A. Hyder, and A. A. Biewener, "Three-dimensional kinematics of hummingbirds flight," *Journal of Experimental Biology*, vol. 210, pp. 2368–2382, 2007.
[528] P. Henningsson, G. R. Spedding, and A. Hedenstrom, "Vortex wake and flight kinematics of a swift in cruising flight in a wind tunnel," *Journal of Experimental Biology*, pp. 717–730, 2008.
[529] D. K. Riskin, J. Iriarte-Diaz, K. M. Middleton, S. M. Swartz, and K. S. Breuer, "The effect of body size on the wing movements of pteropodid bats, with insights into thrust and lift production," *Journal of Experimental Biology*, vol. 213, pp. 4110–4122, 2010.
[530] J.-S. Lee, J.-K. Kim, D.-K. Kim, and J.-H. Han, "Longitudinal flight dynamics of bioinspired ornithopter considering fluid-structure interaction," *Journal of Guidance, Control, and Dynamics*, vol. 34, pp. 667–677, 2011.

[531] A. J. Bergou, L. Ristroph, J. Guckenheimer, I. Cohen, and Z. J. Wang, "Fruit flies modulate passive wing pitching to generate in-flight turns," *Physical Review Letters*, vol. 104, p. 148101, 2010.

[532] S. A. Combes and R. Dudley, "Turbulence-driven instabilities limit insect flight performance," *Proceedings of the National Academy of Sciences*, vol. 106, pp. 9105–9108, 2009.

[533] Z. J. Wang and D. Russell, "Effect of forewing and hindwing interactions on aerodynamic forces and power in hovering dragonfly flight," *Physical Review Letters*, vol. 99, p. 148101, 2007.

[534] J. K. Wang and M. Sun, "A computational study of the aerodynamics and forewing-hindwing interaction of a model dragonfly in forward flight," *Journal of Experimental Biology*, vol. 208, pp. 3785–3804, 2005.

[535] J. H. Brackenbury, "Kinematics of take-off and climbing flight in butterflies," *Journal of Zoology*, vol. 224, pp. 251–270, 1991.

[536] A. Gogulapati, P. Friedmann, E. Kheng, and W. Shyy, "Approximate aeroelastic modeling of flapping wings: Comparisons with CFD and experimental data," in *51st AIAA/ASME/ASCE/AHS/ASC Structures, Structural Dynamics, and Materials Conference*, 2010, pp. AIAA-2010–2707.

[537] C. T. Orlowski and A. R. Girard, "Dynamics, stability, and control analyses of flapping wing micro air vehicles," *Progress in Aerospace Sciences*, vol. 51, pp. 18–30, 2012.

[538] G. Taylor and A. R. Thomas, "Dynamic flight stability in the desert locus Schistocerca gregaria," *Journal of Experimental Biology*, vol. 206, pp. 2803–2829, 2003.

[539] I. Faruque and J. S. Humbert, "Dipteran insect flight dynamics. Part 1: Longitudinal motion about hover," *Journal of Theoretical Biology*, vol. 264, pp. 538–552, 2010.

[540] I. Faruque and J. S. Humbert, "Dipteran insect flight dynamics. Part 2: Lateral-directional motion about hover," *Journal of Theoretical Biology*, vol. 265, pp. 306–313, 2010.

[541] Y.-L. Zhang and M. Sun, "Dynamic flight stability of a hovering model insect: Lateral motion," *Acta Mechanica Sinica*, vol. 26, pp. 175–190, 2010.

[542] M. A. Bolender, "Open-loop stability of flapping flight in hover," in *the AIAA Guidance, Navigation, and Control Conference*, Ontario, 2010.

[543] J. Dietl and E. Garcia, "Stability in hovering ornithopter flight," in *Industrial and Commercial Applications of Smart Structure Technologies*, San Diego, 2008.

[544] J. Dietl and E. Garcia, "Stability in ornithopter longitudinal dynamics," *Journal of Guidance, Control, and Dynamics*, vol. 31, pp. 1157–1162, 2008.

[545] M. Sun, J. Wang, and Y. Xiong, "Dynamic flight stability of hovering insects," *Acta Mechanica Sinica*, vol. 23, pp. 231–246, 2007.

[546] M. Sun and Y. Xiong, "Dynamic flight stability of a hovering bumblebee," *Journal of Experimental Biology*, vol. 208, pp. 447–459, 2005.

[547] X. Deng, L. Schenato, W. Wu, and S. Sastry, "Flapping flight for biomimetic robot insects: Part I – system modeling," *IEEE Transactions on Robotics*, vol. 22, pp. 776–788, 2006.

[548] X. Deng, L. Schenato, and S. Sastry, "Flapping flight for biomimetic robot insects: Part II – flight control design," *IEEE Transactions on Robotics*, vol. 22, pp. 789–803, 2006.

[549] B. Cheng and X. Deng, "Near-hover dynamics and attitude stabilization of an insect model," in *IEEE American Control Conference*, Baltimore, 2010, pp. 39–44.

[550] N. C. Rosenfel and N. M. Wereley, "Time-periodic stability of a flapping insect wing structure in hover," *Journal of Aircraft*, vol. 46, pp. 450–464, 2009.

[551] G. Abate and W. Shyy, "Bio-inspiration of morphing for micro air vehicles," in *Morphing aerospace vehicles and structures*, ed. J. Valasek: Wiley, 2012, pp. 41–53.

[552] M. S. Triantafyllou, F. S. Hover, A. H. Techet, and D. K. P. Yue, "Review of hydrodynamic scaling laws in aquatic locomotion and fishlike swimming," *Transactions of the ASME*, vol. 58, pp. 226–237, 2005.

[553] T. Y. Wu, "Fish swimming and bird/insect flight," *Annual Review of Fluid Mechanics*, vol. 43, pp. 25–58, 2011.

[554] D. Weihs, "A hydrodynamical analysis of fish turning manoeuvres," *Proceedings of Royal Society of London B*, vol. 182, pp. 59–72, 1972.

[555] D. Weihs, "Energetic advantages of burst swimming of fish," *Journal of Theoretical Biology*, pp. 215–229, 1974.

[556] A. L. Douglas, W. B. Dickson, J. T. Vance, S. P. Roberts, and M. H. Dickinson, "Short-amplitude high-frequency wing strokes determine the aerodynamics of honeybee flight," *Proceedings of the National Academy of Sciences*, vol. 102, pp. 18213–18218, 2005.

[557] O. M. Curet, S. M. Swartz, and K. S. Breuer, "An aeroelastic instability provides a possible basis for the transition from gliding to flapping flight," *Journal of the Royal Society Interface*, vol. 10, 20120940, 2013. (doi:10.1098/rsif.2012.0940)

Index

www.ingramcontent.com/pod-product-compliance
Lightning Source LLC
LaVergne TN
LVHW062222200926
848484LV00006B/1456

* 9 7 8 1 1 0 7 0 3 7 2 6 7 *